P9-CJR-805

This book is printed on acid-free paper. ♾

Academic Press, Inc.
A Division of Harcourt Brace & Company
525 B Street, Suite 1900, San Diego, California 92101-4495

United Kingdom Edition published by
Academic Press Limited
24-28 Oval Road, London NW1 7DX

International Standard Serial Number: 0074-7696

International Standard Book Number: 0-12-364561-1

PRINTED IN THE UNITED STATES OF AMERICA
95 96 97 98 99 00 EB 9 8 7 6 5 4 3 2 1

CONTENTS

Ecto-ATPases: Identities and Functions

Liselotte Plesner

Molecular Genetic Approaches to the Study of Human Craniofacial Dysmorphologies

Gudrun E. Moore

Effects of Electromagnetic Fields on Molecules and Cells

Eugene M. Goodman, Ben Greenebaum, and Michael T. Marron

CONTRIBUTORS

Numbers in parentheses indicate the pages on which the authors' contributions begin.

F. C. H. Franklin (1), *Wolfson Laboratory for Plant Molecular Biology, School of Biological Sciences, University of Birmingham, Edgbaston, Birmingham B15 2TT, United Kingdom*

V. E. Franklin-Tong (1), *Wolfson Laboratory for Plant Molecular Biology, School of Biological Sciences, University of Birmingham, Edgbaston, Birmingham B15 2TT, United Kingdom*

Eugene M. Goodman (279), *Biomedical Research Institute, University of Wisconsin—Parkside, Kenosha, Wisconsin 53141*

Ben Greenebaum (279), *Biomedical Research Institute, University of Wisconsin—Parkside, Kenosha, Wisconsin 53141*

M. J. Lawrence (1), *Wolfson Laboratory for Plant Molecular Biology, School of Biological Sciences, University of Birmingham, Edgbaston, Birmingham B15 2TT, United Kingdom*

Michael T. Marron (279), *Office of Naval Research, Arlington, Virginia 22217*

Gudrun E. Moore (215), *The Action Research Laboratory for the Molecular Biology of Fetal Development, Institute of Obstetrics and Gynaecology, Queen Charlotte's and Chelsea Hospital, Royal Postgraduate Medical School, London W6 OXG, United Kingdom*

Vittoria Nuti Ronchi (65), *Institute of Mutagenesis and Differentiation, CNR, 56124 Pisa, Italy*

Liselotte Plesner (141), *Department of Biophysics, University of Aarhus, DK-8000 Aarhus C, Denmark*

Cell and Molecular Biology of Self-Incompatibility in Flowering Plants

F. C. H. Franklin, M. J. Lawrence, and V. E. Franklin-Tong
Wolfson Laboratory for Plant Molecular Biology, School of Biological Sciences, University of Birmingham, Edgbaston, Birmingham B15 2TT, United Kingdom

The potential to exploit self-incompatibility (SI) in plant breeding and its attraction as a system in which to investigate the molecular basis of cell–cell recognition and signaling has resulted in it becoming one of the most intensively studied and exciting topics in plant biology. Much research has encompassed the molecular cloning of genes involved in SI; consequently many reviews of this subject have focused on this area. In this article we have attempted to strike a somewhat different balance, with the objective of giving as broad a picture of the topic of SI as is possible within the confines of a single article. We provide a comprehensive summary of the genetics and population genetics of SI and, in describing progress toward the elucidation of the molecular basis of SI, have surveyed a much wider range of species than has previously been reviewed.

KEY WORDS Self-incompatibility; Cell–cell recognition (higher plants); S gene; S-linked glycoprotein.

I. Introduction

The process of pollination and fertilization in flowering plants involves a series of interactive events between male and female cells, one of the earliest of which is the recognition and consequent acceptance or rejection of pollen grains alighting on the stigma of the recipient plant. Pollen may be rejected either because it is too dissimilar, as when it originates from an individual of another species (interspecific rejection); or because it is too similar, as when it originates from the same plant (intraspecific rejection). This article is concerned with rejection of the second kind, which, following Stout's (1916) proposal, is now referred to as self-incompability (SI).

According to East and Park (1917), Kolreuter was the first to discover SI in *Verbascum phoenicium* in 1764, an observation that was later confirmed by Darwin (1876). Since that time, it has become clear that SI is one of the most common mechanisms preventing self-fertilization in the flowering plants. Thus, East (1940), who examined more than 800 species from 44 orders of monocotyledons and dicotyledons, found that 19 of these orders contained species whose SI had been definitely established and estimated that more than 3000 species of angiosperms were self-incompatible. Brewbaker (1959), noting that SI occurred in at least 71 families and had been recorded in more than 250 of 600 genera, estimated that between one-third and one-half of all species of flowering plants were self-incompatible. Although these estimates must be regarded as provisional, because of the difficulty in establishing that a species is definitely self-incompatible, there is little doubt that SI is broadly comparable, as an outcrossing device, to the sexual dimorphism in the animal kingdom.

In addition to their importance as outcrossing devices, there are several further reasons why SI is of interest to an unusually wide range of biologists, from the mathematical geneticist to the molecular biologist and the biochemist. First, the inheritance of a number of these systems has been elucidated. Second, it is possible to deduce the basic properties of these polymorphisms in natural populations from a knowledge of their inheritance alone because, as is the case with all mating-type polymorphisms, all are maintained by gene frequency-dependent selection. Third, highly specific recognition events take place when a pollen grain alights on a stigma, such that while the germination and growth of one may be inhibited, that of another, immediately adjacent pollen grain is not. Self-incompability, thus, serves as a model system in which to investigate cell–cell recognition and signaling in flowering plants. Fourth, the expression of S genes is subject to both spatial and temporal regulation; hence it provides the opportunity to study gene expression in flowering plants. Last, an understanding of the molecular biology of S genes would open up the possibility of being able to control their expression in self-incompatible crop plant species and of transferring them to self-compatible species, both of which would be of considerable interest to plant breeders in the production of hybrid varieties.

Despite its importance in the flowering plants, SI has, in the past, attracted less attention than it deserves. The subject has been reviewed, however, a number of times (Lewis, 1949a, 1954; Arasu, 1968; Heslop-Harrison, 1975, 1983), including a monograph by de Nettancourt (1977) on the subject. Although these reviews summarize a great deal of valuable information about the genetics and physiology of SI systems, they predate the advances in our knowledge of their molecular biology. More recent reviews (Clarke *et al.,* 1985; Cornish *et al.,* 1988; Ebert *et al.,* 1989; Nasrallah and Nasrallah, 1989; Dickinson, 1990; Haring *et al.,* 1990; Singh and Kao, 1992; Trick and Heizmann, 1992; Hinata *et al.,* 1993) have concentrated on the molecular biology of SI, including the cloning of S genes and the characterization of S gene products, in species of the solanaceous genera and those of the cruciferous genus *Brassica*. The rate at which new knowledge is being acquired about the molecular basis of SI in these and, in particular, in other species that are members of different families of flowering plants, makes it necessary to review this knowledge at frequent intervals. Furthermore, few of these reviews have attempted to cover much about the genetics of SI and none deal with the population genetics of the polymorphisms, despite the fact that their genetic properties determine the context of their molecular properties.

In this article, therefore, we attempt to provide a more comprehensive summary of the genetics and population genetics of SI, together with a broad survey of current state of knowledge on the molecular biology and biochemistry of SI in as wide a range of species as we could obtain information for. It has long been recognized that the identification and characterization of the properties of the S gene products on both the male and female side of the interaction is the obvious first step in unraveling how SI operates. We consider the advances made in identifying components involved in SI and, more critically, the evidence that they function in this role. Information on how SI operates in a range of species is beginning to emerge as work begins to examine how the various components operate in order to bring about the SI response, and we attempt to present models for the operation of SI in various systems where possible. It is a little early to speculate on how SI has evolved in flowering plants, but advances over the last few years have given us a much better idea of the possibilities.

A. Genetics and Distribution of Self-Incompatibility

Although all SI systems depend on an inherited capacity of an individual to reject its own pollen, as well as that of any other individual whose pollen has the same incompatibility phenotype, the majority can be classified on the basis of three criteria, namely, the genetic control of the pollen phenotype, floral morphology, and the number of genes involved. Thus, the incompatibility phenotype of the pollen may be determined either by its own haploid genotype, in which case control is gametophytic; or by the diploid genotype of the pollen parent, in which case control is described as sporophytic. With gametophytic determination, we expect (and observe) segregation of the incompatibility alleles in the pollen; where control is sporophytic, all pollen of an individual has the same phenotype. Where control of the pollen phenotype is sporophytic, the incompatibility alleles may display dominance in either the pollen or stigma or both, whereas these alleles always act independently in the stigma or style of species with gametophytic systems.

The second criterion of classification, that concerning floral morphology, distinguishes two reasonably well-defined groups according to whether incompatibility is associated with floral differences or not, the former being referred to as heteromorphic and the latter as homomorphic species. In most heteromorphic species it is possible to determine the incompatibility phenotype of an individual by examining the form of its flowers. The flowers of most self-incompatible species, however, are homomorphic. It follows, therefore, that the incompatibility phenotype of individuals of homomorphic species cannot be determined by visual inspection; this must be done by crossing each plant with every other of a family or sample taken from a population of the species in question. The classification of the incompatibility phenotype of individuals of a homomorphic species is, thus, much more laborious than that of individuals of heteromorphic species.

Taken together, these two criteria can be used to classify most of the known incompatibility systems into one or the other of the three major groups shown in Table 1. The fourth combination of heteromorphy with gametophytic control of the pollen phenotype does not exist because it would be unworkable. The third criterion, that concerning the number of genes involved, can be used to classify incompatibility systems within each of these three major groups. These are briefly described below.

1. Heteromorphic Systems

There are two kinds of heteromorphic species: those with two forms (distyly) and those with three forms (tristyly) of flowers. Distyly is controlled by a single gene with two alleles, S and s, the genotype of individuals bearing flowers with short styles (thrums) usually being Ss and that of plants bearing long-styled flowers (pins) ss, short usually being dominant to long style (Bateson and Gregory, 1905). Compatible matings are either Ss × ss or ss × Ss, so that the polymorphism is maintained each generation by a recurrent backcross in a way similar to the sexual dimorphism. Distyly occurs in 24 families and more than 164 genera of flowering plants, the most prominent of which are the Primulaceae, the Rubiaceae, and the Plumbaginaceae (Vuilleumier, 1967; Ganders, 1979; Richards, 1986).

Tristyly in most species that have been investigated is controlled by two genes, each of which has two alleles (S,s and M,m), the first of which determines whether the style is short (S) or not short (s) and the second,

TABLE I
Classification of Self-Incompatibility Systems on the Basis of the Control of Pollen Phenotype, Floral Morphology, and Number of Genes Involved[a]

	Control of pollen phenotype	
Floral morphology	Gametophytic	Sporophytic
Heteromorphic		One or two loci; two alleles at each
Homomorphic	One to four loci; many alleles at each	One to three loci; many alleles at each

[a] After Lewis (1954).

given the latter, whether the style is mid (M) or long (m); that is, S is epistatic to M (Barlow, 1913, 1923). Thus, in diploid species long-styled plants are ssmm, mids are ssM–, and short-styled plants are S––––. Each flower has two whorls of stamens that are borne at the levels not occupied by the stigma. Compatible matings involve the transfer of pollen from a given level in one flower to stigmas at the same level in another (e.g., pollen from the middle whorl of stamens of long- and short-styled plants to the stigmas of midstyled plants); all other transfers are incompatible. The determination of incompatibility phenotype in tristylous species is considerably more complicated than in distylous ones, because each plant produces two kinds of pollen, one from each level not occupied by the stigma, and plants of different genotypes produce pollen of the same incompatibility phenotype. Tristylic species are less widely distributed than distylic ones, being confined, apparently, to the Lythraceae, Oxalidaceae, and the monocotyledonous family Pontederiaceae. Heterostyly has been the subject of a major and extensive collaborative monograph edited by Barrett (1992), from which much interesting information can be obtained.

2. Homomorphic Systems with Gametophytic Control of Pollen

Self-incompatibility in *Nicotiana sanderae,* the species in which this system was first discovered, is determined by a single gene, S, with multiple alleles, S_1, S_2, S_3, . . ., S_k (East and Mangelsdorf, 1925). Control of the pollen phenotype is gametophytic and the S alleles are expressed independently in the style or stigma. Incompatibility occurs when the S allele carried by a pollen grain is matched by the same allele in the pistil on which it alights. All individuals in a population are, for this reason, heterozygotes, S_iS_j, so that each produces two kinds of pollen, half of which is expected to carry S_i and the other half S_j. Three kinds of pollination can be recognized with this system of SI, any one of which is either incompatible, when both S alleles in the pollen are matched by the pair in the stigma or style; fully compatible, when no alleles in the pollen are matched; or half-compatible, when while one of the S alleles in the pollen is matched by one in the stigma the other is not. This system probably occurs in more species of flowering plants than any other, being found not only in the Solanaceae, but also the Rosaceae, Leguminoseae, Scrophulariaceae, Onagraceae, Campanulaceae, Papaveraceae, and the monocotyledonous family Commelinaceae.

Self-incompatibility in the grasses is controlled by two, independently inherited, multiallelic loci, S and Z. This system, discovered independently by Lundqvist (1956) in *Festuca pratensis* and by Hayman (1956) in *Phalaris coerulescens,* is similar to the one-locus system, except that it involves the joint effect and segregation of the alleles at two loci rather than one. Thus, whereas the genotype of an individual of a species with a one-locus, gametophytic system is S_iS_j, that of an individual of a species with a two-locus system is $S_iS_jZ_kZ_l$ or $S_{i,j}Z_{k,l}$ for short. Because any pollen grain, being haploid, contains just one S and one Z allele, individuals that are heterozygous at both loci produce four, equally frequent kinds of pollen whose genotypes are S_iZ_k, S_jZ_l, S_jZ_k, and S_jZ_l. The incompatibility phenotype of each pollen grain is determined by the alleles it carries, whose products combine to form a unique specificity; that is, a pollen grain carrying S_i and Z_k alleles has an S_iZ_k phenotype. Similarly, on the female side, because the alleles act independently at each locus, the stigmas of an individual that is heterozygous at both loci form the same four specificites, S_iZ_k, S_iZ_l, S_jZ_k, and S_jZ_l, as in the pollen produced by that individual. Incompatibility occurs when the S–Z pair of alleles carried in the pollen is matched by the same S–Z pair in the stigma on which it alights. In addition to the three kinds of pollination that can be recognized in species with a one-locus system of SI, a fourth, three-quarters compatible, can be observed in the grasses, which occurs when the pair of individuals crossed have one S and one Z allele in common. Unlike those of species with a one-locus system, the individuals of a population of a species with a two-locus system may be homozygous at either the S or Z locus, but not both.

Although the two-locus SI system appears to be confined to the Gramineae, more complex systems have been found in other families. Thus in *Ranunculus acris, R. bulbosus,* and *R. polyanthemos* (Ranunculaceae), and in *Beta vulgaris* (Chenopodiaceae), SI appears to be controlled by four multiallelic genes that are linked in their inheritance (Lundqvist, 1990a; Lundqvist *et al.,* 1973); and Lundqvist (1991) has suggested that in *Lilium martogon* (Liliaceae) at least three loci are involved. The discovery that the genes are linked in these multilocus systems is unexpected, because the efficiency of these systems, in terms of the reduction of cross-incompatibility, is lower than if they were independently inherited. Because the elucidation of these complex systems is much more difficult than when SI is controlled by one or two genes, it is possible that species possessing multilocus, gametophytic systems are more numerous than appears at present.

3. Homomorphic Systems with Sporophytic Control of Pollen

The third, major system of SI, that in which self-incompatibility is determined by a single, multiallelic gene S but in which, unlike *Nicotiana,* the expression of this gene on the male side is sporophytic, was first fully analyzed by Hughes and Babcock (1950) in *Crepis foetida* and by Gerstel (1950) in *Parthenium argentatum.* Bateman (1955) has pointed out, however, that the data of Correns (1912) from *Cardamine pratensis* and that of Riley (1932) from *Capsella bursa-pastoris* can be as easily be explained by a one-locus model as by the two-locus models advanced by these authors. The most extensively investigated species with this system of SI is *Brassica oleracea* (Bateman, 1955; Sampson, 1957; Thompson, 1957). Because determination of the pollen phenotype is sporophytic, any pollination is either incompatible or fully compatible; no partially compatible matings occur. The S alleles may act independently in the stigma and pollen, or display dominance, the latter generally being more common in the the pollen. A consequence of this dominance is that recessive homozygotes occur. For example, if S_1 is dominant to S_2 in the pollen, the cross $S_2S_3 \times S_1S_2$ is compatible and S_2S_2 homozygotes are expected in their progeny.

There are two further points worth making about the dominance of S alleles in this system of SI. First, in the absence of dominance, the incidence of cross-incompatibility both in full-sib families and in populations is considerably higher than in those of species with the one-locus, multiallelic gametophytic system. It is not surprising, therefore, that with one exception (*Cerastium arvense* subsp. *strictum;* Lundqvist, 1990b) the S alleles of all species examined so far display dominance, which has, presumably, evolved to reduce this cross-incompatibility. If so, dominant alleles are more recent than the recessive alleles from which they have evolved. Second, while originally the alleles of each of these species could be arranged into a linear dominance series, those of the highest level being dominant to those of lower levels (Lewis, 1954), this was probably a consequence of the small number of alleles examined. Thus, although S_2, S_5, and S_{15} appear to be recessive to all other alleles in the pollen of

B. oleracea, it is not possible to assign other alleles to a hierarchy of dominance levels (Thompson and Taylor, 1966), because dominance, in general, is nonlinear in both the stigma and pollen. The same appears to be true for other species that have been examined in sufficient detail, such as *Sinapis arvensis* (Ford and Kay, 1985; Stevens and Kay, 1989). There is no obvious reason why dominance should be linear. The only restraint to which the evolution of these alleles must be subject is that the dominance relationships of a pair of alleles in the pollen should not be reversed in the stigma, because if this were the case an individual heterozygous for this pair would be self-compatible. Kowyama *et al.* (1994) have, however, shown that it is possible to arrange 28 S alleles of *Ipomoea trifida* into 6 major levels of dominance that are nearly identical in the stigma and pollen. It is not clear at present whether this remarkable result is due to the fact that *I. trifida* is a self-incompatible species in which it is impossible to obtain seed by selfing, so that the analysis of dominance relationships in this species is not vitiated by the variability and weakness of expression of the S alleles typical of those of *Brassica* species, or to some other as yet unknown reason.

Although the one-locus, multiallelic, sporophytic system appears to be less widely distributed than the one-locus gametophytic system of SI, it has been found not only in the Compositae and Cruciferae, but also in the Convolvulaceae (Martin, 1968; Kowyama *et al.,* 1980), Betulaceae (Thompson, 1979; Germain *et al.,* 1981; Me and Radicati, 1983), Caryophyllaceae (Lundqvist, 1979, 1990b), and the Sterculiaceae (Jacob, 1980).

Several elaborations of one-locus, sporophytic control of SI have been reported in the literature. Verma *et al.* (1977) reported that in the cruciferous species, *Eruca sativa,* SI is determined by three genes and, more recently, Lewis *et al.* (1988) and Zuberi and Lewis (1988) have shown that in *Raphanus sativus* and *Brassica campestris* SI is controlled by two linked genes, one of which is the long-established S gene with sporophytic expression in the pollen and the other, G, is expressed gametophytically in the pollen. Both the S and G genes must be matched for incompatibility. G in both species appears to have a low number of alleles that are fully expressed only in some S allele combinations. They argue that G is an ancestor of the S gene in the gametophytic system of dicotyledons that has been retained in the sporophytic system as an essential component of the incompatibility process. The implications that these extraordinary results and arguments have for both the molecular analysis of SI in *Brassica* and an understanding of the evolution of self-incompatibility in the flowering plants are obvious and it is highly desirable that these results should be independently confirmed as soon as possible. There is one other case in which expression of the genes concerned appears to be both sporophytic and gametophytic; this is the complex system of SI analyzed by Knight and Rogers (1953; 1955) and Cope (1962) in *Theobroma cacao*, in which inhibition of incompatible pollen appears to take place in the ovary, rather than in the pistil, and involves the interaction of gametophytes, rather than that of the male gametophyte with the female sporophyte.

4. Evolution of Self-Incompatibility Systems and Structure of S Genes: Genetic Evidence

Three pieces of evidence from the genetic analysis of SI systems are of particular relevance to an understanding of their molecular biology. First, although Whitehouse (1950) and Crowe (1964) have argued that incompatibility arose only once in the flowering plants and that all known systems have evolved subsequently from a primitive, multiallelic, gametophytic system of SI, the considerable diversity of their genetic determination suggests that self-incompatibility has probably arisen independently at least several times. Thus, with few exceptions, all self-incompatible species of a family appear to possess the same system, which can be more easily accommodated on the assumption that each system arose during the evolution of its family from a self-compatible ancestor, despite our difficulties in understanding how and in what circumstances this might have occurred. Indeed, Lewis (1976) has suggested that the reason why the members of taxonomically quite different families possess what now appears to be genetically the same system is that, although SI has evolved independently in each family, the rigid requirements of the system are such that the same end point is reached by convergence. These arguments lead to the expectation that the molecular biology of SI will be found to differ not only between species possessing different systems, but also between species of different families that appear to possess the same genetic system. This expectation provides, therefore, a critical test of the polyphyletic origin of SI in the flowering plants.

Second, although determined attempts have been made by Lewis (1948, 1949b, 1951) with *Oenothera organensis* and *Prunus avium*, and more recently, by Hayman and Richter (1992), with *Phalaris coerulescens*, to obtain them, nobody has yet unambiguously identified a permanent mutation from one fully functional S allele to another. The reason for this paradoxical result is unknown. These investigations, however, have shown that it is relatively easy to obtain mutants that have lost function, the study of which has provided important evidence about the structure of these loci. Thus, Lewis (1948, 1949b, 1951) and Lewis and Crowe (1954) in a study of spontaneous and X-ray-induced mutants of the S gene of

Oe. organensis and *P. avium* identified two classes of permanent mutants: those in which a previously fully functional allele had lost its activity in the pollen, but not the style, described as pollen part mutants; and those in which this loss of function was confined to the style, described as stylar part mutants. These results led Lewis (1960) to propose that the S gene is a supergene (Darlington and Mather, 1949) consisting of three tightly linked parts that control allelic specificity in both the style and the pollen, a pollen part controlling expression of the gene in the pollen, and a stylar part that controls expression of the gene in the style, respectively. This tripartite model has become the paradigm for most of the current investigations of the molecular biology of SI.

Third, the effect of polyploidy on the expression of SI has revealed some information about the products of incompatibility genes in species with gametophytic systems of SI. Polyploid grasses, both natural and induced, are as self-incompatible as their diploid relatives and parents. Fearon *et al.* (1984a,b,c) have shown that only one S–Z pair of alleles in the diploid pollen of tetraploid *Lolium perenne* must be matched by the same pair in the stigma for incompatibility to occur; that is, the specification of the incompatibility phenotype of the pollen of tetraploids is exactly the same as that in diploids, despite the presence of more than one S allele and more than one Z allele in the former. Autotetraploids of monocotyledonous species in which SI is determined by a single, multiallelic gene, such as *Tradescantia paludosa* (Annerstedt and Lundqvist, 1967) and *Ananus comosus* (Collins, 1961), and those of dicotyledonous species that possess multilocus systems, such as *Ranunculus acris* (Østerbye, 1975, 1977) and *Beta vulgaris* (Larsen, 1977), are also self-incompatible. In contrast, polyploidy in dicotyledonous species that possess a one-locus system causes a partial breakdown of SI in some, but not all, genotypes (Lewis, 1947; Brewbaker, 1954). Analysis of colchicine-induced tetraploids of *Oenothera organensis* and *Trifolium repens* showed that this breakdown was confined to the pollen and that it was caused by the competitive interaction of S alleles in the heterogenic pollen. Lewis (1976) has argued that these results suggest that whereas the polypeptide products of S alleles in the style are monomeric, those in the pollen are multimeric and that the products of the S alleles in heterogenic pollen of some genotypes combine to form a heteromultimer not recognized by the S monomers of the style. In the case of the grasses, however, the products of the incompatibility alleles in both stigmas and pollen must be dimeric polypeptides, one component of which is specified by the S locus and the other by the Z locus, so as to form a unique S–Z product. A similar explanation can be advanced for the products of the alleles of multigenic systems. The immunity of *Tradescantia paludosa* and *Ananus comosus* to polyploidy may be due to their possessing a two-locus system, like the grasses, only one of which is segregating, the other being fixed.

B. Population Genetics of Self-Incompatibility

One of the most attractive features of the self-incompatibility polymorphisms is that it is possible to deduce their properties from a knowledge of their inheritance alone. Among the more interesting of these properties, first deduced by Wright (1939), and listed for the one-locus, gametophytic system by Campbell and Lawrence (1981a) and for the two-locus system by Fearon *et al.* (1994), are the following: (1) the selective advantage of an allele is negatively related to its frequency in the population because the polymorphism is maintained by frequency-dependent selection; hence (2) the number of alleles k in a population and, hence, in the species is potentially large; and (3) provided that the effect of selection on the locus is limited to that associated with incompatibility, the equilibrium frequency of each of the k alleles present in the population is approximately $1/k$; however, (4) the selective advantage of a new allele that appears in the population by mutation or migration is also negatively related to the number of alleles already present.

Deductions 1, 2, and 4 apply to all multiallelic polymorphisms. Deduction 3, however, holds only for the gametophytic polymorphisms. When expression of the S gene in the pollen is sporophytic, the frequency of alleles that are generally recessive to others is expected to be higher at equilibrium than that of alleles which are generally dominant. These deductions can be used as predictions or null hypotheses against which the results obtained from the investigation of the properties of the SI polymorphisms in natural populations can be compared. Because of the considerable labor involved, there are fewer investigations of this sort than is desirable. Among the most extensive of those concerning species with gametophytic systems are the pioneering investigation of Emerson (1939, 1940, 1941) on the rare endemic *Oenothera organensis,* Atwood (1944) on *Trifolium repens,* Williams and Williams (1947) on *Trifolium pratense,* Campbell and Lawrence (1981b) and Lawrence and O'Donnell (1981) on *Papaver rhoeas,* and Fearon *et al.* (1994) on the grass *Lolium perenne*. The two most extensive studies of the polymorphism in natural populations of species with sporophytic systems are those of Ford and Kay (1985) and Stevens and Kay (1989) on *Sinapis arvensis,* and Kowyama *et al.* (1994) on *Ipomoea trifida*. The most thoroughly studied species with a sporophytic system, *Brassica oleracea* and its allies, is of limited interest for present purposes because nearly all of the information concerning the number, frequency, and dominance relationships of S alleles comes from the investigation of commercial, rather than natural, populations of these species (Ockendon, 1974, 1985).

1. Allele Frequencies

Whereas the frequencies of the incompatibility alleles in the *Oenothera organensis* population (Emerson, 1939; Campbell and Lawrence, 1981b) and probably also in the clover populations (Atwood, 1944; Williams and Williams, 1947) appeared to be approximately equal, as the theory suggests, those in three populations of poppies (Campbell and Lawrence, 1981b; Lawrence and O'Donnell, 1981) and in a population of *Lolium perenne* (Fearon *et al.,* 1994) were found to be significantly unequal. Although the possibility that the unequal allele frequencies in these poppy populations was caused by sampling effects over and above that due to drift cannot be ruled out (Lawrence *et al.,* 1994), the results obtained from an analysis of disturbed segregation ratios in a number of full-sib families suggested that these inequalities were more likely to be caused by an extra effect of selection acting on the S locus via linkage to other genes that were the chief target of this additional selection, among which were those determining seed dormancy (Lawrence and Franklin-Tong, 1994; Lane and Lawrence, 1995). The disturbances in the majority of these families indicated that the extra effect of selection was of the gametic, rather than the zygotic, type and that this selection was confined to the female side of the cross. The S alleles in this species appear, therefore, to be subject to several kinds of selection in addition to the frequency-dependent selection that maintains the polymorphism. The cause of the unequal allele frequencies at both the S and the Z loci in the ryegrass population is unknown, although it is possible that these are grossly out of equilibrium in a population that persists by vegetative, rather than sexual, reproduction (Fearon *et al.,* 1994). These results suggests that unequal S allele frequencies may be more common than has hitherto been supposed.

In the absence of a detailed knowledge of the dominance relationships between alleles both in the stigma and pollen, it is not possible to apply a statistical test to the frequencies of these alleles in species with a sporophytic system of SI. Nevertheless, as expected, the frequencies of alleles that were generally dominant in the pollen of *Sinapis arvensis* appeared to be lower than those that were generally recessive (Stevens and Kay, 1989). This effect was much more marked in the six populations of *Ipomoea trifida* investigated by Kowyama *et al.* (1994), in all but one of which the allele that is recessive to all others, both in the stigma and pollen, S_3, was the most frequent.

2. The Number of Alleles in a Population

The number of alleles in a population of a self-incompatible species depends on the number of alleles in the species and its effective size (Wright, 1939), neither of which will be known. The number of alleles estimated to be present in a population will also depend on the kind and size of the sample drawn from it. Comparisons between the results obtained from the investigation of the number of alleles in populations of different species are, for these reasons, not easy. It is surprising, therefore, that the populations of most of the gametophytic species that have been investigated appear to contain a similar number of alleles. Thus, Emerson (1940) ultimately found 45 different S alleles in the *Oenothera organenesis* population, each of 3 *Papaver rhoeas* populations were estimated to contain between 40 and 45 alleles (O'Donnell and Lawrence, 1984; Lawrence *et al.*, 1993), and *Lolium perenne* population was estimated to contain not less than 40 alleles at each locus (Fearon *et al.*, 1994). Indeed, a population of at least one species with a sporophytic system of SI also appeared to contain a similar number of alleles, because Stevens and Kay (1989) found 35 different S alleles in 35 plants raised from seed taken at random from a population of *Sinapis arvensis* in South Wales. Populations of *Ipomoea trifida,* however, contain only about half this number of alleles, because even the 4 central populations examined by Kowyama *et al.* (1994) contained only between 15 and 21 different S alleles in 38 to 41 plants, despite the fact that the thoroughness of these studies, measured by repeatability (Campbell and Lawrence, 1981a), was high.

Populations of clover species, however, appear to contain many more alleles than those of other species. Atwood (1944) identified 36 different alleles, of a total of 49 examined in a sample taken from one natural population of *Trifolium repens,* and 39 of 49 from another population; and Williams and Williams (1947) found 41 of 48 and 35 of 38 different alleles in samples taken from two nonpedigree strains of *Trifolium pratense*. Although the repeatability of these surveys is much lower than others and their data have yet to be fully analyzed, estimates of the number of alleles in these populations are at least three times those contained in populations of other species. O'Donnell and Lawrence (1984), for example, have shown that the maximum likelihood estimates of the number of alleles in the *T. pratense* populations are 156 and 294, respectively. Why populations of clovers appear to contain many more S alleles than those of other species is not known. It is worth pointing out, however, that the widespread belief that populations of self-incompatible species contain "hundreds" of S alleles is based solely on the evidence from these clovers and that the evidence from the more thoroughly investigated species suggests that populations of these are unlikely to contain more than 50 alleles.

3. The Distribution of Alleles between Populations

If deduction 2 (above) is true, any population, unless it is long established and large, is expected to contain only a subset of these alleles. It follows, therefore, that different populations are expected to contain different subsets of these alleles. To test this prediction, it is necessary to analyze samples from two or more populations, to cross-classify the alleles identified in each against those of the other(s), and to apply a statistical procedure to these data in order to determine the overlap between the complements of alleles these populations contain. To date, this procedure has been used on two species only, *Papaver rhoeas* and *Ipomoea trifida*. The cross-classification of the S alleles of samples taken from three British populations of poppies revealed that each contained essentially the same complement of alleles (Lawrence *et al.*, 1993; O'Donnell *et al.*, 1993). In a discussion of these unexpected results, it was argued that this outcome was due to there being only a limited number of alleles in the species as a whole, rather than to any local limitation of this number in the peripheral populations of the British Isles, and that the chief cause of this limitation was the attenuation of the strength of frequency-dependent selection which Lawrence *et al.* (1994) later showed became weak once a population contained 40 alleles. This argument received some support from the results obtained from the partial cross-classification of the alleles from one of the British populations of poppies against those of a sample taken from a population in Spain, which showed that an estimated 53% of the alleles of the latter also occurred in the former (Lane and Lawrence, 1993). The results obtained from the complete cross-classification of the alleles of the six populations of *Ipomoea trifida* investigated by Kowyama *et al.* (1994), in contrast, indicate considerable differentiation between populations with respect to the alleles they contain. None of the samples taken from these populations contained more than 21 of a total of 49 alleles identified overall and each contained alleles that occurred in no other sample. Because of the difficulty of devising a procedure that would allow the estimation of the number of alleles in populations of species with a sporophytic systems of SI, it is not possible, as with the poppy data, to estimate the overlap between the complements of alleles that these populations of *I. trifida* contain. Nevertheless, one of the most striking features of these results is that the samples from the two peripheral populations, whose sizes were only marginally smaller than the others, contained only one-third the number of alleles contained by those from the four central populations.

4. The Number of S Alleles in the Species

Our discussion of the number of alleles in a population has provided some information on the number of alleles in the species, because the latter cannot be smaller than the former. Indeed, because alleles have been systematically cross-classified over two or more populations only in *Papaver rhoeas* and *Ipomoea trifida*, the best estimate of the number of alleles in most species is the same as that of the number in their populations. In the case of *P. rhoeas*, the data obtained from the cross-classification of the alleles of a sample taken from a Spanish population against those of a British population were also used to obtain an approximate estimate of the number of alleles in the species. This estimate of 66 alleles is, for reasons given by Lane and Lawrence (1993), almost certainly biased upward, by an unknown amount; the three British samples contained among them 45 different alleles. As mentioned previously, the samples taken from the six *I. trifida* populations contained a total of 49 different alleles (Kowyama *et al.*, 1994). On the present evidence, then, the species that have been investigated appear to fall into two groups with respect to the number of alleles they contain: the clovers, whose species appear to possess several hundred alleles, falling into the first group and all of the other species, which possess 40–70 alleles, into the second. *Brassica oleracea* can also be placed in the second group, because Ockendon (1985) states a total of 49 different alleles have been identified in a wide range of cultivars. Two chief points emerge from this comparison. First, the classification of species in terms of the number of incompatibility alleles they appear to possess cuts across the gametophytic/sporophytic divide. Second, most self-incompatible species, despite the fact that they differ widely in their life histories, reproductive biology, and ecology, appear to possess a similar number of alleles. This suggests that the extent of these polymorphisms may be subject to a dynamic restraint, because of the negative relationship between the selective advantage of a new allele and the number of alleles already present in a population (Lawrence *et al.*, 1994).

C. Self-Incompatibility: A Model System for Studying Cell–Cell Recognition and Signaling

It has been long assumed that pollination and self-incompatibility involve cell–cell recognition and signaling to mediate the interaction between the stigma and pollen that lands on it. Clearly the precise nature of the interaction during both fertilization and self-incompatibility requires sophisticated control. In addition, given the infrequency of the breakdown of SI, it must be robust. One might anticipate that for each S allele there is a specific receptor-binding site, which only recognizes and interacts with a molecule produced by that particular S allele. However, until recently, there was little evidence to substantiate this concept. Part of the reason for this has been the lack of evidence for signal transduction pathways operating in plant cells. Although many of the components of signaling pathways that have been found in mammalian cells have now also been detected in plant cells, it is not straightforward to show that the components actually function in the same way. Unequivocal identification of a receptor in a plant cell, together with a demonstration that it acts in concert with its ligand, is still elusive. Thus, although the concept still stands, we are so far no nearer to identifying the pollen "receptor" involved in the SI response.

Nevertheless, during the last few years strong evidence has been gathered that suggests that signal transduction plays an important role in cell–cell signaling in plant cells. One major mechanism for signal transduction in animal cells, the "phosphatidylinositol response" (Michell, 1975), involves hydrolysis of phosphoinositides into diacylglycerol and inositol trisphosphate, which act as second messengers (Berridge and Irvine, 1989). This pathway also operates in plants, although there are relatively few examples to date, owing mainly to difficulties in the detection of intermediates of the pathway. Another major mechanism for signal transduction in both animal and plant cells involves the transient release of Ca^{2+} into the cytosol (Hepler and Wayne, 1985; Trewavas and Gilroy, 1991). Cytosolic free Ca^{2+} ($[Ca^{2+}]_i$) acts as a second messenger, activating Ca^{2+}-dependent protein kinases, which causes substrate alteration by phosphorylation, eliciting a "response." This has attracted much attention because it is well established that Ca^{2+} plays a role in pollen germination and pollen tube growth (Brewbaker and Kwack, 1963; Jaffe *et al.*, 1975; Mascarenhas and Lafountain, 1972; Reiss and Herth, 1978; Picton and Steer, 1983). It has been suggested that Ca^{2+} channels function in pollen tubes because Ca^{2+} is known to be taken up from the stigma by pollen grains during germination (Bednarska, 1989) and inhibition of this uptake inhibits pollen tube growth (Picton and Steer, 1985; Reiss and Herth, 1985). It has been established that growing, but not nongrowing, pollen tubes have a gradient of $[Ca^{2+}]_i$ in their tips (Obermeyer and Weisenseel, 1991; Rathore *et al.*, 1991; Miller *et al.*, 1992), which suggests that Ca^{2+} signaling may be involved in pollen tube growth, because maintenance of this gradient appears to be essential for pollen tube growth. Calcium-dependent protein kinases have been identified in germinated pollen of *Nicotiana alata* (Polya *et al.*, 1986), which further suggests that a Ca^{2+}-mediated signal transduction pathway may operate during pollination. Nevertheless, much of the evidence for a Ca^{2+}-mediated signaling system remains circumstantial, because interference with Ca^{2+} levels in the cell usually results in pollen tube inhibition.

Information about the various signal transduction components involved in SI has in the past been limited. The levels of calcium in incompatible and compatible pollen grains of *Brassica oleracea* have been investigated using chlorotetracycline, which reports membrane-associated calcium, and energy-dispersive analysis of X rays, which measures total calcium (Singh *et al.*, 1989). However, although incompatible pollen grains were found to contain higher levels of calcium than compatible grains, these data do not provide evidence that calcium is acting as a second messenger because $[Ca^{2+}]_i$ levels were not measured. The cloning of an S receptor kinase (SRK) gene from *B. oleracea* that resides at the S locus (Stein *et al.*, 1991) provided the first firm evidence implicating the involvement of a signal transduction pathway. The finding that this gene is required for the SI response (Nasrallah *et al.*, 1994) provides good evidence for its involvement in SI. This is discussed in more detail in Section III. Further evidence that signaling is involved in the SI response comes from work on *Papaver* and, more recently, on *Secale*. Rapid, transient phosphorylation of pollen proteins occurring as a result of an SI reaction (Franklin-Tong *et al.*, 1992, 1993; Wehling *et al.*, 1994a) suggests a possible role for protein kinase activity in this reaction. Other experiments have provided evidence that the SI response in *Papaver rhoeas* is mediated by $[Ca^{2+}]_i$ acting as a second messenger (Franklin-Tong *et al.*, 1993). Further details are presented in Section IV. These exciting finds will doubtless stimulate further studies that should establish the role of signal transduction in the SI response, together with the pathways involved.

II Physiology of Self-Incompatibility

Although the general developmental and tissue-specific control of SI is well established, the basis for this is still far from clear. In most self-incompatible species, only after the pistil has reached a particular developmental stage (usually 1–2 days before floral maturation and anthesis) does SI begin to operate. This knowledge has had practical applications; for example, plant breeders commonly use bud pollination to obtain self-seed set, because SI is not expressed in the bud stage. With some of the other methods utilized for overcoming SI, the physiological basis of why they are effective is unclear. For example, exposing plants to CO_2 or high temperatures is often used to promote self-seed set in the breeding programs with *Brassica* and other crop plants.

There is much still to be discovered about the physiology of the SI reaction and response in the pollen and stigmatic cells. Most of the information to date has been derived from careful ultrastructural studies. One approach, which may complement these studies of *in vivo* pollinations, is to reproduce the SI response *in vitro* or *semi-vivo*. This provides a simplified way in which to investigate the components involved in the SI reactions, which could be helpful, because there is no possibility of the results being confounded by the presence of pistil and pollen tissue. Another approach, involving the study of the effect of metabolic inhibitors, may help to elucidate some of the mechanisms that are required for pollen germination and tube growth, as well as the SI response. Compatible pollen tubes from most species, whether they are from plants with sporophytic or gametophytic SI, appear to be similar. Investigations into the physiological requirements for normal pollen germination and tube growth may provide useful information for investigations into SI.

For some time now, it has been convenient to neatly categorize SI into "sporophytic" and "gametophytic" systems, and this is the way we have subdivided the SI physiological responses. The current knowledge about the physiology of the SI response in both the stigma and the pollen is outlined below. Some of the work that relates more directly to the molecules involved in the SI response is dealt with in detail below (Sections III and IV). It will begin to emerge that, although in the main this grouping of types of SI stands up to scrutiny, there are clear exceptions to the rule. This theme is elaborated on toward the end of this article.

A. The Self-Incompatibility Response in Sporophytic Systems

The physiology of the SI reactions of all the sporophytic SI species so far examined (*Brassica* and *Ipomoea*) appears to be similar. The stigma surface is "dry" and the pollen trinucleate. Self-pollen is generally inhibited early, either before or just after germination; indeed, there is evidence that recognition events in *Brassica* may occur before the pollen grain hydrates (Sarker *et al.*, 1988). It also appears that this initial blocking of germination may not be irreversible, because self-pollen could be "rescued" by transfer to a compatible stigma several hours after pollination (Singh *et al.*, 1989; H. G. Dickinson, personal communication). Self-pollen tubes, if produced, penetrate the cuticle, but not the stigmatic papillar cells. There is much debate about the production of callose (a $\beta \rightarrow 1,3$ glucan) in the stigmatic papillar cells, which is localized at the point where incompatible pollen grains land. Although this is characteristic of an SI response, whether this deposition occurs as a result of an incompatible reaction or plays a part in causing the SI response is still open to question. Experiments by Singh and Paolillo (1989), using stigmas pretreated with 2-deoxy-D-glucose, which inhibits callose production in the stigma, suggest that the callose response in the stigmatic papillae is not essential for inhibition of incompatible pollen.

One of the major obstacles to the development of a reliable bioassay for *Brassica* has been the lack of a reliable germination medium for its pollen, which is reported to be difficult to grow (Roberts *et al.*, 1983; Ferrari *et al.*, 1981), reputedly because it is trinucleate. As a result, this approach has not often been used to investigate the physiological basis of SI for sporophytic species. Ferrari and co-workers (Ferrari and Wallace, 1975; Ferrari *et al.*, 1981) appeared to have established a good bioassay, but subsequent reports of bioassays for *Brassica* have suggested, until recently, that they have not been reliable. However, there are isolated reports that indicate that if the right conditions are found, this may be a useful technique. For instance, Roberts *et al.* (1983) reported that although addition of crude stigma extracts to *Brassica* pollen *in vitro* affected the rate of hydration of the pollen, they did not obtain any evidence for specific inhibition of self-pollen. More recently, Singh and Paolillo (1989) have performed detailed work aimed at perfecting an *in vitro* system for *Brassica*. They found that by pretreating stigmas of *Brassica* with hexane, biologically active stigmata S glycoproteins could be eluted that had discriminatory activity against self- and cross-pollen germination.

B. The Self-Incompatibility Response in Gametophytic Systems

It has been generally supposed that species with gametophytic SI systems, in contrast to those with sporophytic SI systems, have a "wet" stigma and stylar inhibition of pollen tube growth. The idea that all species with gametophytic SI have this type of morphology and physiology is a gross oversimplification, however, as there appear to be many different types of floral morphology found in plants that have gametophytic control of SI.

The main group, which is generally regarded as the "model" example for gametophytic SI (because it has been studied most intensively) is the group of solanaceous plants (*Nicotiana alata, Lycopersicon peruvianum, Petunia inflata,* and *Solanum tuberosum*) for which this description of a wet stigma and stylar inhibition is correct. This description also fits the morphology and physiology of SI for the Liliaceae (*Lilium longiflorum*) and Rosaceae (*Prunus avium, Malus domestica,* and *Pyrus serotina*), which also have wet stigmas. Pollen from this group is generally binucleate. The stigma is wet owing to exudates produced by the stylar transmitting tract, which includes arabinogalactans; arabinogalactans probably play a role in promoting pollen tube growth through the style (Clarke *et al.*, 1979; Mau *et al.*, 1982). On landing on the stigma, the pollen grain hydrates, germinates, and grows through the stigmatic papillae and intercellularly in the transmitting tract. Growth of incompatible pollen tubes is arrested by the time they have grown through a third of the stylar canal; the tip of the pollen tube often swells or bursts at this point. Thus, it appears that inhibition is apparently not rapid. In contrast, inhibition of incompatible pollen in the Papaveraceae, Poaceae, and Oenotheraceae occurs on or just beneath the stigma. In both the Papaveraceae and the Poaceae, the stigmatic surface is also "dry." Although the floral morphology of *Papaver rhoeas* is different from other self-incompatible species so far described, because there is no style, the flowers of other species in this group with stigmatic inhibition possess styles. It is likely that other genera, if examined with sufficient care, will exhibit stigmatic inhibition of incompatible pollen. Thus, these species should not be regarded as mere "curiosities," which do not conform to the "norm." In this group, inhibition of incompatible pollen appears to be rapid, occurring before tube emergence, during or just after pollen germination. A localized deposit of callose is deposited inside the pollen grain, at what appears to be the colpal aperture from which the pollen tube would have emerged. Pollen tubes that do emerge may vary in length, but are typically short and distorted, with a heavy deposition of callose at the pollen tube tip. It has already been suggested that this group with dry stigmas differs physiologically from the wet stigma gametophytic SI species (Heslop-Harrison and Shivanna, 1977). Further studies should help reveal the basis of this apparent difference.

Attempts to develop an *in vitro* bioassay for the SI response in species with gametophytic SI systems have generally met with more success than with those possessing sporophytic systems, which is attributed to the fact that binucleate pollen is much easier to grow in an artificial medium. Many of the earlier attempts reported S-specific effects of pistil extracts on pollen tube growth *in vitro* in *Petunia inflata, P. hybrida, Primula vulgaris, Prunus avium,* and *Lilium longiflorum* (Brewbaker and Majumder, 1961; Sharma and Shivanna, 1983; Williams *et al.*, 1982; Dickinson *et al.*, 1982); however, much of this work was not exploited further because reproducibility was low. Later work reported many problems with obtaining S-specific effects, especially with *Nicotiana alata;* many different bioassays have been tested for *N. alata,* although none has been entirely satisfactory (Sharma and Shivanna, 1986; Clarke *et al.*, 1985; Harris *et al.*, 1989; Jahnen *et al.*, 1989; Gray *et al.*, 1991) because although growth of the pollen of some genotypes was inhibited by the addition of self-S glycoproteins, those of others were not. Harris *et al.* (1987) tried a different approach to the problem of finding a reliable assay for the S glycoprotein, by using an enzyme-linked immunosorbent assay. They showed that the inhibition of *in vitro* growth of *N. alata* pollen was accompanied by a reduction in binding of a monoclonal antibody, whereas treatment with the S_2 glycoprotein caused an increase in binding. More successfully, an *in vitro* system for pollen has been developed for *Papaver rhoeas*

(Franklin-Tong *et al.*, 1988). Addition of either crude stigma extracts, purified stigmatic S proteins, or recombinant S gene product elicits the expected genotype-specific inhibition of pollen tube growth (Franklin-Tong *et al.*, 1988; 1989; Foote *et al.*, 1994). Although this is not a perfect system, as there are some nonspecific effects, the S-specific effects are clearly detectable over and above these; differences are statistically significant and results are repeatable with any genotypic combination examined. This *in vitro* system is robust and has been adapted and modified in order to study various physiological aspects of the SI reaction. Not only has it allowed the isolation and characterization of the stigmatic components involved in SI by the monitoring of fractions for S activity during the course of purification, but it has also permitted investigations into the pollen side of the reaction, which has hitherto been difficult to study. The system has also been used to analyze the effects of certain metabolic inhibitors, changes in gene expression, cytosolic Ca^{2+} levels, and protein phosphorylation, which occur in the pollen as a result of the SI reaction (see Section IV,C for further details).

It is possible that further information about the physiology of the SI responses in these species may help to explain some of the anomalies outlined above. It seems possible that the problems associated with obtaining clear-cut, reproducible SI reactions *in vitro* for *Nicotiana alata* may be due to the rapid response that is found *in vivo*. Thus, the difference between the type of SI response in *P. rhoeas* and *N. alata* may be crucial.

III. Sporophytic Self-Incompatibility

A limited number of species with sporophytic SI have been studied in any detail. Most studies have concentrated on species in the Brassicaceae: *Brassica oleracea* (cabbage, broccoli, cauliflower) and *B. campestris* (Chinese cabbage, oil seed rape) and *Raphanus sativus* (radish). More recently, a member of the Convolvulaceae, *Ipomoea trifida,* a wild relative of *I. batatas* (sweet potato), has also been studied at the molecular level. In this section, we describe the biochemical and molecular cloning data with respect to the S genes and their products, together with the evidence that they operate in the SI response.

A. Cell and Molecular Biology: *Brassica oleracea* and *Brassica campestris*

1. Identification of S-Linked Glycoproteins from the Pistil

The first S-associated proteins to be identified were from *Brassica oleracea* (Nasrallah and Wallace, 1967a,b; Nasrallah *et al.*, 1970, 1972), which has a sporophytic self-incompatibility system. These early studies were based on immunological evidence. Using antigens and antisera raised to stigma extracts of *B. oleracea* for immunodiffusion tests, specific patterns attributable to S alleles were detected. These S genotype-specific antigens are genetically determined because parental S antigens were found in their F_1 and F_2 progenies; they were also shown to be tissue specific and the quantity of S antigen correlated with both the development of self-incompatibility and with the strength of the self-incompatibility allele in question (Nasrallah and Wallace, 1967a,b). A specific protein was first correlated with the S_2 allele of *B. oleracea* (Nasrallah *et al.*, 1972) and further S alleles were identified by Sedgley (1974a,b) and Nishio and Hinata (1977a,b, 1978a,b). These bands segregating with S alleles in *Brassica* were found to be glycoproteins (Nasrallah *et al.*, 1970; Nishio and Hinata, 1978b). Although there was no evidence at the time that these S-linked molecules were involved in the SI reaction, the information about them implicated a role in SI, making them strong candidates for the products of the S gene. It is interesting to note that the identification of S-linked glycoproteins is not always easy, because some S genotypes produce more than one S-specific band (Nishio and Hinata, 1977a, 1978b), whereas no S-linked proteins could be detected in other S genotypes (Nishio and Hinata, 1978b). This phenomenon may be related to the strength of the S allele; Sedgley (1974a,b) has observed that antisera can apparently be raised only from S alleles that are high in the dominance series and has suggested that these were either more antigenic or more abundant.

Purification of the first S-specific glycoprotein was achieved in *Brassica campestris* by Nishio and Hinata (1979), using concanavalin A (ConA)–Sepharose affinity chromatography and isoelectric focusing. The resultant glycoprotein had an apparent molecular mass of 57 kDa and was postulated to be the specific product of the S_7 allele, because it was found only in stigmatic tissue and cosegregated with the S_7 allele. Evidence that the S-specific glycoproteins might actually be involved in the SI reaction was provided by Ferrari *et al.* (1981), who purified the S_2 glycoprotein from stigmas of *B. oleracea* which had characteristics similar to those of the S glycoprotein from *B. campestris*. This S_2 glycoprotein, when applied to S_2S_2 pollen, prevented it from germinating on three types of compatible stigmas, whereas pretreatment of S_3S_3 and S_8S_8 pollen with the S_2 glycoprotein did not interfere with their compatibility. Thus, this molecule was shown to have S-specific inhibitory biological activity. Other S-associated proteins have since been identified or isolated from *Brassica* (Roberts *et al.*, 1979; Nishio and Hinata, 1982; Takayama *et al.*, 1986a; Isogai *et al.*, 1987). In 1980, a method for quickly identifying S glycoproteins using ConA–fluorescein isothiocyanate (FITC) to visualize glycoproteins by sodium dodecyl sulfate-polyacrylamide gel electrophoresis (SDS-PAGE) was published (Nishio and Hinata, 1980). Since then, similar methods have been extensively used to identify ConA-binding glycoproteins associated with the S genotype. The main characteristics of the S-associated proteins appear to be that they are all glycoproteins and have similar molecular weights and unusually high isoelectric points (p*I*s) that vary allelicly. The biosynthesis of these molecules during stigma development using ^{14}C-labeled amino acids and [^{35}S] methionine, has been investigated by Nasrallah *et al.* (1985) and the accumulation of S glycoproteins in the stigmatic pappillae has been shown to coincide with the acquisition of SI.

2. Cloning and Characterization of the S-Linked Glycoprotein Genes

Cloning of a gene encoding an S glycoprotein from *Brassica oleracea* was first reported by Nasrallah *et al.* (1985). A clone (pBOS5) was obtained by differential screening of a cDNA library made from stigma tissue. The identity of the clone was established using antiserum raised against the S_6 glycoprotein, which reacted with a Western blot of a polypeptide synthesized in *Escherichia coli* carrying a β-galactosidase fusion to the open reading frame of the cloned sequence. Evidence that pBOS5 encoded the product of the S gene was provided by linkage analysis, using Southern blot analysis. Restriction fragment length polymorphisms (RFLPs) in the *Brassica* genome detected with pBOS5 cosegregated with the S locus in P_1, P_2, F_1, and F_2 populations (Nasrallah *et al.*, 1985), providing evidence that the gene encoded by this clone is linked to the S locus. More evidence that this clone might encode the S gene came from Northern hybridizations, which revealed that it exhibited the anticipated spatial and temporal pattern of expression (Nasrallah *et al.*, 1985); more recently, *in situ* hybridizations have shown that it is highly expressed in the stigmatic papillae, which is the site of pollen inhibition (Nasrallah *et al.*, 1988; Kandasamy *et al.*, 1989). Evidence has also emerged that indicates that the S-linked glycoprotein (SLG) gene is expressed in the male reproductive tissues (Nasrallah and Nasrallah, 1989; Guilluy *et al.*, 1991; Sato *et al.*, 1991); this is discussed in Section III,A,6.

The nucleotide and/or amino acid sequences of a considerable range of SLG alleles from *Brassica oleracea* are now available, including those of S_{2A}, S_6, S_{13}, S_{14}, S_{22}, and S_{29} (Trick and Flavell, 1989; Chen and Nasrallah, 1990; Dwyer *et al.*, 1991). Amino acid sequences have also been obtained for *Brassica campestris* alleles S_8, S_9, and S_{12} (Isogai *et al.*, 1987; Takayama *et al.*, 1987). The relationship between these alleles has been reviewed in detail (Trick and Heizmann, 1992; Dickinson *et al.*, 1992), so that only a brief overview is presented here. Sequence analysis has shown that the SLG gene has a signal peptide, which is in keeping with the idea that the product of the gene is secreted. A notable feature that is common to virtually all of the *Brassica* SLG genes is 12 conserved cysteine residues, which are located in the C-terminal region of the proteins and thought to form disulfide bridges necessary for correct folding of the polypeptide (Nasrallah *et al.,* 1987). Comparison of 9 SLG sequences reveals a total of 13 potential N-glycosylation sites, of which four are conserved among all the alleles compared (Dickinson *et al.,* 1992). It seems likely that not all these potential sites are glycosylated, because there is evidence that of the nine potential sites in the S_8 allele from *B. campestris* only seven are actually glycosylated (Takayama *et al.,* 1987). Analysis of the carbohydrate chains of three SLG genes from *B. campestris* revealed that the major N-glycosidic oligosaccharide chains A and B were identical and that they are not unique to SLG genes, because they are commonly found in other plant glycoproteins (Takayama *et al.,* 1986a,b). Umbach *et al.* (1990) confirmed that this carbohydrate backbone is a substrate for a typical glycosylation pathway in plants, rather than one specific to *Brassica* stigmas. Thus, these differences in the composition of glycan side chains appear unlikely to be involved in determination of allelic specificity, although this possibility cannot be ruled out. However, the precise role that glycosylation plays remains to be determined. That glycosylation does have a functional role is indicated by the observation that stigmatic papillae treated with the glycosylation inhibitor tunicamycin no longer exhibit an SI phenotype (Sarker *et al.,* 1988).

Overall sequence comparisons of a range of S alleles reveal that the alleles can be placed into two groups. Of particular interest is that these groupings coincide with allelic dominance. The class I alleles, which include dominant S alleles (e.g., S_{29}), exhibit around 80% homology at the protein level with other members of this class. This decreases to about 65% when compared with members of class II, which is made up of the recessive alleles, (e.g., S_2), although within class II the level of homology is again on the order of 80%. From the analysis of SLG_6, SLG_{13}, and SLG_{14} Nasrallah *et al.* (1987) divided the protein into four regions based on homology between the alleles. Regions A (residues 1–182) and C (residues 275–378), which contain the conserved cysteines, exhibit around 80% homology, whereas regions B (residues 183–274) and D (residues 379–405) are highly polymorphic with only a 44% homology among alleles, suggesting that these domains are likely to have a key role in allelic specificity.

3. Identification of S Locus-Related Genes

Southern analysis provided evidence that the SLG gene is a member of a multigene family. Scutt *et al.* (1990) reported that there may be as many as 11 SLG-related sequences in the *Brassica* genome. However, it is clear that not all of these could encode functional genes, because they contain a number of sequence deletions and insertions (Nasrallah *et al.*, 1988). The S locus-related (SLR) genes have been cloned (Lalonde *et al.*, 1989; Trick and Flavell, 1989; Trick, 1990; Scutt *et al.*, 1990; Boyes *et al.*, 1991). The pattern of expression of these SLR genes (SLR-1 and SLR-2) is identical to that of the SLG gene with respect to its tissue specificity and developmental timing. However, in contrast to the SLG gene, SLR genes are not linked to the S locus, and consequently do not contribute to S allele specificity. Genetic analysis has revealed that although the SLR-1 and SLR-2 loci are distinct from the SLG they are themselves genetically linked (Boyes *et al.*, 1991). The SLR-1 genes share about 70% DNA sequence homology with the SLG class I genes (Lalonde *et al.*, 1989; Trick and Flavell, 1989), whereas the SLR-2 gene has over 90% homology to the S_2 gene (Boyes *et al.*, 1991). The three genes are clearly related because they have a number of features in common, such as the putative signal sequence and the cysteine-rich region. This suggests that the SLG and SLR loci are diverged products of gene duplication, sharing a common lineage, and that their products may have evolutionarily related functions.

In contrast to SLG, the SLR-1 locus appears to be a single, highly conserved gene. Analysis of the SLR-1 and SLR-2 genes isolated from different *Brassica* genotypes confirmed that the SLR gene is characterized by an extremely low level of allelic variation, with different S alleles being 99–100% conserved (Lalonde *et al.*, 1989; Trick, 1990; Scutt *et al.*, 1990). The detection of SLR-1 genes in both self-incompatible *Brassica* and *Raphanus* and self-compatible *Brassica* and *Arabidopsis* suggests that there is widespread conservation of this gene. As SLR-1 is highly expressed in both self-incompatible and self-compatible strains of *Brassica* and is highly conserved, these genes may play an important general role in pollination.

4. Identification and Characterization of the S Receptor Kinase Gene

A major development in the study of SI in *Brassica* has been the discovery of the *S* receptor kinase (SRK) gene, which was identified following the analysis of a number of SLG-homologous genomic clones from a *Brassica oleraclea* S_6S_6 homozygote. Nucleotide sequencing of the regions that flanked the SLG region within these clones resulted in the identification of sequences similar to those encoding protein kinases (Stein *et al.*, 1991). The most notable homology was found with ZmPK1, a serine/threonine receptor kinase of unknown function from maize (Walker and Zhang, 1990). More recently a vegetatively expressed putative receptor kinase gene from *Arabidopsis thaliana,* ARK1, has been identified (Tobias *et al.*, 1992). The extracellular domain of this gene exhibits around 60% homology with the corresponding SRK domain, which rises to about 76% homology within the kinase domain. These observations suggest that the S receptor kinase family is extremely ancient and that members of this family may have a diverse range of roles in plant signaling systems.

Sequencing of the SRK_6 genomic clone and cDNA clones predicts a protein of 857 amino acid (aa) residues (98 kDa) (Stein *et al.*, 1991). These comprise an amino-terminal putative signal peptide, which precedes an S domain of 406 residues that exhibits 89% identity to SLG_6 and contains the 12 conserved cysteine residues found in all members of the S gene family. The nucleotide sequences encoding the S domain are separated by an 896-bp intron from exon 2, which encodes within it a 20-aa membrane-spanning helix. The remainder of the protein is encoded by exons 3–7 and contains the 11 subdomains that are conserved among the protein kinase family. The sequences found within subdomains VI (DLKVSN) and VIII (GTYGYMSPE) suggest that SRK_6 is a serine/threonine protein kinase. Thus SRK_6 would appear to be a membrane-bound receptor kinase, with an extracellular domain that is highly homologous to SLG_6. A previously identified S gene, SLG_{2B} (Chen and Nasrallah, 1990), was found during this same study to be in fact the S domain of an SRK_2 gene. In this case the S domain of SRK_2 exhibits 90% identity with its SLG_2 partner. Comparison with SRK_6 revealed that the two receptor kinases possess an overall homology of 68%.

The high degree of similarity between the SLG gene and the extracellular domain of its SRK partner suggests that the former arose as a result of gene duplication, with sequence similarity being maintained by a process such as gene conversion (Tantikanjana *et al.*, 1993). Although the initial study revealed that both SRK and SLG are genetically linked to the *S* locus, their physical organization was more recently established using pulsed-field gel electrophoresis, which showed that they reside within a region on the order 220–350 kb (Boyes and Nasrallah, 1993). Whether this region delineates the entire S locus remains to be established. In view of the complexity of this locus, however, it has now been proposed that classic "S alleles" should be more correctly referred to as "S haplotypes" (Boyes and Nasrallah, 1993).

Direct evidence that the SRK gene encodes a serine/threonine protein kinase has been obtained (Goring and Rothstein, 1992) from the clone SRK-910 from a self-incompatible *Brassica napus* line. The S domain of this gene exhibits 75.4 and 67.2% amino acid homology to the corresponding domain in SRK_6 and SRK_2, respectively. The kinase domain from SRK-910, and a mutant derivative in which the conserved lysine residue in subdomain II was replaced with an alanine residue, were expressed in *E. coli* as fusion proteins with glutathione *S*-transferase. When tested for kinase activity, only the fusion containing the wild-type kinase domain was active; phosphoamino acid analysis revealed that the target for this activity was serine and threonine residues.

SRK exhibits the same developmental and spatial pattern of expression as the SLG gene. Expression studies with SRK_6 revealed expression to be maximal in mature pistils (Stein *et al.*, 1991). A range of transcripts from 1.6 to 5.9 kb was detected, of which three (2.3, 3.0, and 4.1 kb) were ascribed to the SRK gene. The 3.0-kb transcript is the predicted size for the fully spliced gene product; the larger transcript represents the unspliced transcript, whereas the 2.3-kb product is thought to be the result of alternative splicing or transcription from a different promoter, because it hybridizes to the kinase domain and introns therein. In addition to expression in the female tissue, SRK transcripts were also detected in developing anthers at the uninucleate stage and maximally at the binucleate stage. However, the level of expression of SRK is considerably lower than that of the SLG gene by a factor of some 140- to 180-fold. This is probably not surprising given the fact that SRK is a membrane protein and that SLG is expressed at extremely high levels in pistil tissue (Nasrallah and Nasrallah, 1989).

Evidence has emerged that indicates that the regulation of SRK expression is, to some extent at least, independent from that of other members of the S gene family. Nasrallah *et al.* (1992) report the identification of a locus SCF1 that positively regulates the SLG, SLR-1, and SLR-2 members of the S gene family. A spontaneous self-fertile, *scf1* mutant of *Brassica campestris* expresses these genes at much reduced levels, whereas expression of SRK is unaffected, providing a clear indiciation of the crucial role of the S genes in the SI reaction. Because the pollen in the *scf1* mutant is phenotypically normal it would appear that expression of the S locus in the male tissue involves a regulatory element (or elements) distinct from SCF1.

5. Functional Analysis of *Brassica* S Genes

Assuming that the SLG and SRK genes, either independently or together, control S-specific recognition, it should be possible to demonstrate this by introducing them by transformation into a recipient plant and demonstrating that the transgenic plant acquires the appropriate S phenotype. When the SLG_8 gene from *Brassica campestris* was transformed into a *Brassica oleracea* recipient, however, the transgenic plants did not exhibit a new allelic specificity (Toriyama *et al.*, 1991); indeed, the plants became self-compatible, owing to apparent silencing of the endogenous S genes by cosuppression. Introduction of the SRK_6 gene into *B. oleracea* carrying the S_2 haplotype revealed that, despite expression of the transgene, no change in S specificity was obtained (Stein *et al.*, 1991). As a consequence of this result the authors suggested that the SLG/SRK gene pair work in conjunction and therefore both elements would be expected to be required in order to specify a functional S phenotype. To date there are no reports of transformation experiments that have directly tested this hypothesis. Nevertheless, this is an attractive proposition, which is certainly consistent with available evidence. The decreased expression of SLG in the self-compatible *scf1* mutant has already been mentioned. Evidence has also been obtained that demonstrates that plants defective in *SRK* expression are also self-compatible. Thus, Goring *et al.* (1993) have identified a self-compatible *Brassica napus* line that produces a truncated SRK protein as a result of a 1-bp deletion within the gene. A similar situation has also been described in *Brassica oleracea* (Nasrallah *et al.*, 1994). In this case a self-compatible deletion mutant (S_{f1}) has been identified that lacks the SRK promoter and the first two exons of the gene. The authors speculate that this deletion may have arisen as a result of an unequal cross-over event between the SLG/SRK gene pair in S_{f1}. Expression of the SLG gene is not affected in the S_{f1} haplotype.

Detailed analysis of the SLG gene from the S_2 haplotype has revealed that, in addition to producing a 1.6-kb RNA transcript that encodes the secreted SLG protein, it also encodes a 1.8-kb transcript (Tantikanjana *et al.*, 1993). The protein encoded by this longer transcript differs from SLG in that it possesses a hydrophobic membrane-anchoring helix at its C terminus. It has been suggested that this has arisen as a result of the initial duplication and possible subsequent deletion events that generated the SLG/SRK gene pair. But unlike the S_6 haplotype, in which only the region encoding the extracellular domain was duplicated, in the case of S_2 a larger segment encompassing both this and the transmembrane domain was duplicated, enabling two products to be generated by differential splicing. Thus, three S proteins are produced by the S_2 haplotype: SRK, SLG, and what is effectively an SRK protein lacking the kinase domain. Tantikanjana *et al.* (1993) suggested that this may account for the recessivity of the class II haplotypes, because it may be anticipated that the truncated protein would interfere with the proposed SLG/SRK functional pairing.

6. The Pollen S Gene Component and Mechanism of Pollen Inhibition: Current Progress

In contrast to the considerable progress that has been made in the analysis of the stigmatic S gene products, little is known regarding the male side of the interaction. From genetic and physiological studies it may be predicted that the pollen component (or components) are expressed and incorporated into the grain as it matures within the tapetum. Attempts to identify proteins from pollen that segregate with known S alleles (haplotypes) have been unsuccessful (Nishio and Hinata, 1978b).

As previously mentioned, evidence has emerged that indicates that transcripts with homology to SLG and SRK genes are expressed in the male reproductive tissues. A low level of expression has been detected (several hundredfold lower than in stigmas) in anthers that are postmeiotic 3–4 days before anthesis, but not in premeiotic, later postmeiotic anthers, or mature pollen (Nasrallah and Nasrallah, 1989; Guilluy *et al.*, 1991). This is supported by expression studies in transgenic *Brassica* plants, using a GUS reporter gene fused to the SLG promoter (Sato *et al.*, 1991). Although these findings are compatible with the dimer hypothesis of Lewis (1954), which predicts that an incompatible reaction arises from the interaction of near identical male and female S gene products, the low levels of these products suggest it would be prudent to reserve judgment on the significance of this finding until a more detailed analysis of S locus expression within the developing anthers has taken place.

A particularly interesting development has come from Dickinson and colleagues, who reported the identification of a 7-kDa pollen coat protein (PCP) that interacts with stigmatic extracts to form a complex (Doughty *et al.*, 1993). Analysis of the interaction product revealed that the PCP had specifically interacted with SLG protein. On this evidence the PCP must be considered as a candidate for the male S determinant, although further studies are required to confirm this possibility. In the absence of a confirmed pollen SI determinant it is of course difficult to propose with certainty a model for inhibition of incompatible pollen. Nevertheless, a model proposed by Dickinson (1994) would seem to be a reasonable interpretation of current evidence, and is illustrated in Fig. 1. Dickinson proposes that an as yet unidentified male S gene product present in the pollen coat interacts with the corresponding female SLG and SRK gene products to form an activated receptor complex within the stigma plasma membrane. The activated receptor kinase then initiates the phosphorylation of one or more intermediates that ultimately produce a localized response within the stigmatic papilla cell that prevents pollen growth. What is intriguing about this model is that the inhibition reaction occurs within the stigma tissue rather than the pollen, which appears to be the case in the gametophytic SI systems. That phosphorylation of stigma proteins is important for a functional SI system is supported by studies that have investigated the effects of inhibitors of protein phosphatases (Scutt *et al.*, 1993; Rundle *et al.*, 1993), both of which revealed that pretreatment of pistil tissue with inhibitors of type 1 and 2A protein phosphatases resulted in breakdown of the SI response. A more detailed understanding of the *Brassica* SI response awaits the definite identification of

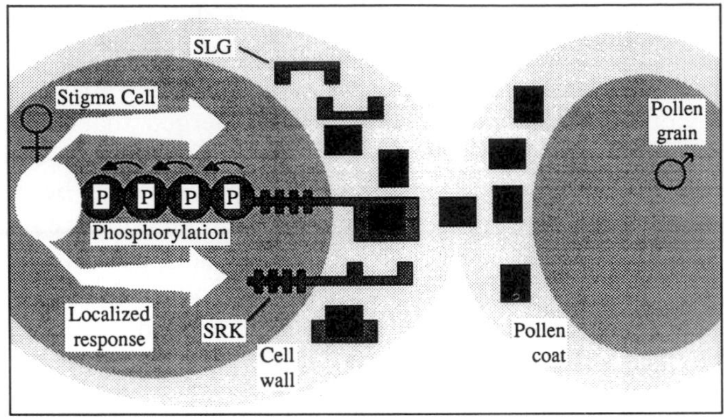

FIG. 1 Proposed model for inhibition of incompatible pollen in *Brassica* (based on Dickinson, 1994). An as yet unidentified male S gene product located in the pollen coat specifically interacts with cognate SRK/SLG molecules present in the stigmatic tissue to form an activated receptor complex. This stimulates phosphorylation of one or more intermediates which in turn bring about a localized response in the stigmatic papilla cell that prevents pollen growth. [Adapted with permission from Dickinson, H. G. (1994). *Nature* (*London*) **367,** 517–518. Copyright (1994) Macmillan Magazines Limited.]

the male S gene product and the identification of the components of the signal transduction pathway that mediates arrest of pollen growth.

B. Cell and Molecular Biology: *Ipomoea* (Convolvulaceae)

Ipomoea trifida has an SI system that appears to be similar to that found in *Brassica* (Kowyama *et al.*, 1980). Like *Brassica,* it has a single-locus sporophytic SI system, with multiple S alleles that display dominance. Molecular studies on SI in *Ipomoea* have been initiated by Kowyama's group (Mie, Japan).

1. Cloning of a S Receptor-like Kinase Gene in *Ipomoea*

Using conserved sequences of the receptor domain encoded by the *Brassica* SRK_6 gene as polymerase chain reaction (PCR) primers, PCR products have been amplified from cDNAs from stigma, anther, and pollen in *Ipomoea* (Y. Kowyama, personal communication). Four distinct classes of clones were obtained and characterized. Three of these classes of clones (IPK A, IPK B, and IPK D) were found to have approximately 70% homology with the *Brassica* SRK_6 gene. Northern blots have determined that these clones have a pattern of developmental expression that correlates with the acquisition of SI and expression is restricted to floral tissue (stigma, anther, and petal). IPK A was subsequently used as a probe to obtain a 2.8-kb clone from a stigma cDNA library from *Ipomoea*. This gene encodes a kinase and receptor domain and has high homology to the kinase genes: SRK from *Brassica* (Stein *et al.*, 1991), ARK1 from *Arabidopsis* (Tobias *et al.*, 1992), and the gene encoding ZmPK1 from *Zea mays* (Walker and Zhang, 1990; Y. Kowyama, personal communication). Thus, it is clear that a receptor kinase from *Ipomoea* has been identified.

Because the SRK gene from *Brassica* has both homology to the SLG gene and is linked to the S locus, RFLP analysis has been used to attempt to see if these genes from *Ipomoea* are also linked to the S locus. However, although these four classes of clones are highly homologous to *SRK*, surprisingly, no tight linkage to the S locus was detected. Nevertheless, because these genes are expressed in floral tissue it is likely that they play a role in pollination. Thus, for the moment, the identity of the S gene(s) from *Ipomoea* remains to be resolved and may, in fact, be distinct from those found in *Brassica*.

IV. Gametophytic Self-Incompatibility

A number of different species with gametophytic control of self-incompatibility has been characterized. Studies in the mid-1970s and early 1980s initiated characterization of pistil components of self-incompatibility in *Nicotiana alata* (Bredemeijer and Blaas, 1981) and *Lilium longiflorum* (Dickinson *et al.*, 1982). The first S-associated protein to be found in a species with a gametophytic self-incompatibility system was a glycoprotein from *Prunus avium* (Raff *et al.*, 1981; Mau *et al.*, 1982). The mid-1980s saw the cloning and characterization of several *S* genes and their products from solanaceous species, including *Nicotiana alata* (Anderson *et al.*, 1986). *Petunia hybrida* (Broothaerts *et al.*, 1989; Clark *et al.*, 1990) and *Petunia inflata* (Ai *et al.*, 1990; Singh *et al.*, 1991), and *Solanum tuberosum* (Kirch *et al.*, 1989) and *Solanum chacoense* (Xu *et al.*, 1990a,b). More recently studies have widened to include *Papaver rhoeas* (Franklin-Tong *et al.*, 1989, 1991, 1993; Foote *et al.*, 1994), *Pyrus serotina* (Hiratsuka, 1992a; F. Sakiyama, personal communication), *Secale cereale* (Wehling *et al.*, 1994a,b), and *Phalaris* coerulescens (Li *et al.*, submitted). We review the main findings from these molecular studies here.

A. Cell and Molecular Biology: *Nicotiana alata, Petunia, Solanum chacoense, Lycopersicon peruvianum* (Solanaceae)

1. Identification of S-Linked Glycoproteins from the Pistil

Although Bredemeijer and Blaas (1981) correlated banding patterns of S-specific proteins from stylar extracts of *Nicotiana alata*, these molecules were not isolated. A few years later, a glycoprotein from styles, which cosegregated with S genotype was isolated and characterized (Clarke *et al.*, 1985; Anderson *et al.*, 1986). S-Associated glycoproteins have since been identified and isolated in *Lycopersicon peruvianum* (Mau *et al.*, 1986), *Petunia hybrida* (Kamboj and Jackson, 1986), *Petunia inflata* (Ai *et al.*, 1990), *Solanum tuberosum* (Kirch *et al.*, 1989), and *Solanum chacoense* (Xu *et al.*, 1990a). All of these species are members of the Solanaceae. Generally, these molecules are not present in pistils of self-compatible relatives. However, the S glycoproteins from *Solanum* and *Petunia* are also found in self-compatible lines (Kirch *et al.*, 1989; Clark *et al.*, 1990), which suggests that self-compatibility is not entirely due to the loss of these major style-specific glycoproteins in these cases.

All of the S proteins from the solanceous species examined to date are glycosylated. Analysis of the structure of the N-linked glycan chains by fast atom bombardment mass spectrometry and ^{1}H nuclear magnetic resonance has revealed heterogeneity in the chain structure and that some of the S glycoproteins consist of a number of ''glycoforms'' (Woodward *et al.*, 1989). Although there is more than sufficient variability in the glycan chains to encode allelic specificity, it is not known whether these carbohydrate domains are involved in determining S allele specificity. Details of the amino acid sequence information of the S glycoproteins are described in the next section, because much of this has been derived from nucleotide sequence data.

2. Cloning of the Pistil S Gene

The first S gene to be cloned from a plant with a gametophytic system of SI was from *Nicotiana alata* (Anderson *et al.*, 1986). The putative S gene clone isolated from *N. alata* was identified by homology with sequence from the S_2 glycoprotein. Stylar-expressed cDNA clones were screened with an oligonucleotide complementary to the N-terminal end of the purified *S* glycoprotein. The *N. alata* S_2 clone contains an open reading frame of 642 nucleotides, which is sufficient to encode a protein of 24.8 kDa, including a putative signal peptide of 22 amino acids. Because the gene cosegregates with the S locus, it appears to be a good candidate for the S gene. In contrast to the *Brassica S* gene family, the solanaceous gene is present as a single copy within the genome (Anderson *et al.*, 1986).

The expression of the *Nicotiana S* genes is limited to mature styles, the concentration of their glycoproteins increasing during floral development and, hence, coinciding with the acquisition of self-incompatibility. *In situ* hybridization and immunocytochemistry, using mRNA and antibody to the S glycoprotein from *Nicotiana* has revealed that their localization is restricted to the surface of stigmatic papillae, the extracellular matrix secreted by the transmitting tract tissues, and the inner layer of the placenta in the ovary (Cornish *et al.*, 1987; Anderson *et al.*, 1989). Thus, expression is detected only in female reproductive tissue where the pollen tubes grow, so the pollen tubes are in direct contact with the S glycoproteins throughout pollination. Interestingly, it has been found that the S gene in *Nicotiana* is also expressed in developing pollen (Dodds *et al.*, 1993). A transcript homologous to the cDNA of the S_2 gene was found in RNA from S_2S_2 postmeiotic anthers during late development, but not in mature pollen. However, the level of expression is only 1% of that detected in the style. The S protein also appears to be present in the intine of mature pollen. Although these results are consistent with the hypothesis that the same S gene encodes both the stylar and the pollen components of the SI interaction, it is difficult to reconcile them with the proposed model for inhibition of incompatible pollen.

3. Structure and Sequence Comparisons of the Solanaceous S Genes

A number of other S alleles have now been isolated from *Nicotiana alata* (Anderson *et al.*, 1989), together with a considerable number from other solanaceous species. Extensive homology was observed at the N-terminal sequences of the S_2 glycoprotein from *N. alata* and S_1 and S_3 from *Lycopersicon peruvianum* (Anderson *et al.*, 1986; Mau *et al.*, 1986). cDNA clones corresponding to three S alleles from *Petunia hybrida* were isolated using the highly conserved 15 amino acids from the *N. alata* and *L. peruvianum* sequences (Clark *et al.*, 1990) and cDNA clones for three S alleles (S_1, S_2, and S_3) were obtained from *Petunia inflata* (Ai *et al.*, 1990). Two clones for S_2 and S_3 have been obtained from *Solanum chacoense* (Xu *et al.*, 1990b) and clones for three S alleles from *Solanum tuberosum* have been isolated using homology with *N. alata* sequences (Kaufmann *et al.*, 1991).

Amino acid sequence alignments have revealed important features of the S glycoproteins (Anderson *et al.*, 1989; Haring *et al.*, 1990; Ioerger *et al.*, 1990, 1991). First, all possess a classic signal peptide, which is consistent with the observation that they are secreted into the stylar canal. Second, they appear to be highly divergent (Ioerger *et al.*, 1990; Kheyr-Pour *et al.*, 1990); comparison of S_1, S_2, S_3, and S_6 from *N. alata,* for example, reveal an overall sequence identity of only 51.5% (Haring *et al.*, 1990), which is unusual for allelic gene products. This diversity is due to several hypervariable regions within the molecule, which have been suggested to be prime candidates for a functional role in determining the allelic specificity (Ioerger *et al.*, 1990). There are also five regions that are highly conserved, suggesting that they are structurally important, three of which are predicted to be hydrophobic and likely to form a core, while eight conserved cysteines potentially form disulfide bridges. Other regions of the molecule have proved to be of critical importance to the function of the S protein. One of the most significant findings with respect to the solanaceous S genes, discovered by Sakiyama and co-workers, was that two conserved regions of the *Nicotiana* S gene amino acid sequence are homologous to regions conserved in the fungal ribonucleases, RNases T_2 and Rh (McClure *et al.*, 1989). These regions are close to the catalytic domain of these RNases and include histidine residues which are critical for enzymatic activity. At the time this was discovered, only the S_2 *Nicotiana S* gene sequence had been published; subsequently, all of the solanaceous S genes identified have been found to have these conserved regions. This suggested that RNase activity could have a biologically important function in this SI system. This is discussed in the next section.

Sequence comparison has also permitted investigation of the evolutionary relationships of the S alleles to be investigated (Ioerger *et al.*, 1990; Clark and Kao, 1991). This has shown that the S locus is highly polymorphic, with amino acid similarities within species as low as 40%, and most interestingly has revealed that some interspecific similarities are greater than intraspecific similarities. For example, the SF11 and Sz alleles from *Nicotiana alata* exhibit a higher degree of sequence similarity to S alleles from *Petunia inflata* than to other *N. alata* alleles. This suggests that the S polymorphism predates divergence of these species and, as this is thought to have occurred about 27 million years ago, must be extremely ancient.

4. Identification of a Function for the S Gene Product: RNase Activity

Finding a function for the S glycoproteins was of primary importance, because this would shed light on the mechanism involved in this SI system. A series of studies showed that all the S glycoproteins from the solanaceous species were functional RNases and had high RNase activities (McClure *et al.*, 1989; Clark *et al.*, 1990; Singh *et al.*, 1991; Xu *et al.*, 1990b) and led to these S glycoproteins being called S RNases.

A major step forward was the finding that pollen rRNA was degraded in incompatible but not compatible pollinations *in vivo* (McClure *et al.*, 1990), which suggested that the S RNases functioned enzymatically in the SI reaction. S RNase uptake into the pollen tube cytoplasm has also been demonstrated and these molecules will inhibit translation of pollen RNA. Importantly, not only could the S RNases degrade pollen RNA, but their action was not S specific and they did not act on specific pollen mRNAs (Gray *et al.*, 1991; McClure *et al.*, 1990).

5. Direct Evidence That S RNases Are Required for Pollen Inhibition

It should be noted that all the data proposing that the putative S genes and their products are involved in the control of self-incompatibility that have been discussed so far are based entirely on indirect, correlative evidence. Experiments involving transformation of plants with the putative S gene would provide definitive evidence that these molecules are, indeed, S genes, because a change in incompatibility phenotype should be detected, showing that the sequences encode a functional S gene. However, until recently, there have been difficulties with transformation of *Nicotiana alata*. Thus, to date there have been few reports of successful transformation of the putative S genes. However, Lee *et al.* (1994) and Murfett *et al.* (1994) have performed elegant experiments using transformation in *Petunia* and *Nicotiana,* respectively, providing the first direct evidence that the cloned S genes and their products from the Solanaceae do indeed function in the self-incompatibility reaction.

Transformation of *Petunia* plants with a 2-kb antisense S_3 gene resulted in transgenic plants that, when self-pollinated, set seed with variable ability to reject self pollen. This loss of SI behavior has been analyzed (Lee *et al.,* 1994). Northern blots showed that the appropriate transcripts were missing in those plants that failed to reject pollen. Using fast protein liquid chromatography (FPLC) to examine the levels of *S* proteins in the pistils, it was found that the transgenic plants that failed to reject S_3 pollen were missing the S_3 protein. The antisense gene, therefore, appears to specifically knock out the S protein, with the result that the ability of the pistil to reject that pollen is specifically lost. The pollen phenotype was not changed, which suggests that the pollen S gene is likely to be separate from the pistil S gene. Using a gain-of-function approach, the transformation of S_1S_2 plants with the S_3 gene resulted in transgenic plants that, when pollinated with S_3 pollen, completely rejected it. Analysis of the protein levels in these plants, using FPLC, revealed that transgenic plants that rejected S_3 pollen had the S_3 protein in addition to the S_1 and S_2 proteins, at comparable levels. Other transgenics displayed partial seed set; these plants had a smaller amount of S_3 protein. Thus, the level of protein expressed in the pistil appears to be proportional to the level of seed set (Lee *et al.,* 1994).

Murfett *et al.* (1994) have also used transformation, with the *Nicotiana* S_{A2} allele, to demonstrate that the S RNases play an essential role in the S allele-specific rejection of pollen. Because of problems with both the transformation of *Nicotiana alata* and expression of the intact S genes, they developed a new hybrid recipient plant (*N. alata* × *N. langsdorfii*) and a heterologous promoter that is active in the pistil. Plants transformed with the S_{A2} gene had high levels of the S_{A2} transcript, high levels of the S_{A2} RNase, and rejected S_{A2} pollen, demonstrating directly that this gene alone can confer self-incompatibility.

These gain-of-function and loss-of-function experiments provide the most conclusive evidence to date that the S RNases actually function in SI. They lead the way in demonstrating an elegant way in which to definitively and directly attribute S function to S allele-specific S RNases. Not only do these results provide the first direct *in vivo* evidence that these RNases play an essential role in self-incompatibility, but they also demonstrate the feasibility of using both sense and antisense approaches to dissect the functional components involved in the SI response. However, the question remains as to how this type of SI mechanism operates; do these proteins function as RNases and how do they inhibit pollen tube growth in an S allele-specific manner?

6. A Working Model for Self-Incompatibility in Solanaceae

As a result of the finding that the S glycoproteins from *Nicotiana* were RNases, a model for the operation of the *Nicotiana* SI system was proposed (Gray *et al.*, 1991; McClure *et al.*, 1989, 1990; Haring *et al.*, 1990; Thompson and Kirch, 1992). This hypothesized that the S RNases function cytotoxically, with S allele specificity either achieved by S-specific uptake of the RNase into the pollen tube or S-specific inactivation of S RNases after nonspecific uptake. The degradation of pollen rRNA and interference with protein biosynthesis would result in the arrest of pollen tube growth. This model is outlined in Fig. 2. The transformation of both *Petunia* and *Nicotiana* with the S RNases gives good evidence that the RNases are essential for the SI response.

Although this is an attractive model, a few points require elucidation before it can be accepted as a complete description of inhibition of pollen tube growth in this SI system. For instance, the mechanism by which specificity is controlled and the nature of the pollen component controlling uptake of the S RNases is unclear. A clue may be provided by the observation that S RNases with levels of RNase activity comparable to those exhibited by incompatible S RNases are present in self-compatible *Petunia* styles (Clark *et al.*, 1990; Singh *et al.*, 1991). A possible explanation could be that these S RNases, although present and active, cannot recognize self-pollen and therefore an SI phenotype is not expressed. Analysis of the breakdown of SI may well provide important insights into how this SI system operates. Although the possibility that this RNase activity was recruited during evolution and that its catalytic activity is unrelated to the SI function has been considered (Haring *et al.*, 1990), this seems unlikely at present. The RNase hypothesis is the only feasible model

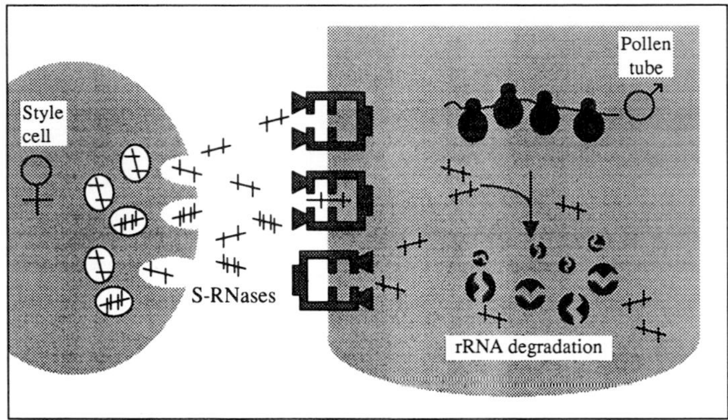

FIG. 2 Proposed mechanism for the self-incompatibility reaction in Solanaceae. (Based on Dickinson, 1994; derived from work of Clarke and colleagues. See text for details.) S RNases secreted by the stylar cells are taken up by the corresponding male S gene component in the pollen tube. The internalized S RNase then degrades the rRNA within the pollen tube, leading to eventual cessation of protein synthesis and arrest of growth. [Adapted with permission from Dickinson, H. G. (1994). *Nature (London)* **367,** 517–518. Copyright (1994) Macmillan Magazines Limited.]

postulated for this SI system on the information available at present and all the features of this gene are consistent with it being the S gene. All that remains to be elucidated is the key to the model; namely, how the specificity of the interaction is controlled.

B. Identification of RNases in the Japanese Pear *Pyrus serotina* (Rosaceae)

The detection of RNase activity associated with self-incompatibility in the Solanaceae has led to a number of groups attempting to identify S RNases associated with self-incompatible genotypes in other species. Two Japanese groups are currently investigating the role of RNases in *Pyrus serotina* (Japanese pear). Using a self-incompatible strain Nijisseiki (S_2S_4) and its self-compatible mutant Osa-Nijisseiki (also S_2S_4), Hiratsuka (1992a,b) has identified an S_2 allele-associated protein. Sakiyama's group (Osaka, Japan), since discovering that the *Nicotiana* S gene has homology to fungal RNases, has also initiated studies aimed at isolating S RNases from *P. serotina*. Using the same S_2S_4 strains as previously described, together with two other self-incompatible strains (genotypes S_2S_3 and S_4S_5), they have identified and purified two RNases from the style of each species. The RNase activities measured in the crude extracts were found to be at levels comparable to those found in *Nicotiana alata* and are developmentally expressed with a temporal pattern that correlates with the acquisition of SI. However, both groups have found that the self-compatible mutant had levels virtually identical to those of its self-incompatible progenitor. Thus, analysis of these RNases so far has shown that expression of self-incompatibility and self-compatibility is not associated with a simple absence and appearance, respectively, of these RNases in the style.

Because it appears that although there are stylar RNases associated with self-incompatibility, they are present in both self-incompatible and self-compatible lines, RNase activity alone is not sufficient to explain the expression of self-incompatibility in these plants. Thus, it remains to be seen if these RNases are responsible for the control of self-incompatibility in this species. It is possible that there are subtle differences in the structures of the RNases found in the self-incompatible and self-compatible strains; however, the involvement of factors other than RNases cannot, at this stage, be ruled out. Nevertheless, it may be borne in mind that there may be a number of RNases occurring as a multigene family, only some of which are responsible for self-incompatibility. A molecular approach, using the solanaceous RNases as a probe, may reveal that there are other RNases that are more closely related to those already identified as being involved in SI.

C. Cell and Molecular Biology: *Papaver rhoeas* (Papaveraceae)

1. Identification and Characterization of S-Linked Proteins in the Stigma

An *in vitro* system for *Papaver rhoeas,* which has enabled the detection of biological activity of the stigmatic S gene products by testing them on pollen grown *in vitro* (Franklin-Tong *et al.,* 1988), has already been described (Section II). Using this system to monitor S activity, the stigmatic S gene products from several S genotypes of *P. rhoeas* have been isolated and characterized (Franklin-Tong *et al.,* 1989; Walker, 1994). Preparative isoelectric focusing (IEF) has been used to separate biologically active proteins (Franklin-Tong *et al.,* 1989). The stigmatic S gene products for the S_1 and S_3 alleles have been completely purified (Foote *et al.,* 1994; Walker, 1994). Segregation analysis of stigmatic proteins from over 250 individual plants using analytical IEF has resulted in the detection of proteins that cosegregate with the S_1 allele. These putative S_1 products consist of two proteins, one with a p*I* of 7.5, the other with a p*I* of 6.9, which corresponds closely with the p*I* assigned using biological activity. These two forms of the S_1 protein have been designated S_1a and S_1b, respectively. Using SDS-PAGE, S_1a and S_1b were further analyzed. Each isoform separated into two proteins with apparent molecular masses of 16.7 and 14.7 kDa for S_1a products; and 16.8 and 14.8 kDa for S_1b products (Foote *et al.,* 1994). The molecular weights of the S_1 proteins are lower than the previous estimate of 22kDa (Franklin-Tong *et al.,* 1989), which was based on gel filtration information; it is now apparent that the proteins behave abberantly during many forms of chromatography. The observation of more than one S_1-specific band was of considerable interest because it raised questions as to the relationship between the forms and the implications for biological activity. There is good supportive evidence that the proteins characterized are genuine S gene products. Not only are these molecules completely associated with the S locus and are developmental expression correlates with the expression of self-incompatibility, but they also have the S-specific biological activity expected of them. In contrast with other SLG genes described to date, which are major pistil components (e.g., the SLG genes from *Petunia hybrida* comprise 16–18% of total protein content; Broothaerts *et al.,* 1989), the abundance of the *Papaver* S proteins is estimated to be around 0.5–1% of total protein in the stigmatic papillae.

What then, is this S protein? Because the S glycoproteins of the solanaceous self-incompatible species all appeared to be ribonucleases (McClure *et al.,* 1989; Clark *et al.,* 1990; Ioerger *et al.,* 1991) and as *Papaver rhoeas* has an SI system genetically identical to these species, it might be expected that the stigmatic S glycoproteins may also be RNases. Surprisingly, the S gene product in *Papaver* is not a ribonuclease. Not only do *P. rhoeas* stigmas have levels of ribonuclease activity at least several hundredfold lower than that found in *Nicotiana alata* and immature stigmas exhibit ribonuclease activity at a level indistinguishable from that of the mature self-incompatible stigmas, but more importantly, there is no detectable RNase activity that correlates with the presence of the functional stigmatic S gene product in *P. rhoeas* and pollen from *P. rhoeas* is insensitive to ribonuclease activity (Franklin-Tong *et al.,* 1991). All this suggests that the involvement of RNases in the SI response in *Papaver* is unlikely; indeed, transcription of pollen genes is required in order for the SI response to operate in *P. rhoeas* (Franklin-Tong *et al.,* 1990). These conclusions were confirmed by the cloning of the S gene (Foote *et al.,* 1994); this evidence is discussed in the next section. Because it is clear that the S proteins from *Papaver* are not ribonucleases, an alternative function for this molecule has been proposed, based on evidence discussed in Section IV,C,4.

2. Cloning of the Stigmatic S Gene from *Papaver* and Evidence That This Is the S Gene

The S_1 allele of the S gene from *Papaver rhoeas* has been cloned (Foote *et al.*, 1994) by screening a cDNA library using an oligonucleotide based on the N-terminal nucleotide sequence of the S_1 protein. Thus, the evidence that will be described for this being the S gene also relates to the proteins characterized. The gene has an open reading frame (ORF) of 417 bp, which is sufficient to encode a mature 14.5-kDa polypeptide comprising 120 amino acids. The ORF is preceded by a 19-residue putative signal peptide sequence, which has a relatively hydrophilic C terminus and a central hydrophobic region. The sequence also predicts the position of a single potential N-glycosylation site. The mature peptide is largely hydrophilic, except for a short N-terminal hydrophobic region; the protein is predicted to be relatively rich in charged amino acids.

Expression of the S_1allele appears to be confined to stigmatic tissue; the S_1 clone hybridizes to a transcript of ~1.0 kb. Developmental expression is as expected for an SI gene: transcripts are first detected in stigmas at extremely low levels in flowers 2 days preanthesis and there is a rapid burst of activity 1 day before anthesis, which lasts for several days after flower opening. There is no detectable expression in the pollen. Southern analysis indicates that the gene is single copy and does not readily cross-hybridize to a variety of other S alleles that have been tested, indicating that in common with their solanaceous counterparts the *Papaver* S genes may exhibit a high degree of sequence polymorphism. Comparison of the amino acid sequence of the S_1 allele with that of the more recently cloned S_3 endorses this view, as it reveals they share only 55% sequence identity (Walker, 1994). However, despite this high degree of polymorphism, secondary structure predictions reveal that the two polypeptides may adopt remarkably similar conformations, indicating there may be topological constraints on these molecules relating to their interaction with the pollen S gene product.

The cloned S_1 sequence was expressed in *E. coli* to obtain recombinant protein (denoted S_1e). This had an apparent molecular mass around 0.8–0.9 kDa larger than the nonglycosylated forms of S_1 found in the plant, suggesting that posttranslational processing events, in addition to glycosylation, occur in stigmatic papillae. Thus, these data would appear to back the explanation for the several forms of S_1 detected in the stigma. Use of recombinant S_1 protein (S_1e) has clearly demonstrated that the sequence cloned is indeed the S_1 allele, because S_1e is biologically active specifically against pollen carrying the S_1 allele, providing a direct demonstration of the specific biological activity of this cloned S gene (Foote *et al.*, 1994). In addition, these studies also show that glycosylation and the presumed processing of the S_1 product are not absolute prerequisites for either biological activity or specificity. However, it is considered that this post-translational processing has some role, possibly in the modification of the activity or stability of the S_1 protein.

Most significantly, searches of protein and DNA databases using the S_1 and S_3 sequences have failed to detect any significant homology with any other genes, including those that encode the S locus glycoproteins from *Nicotiana* and other solanaceous species, and those from *Brassica*. This indicates that the *Papaver* S gene is different from any of the other S genes cloned, providing a new class of S gene (Foote *et al.*, 1994), and confirms the earlier finding that the stigmatic S gene from *Papaver rhoeas* is not a ribonuclease (Franklin-Tong *et al.*, 1991). Not only does this suggest that the SI mechanism operating in this species is different from other S genes cloned to date in different species, but it also has important implications for the evolutionary relationships between the S genes, which is discussed below in Section V.

3. Cellular and Molecular Studies on the Pollen–Stigma Interaction

The *in vitro* system has enabled a study of pollen metabolism, by testing the effects of various metabolic inhibitors on pollen tube growth and the SI reaction. The system has also been adapted to allow analysis of changes in gene expression (Franklin-Tong *et al.*, 1990), protein phosphorylation (Franklin-Tong *et al.*, 1992), and cytosolic calcium levels (Franklin-Tong *et al.*, 1993), which occur in the pollen as a result of the SI reaction. Some of these studies are outlined below.

a. De Novo RNA Synthesis in Pollen Occurs as a Self-Incompatible Response In *Papaver rhoeas*, as with many species, no RNA synthesis occurs during the first stage of pollen tube growth, because two-dimensional (2D) gels have shown that the mRNA species in poppy pollen prior to, and following, germination are indistinguishable (Franklin-Tong *et al.*, 1990). It is, therefore, not surprising that the transcription inhibitor actinomycin D has no detectable effect on pollen tube growth. Presumably mature pollen contains the full complement of mRNA species that are required to specify the proteins that will be needed for pollen tube extension. However, when pollen from *P. rhoeas* is challenged with the stigmatic S protein in the presence of actinomycin D the inhibition of pollen tube growth is partially alleviated (Franklin-Tong *et al.*, 1990; Franklin *et al.*, 1991). Although the observation that transcription inhibitors can overcome inhibition of incompatible pollen is not new, it has not previously been known in which tissues differential RNA synthesis occurs; in this case it is clearly in the pollen. These observations suggest that the induction of *de novo* pollen gene expression is required for a full SI response and that the product(s) of these pollen genes have a role in the SI response.

Proteins associated with the incompatibility response have been identified on 2D gels. Translation products from the RNA isolated from incompatible pollen revealed several new proteins, which correspond to new RNA transcripts produced in pollen as a result of the SI response; their appearance is specific to an incompatible reaction (Franklin-Tong *et al.*, 1990). There is also an indication of posttranslational "downregulation" of some transcripts in an incompatible response. Although the differential synthesis of RNA in self- and cross-pollinated styles has been previously reported (van der Donk, 1974), this is the first report to date showing that there are specific changes in RNA synthesis in the pollen as a consequence of the SI response. Several clones have been identified from a pollen cDNA library, whose expression appears to be associated with the SI response in pollen (Franklin-Tong and Franklin, 1992). Preliminary experiments have shown that at least one of the upregulated clones shows upregulation in an incompatible *in vivo* pollination, lending confidence to the idea that these genes, which have been identified in *in vitro*-challenged pollen, are expressed in pollen involved in the pollen–stigma interaction in the plant (Franklin-Tong and Franklin, 1992). However, whether these pollen "response" genes are directly involved in the inhibition of incompatible pollen or whether they are expressed as a consequence of the SI response (i.e., they may characterize, rather than be involved in, the SI response) is not yet clear. This response is currently being studied in more detail, because characterization of these genes should help establish how SI operates in this species.

Comparisons between pollen–pistil interactions and host–pathogen interactions have often been made, because they both involve specific recognition processes and the subsequent inhibition or death of specific cells (Teasdale *et al.*, 1974; Bushnell, 1979; D. H. Lewis, 1980). It is already established that the host–pathogen response is an active process, involving *de novo* mRNA and protein synthesis (Dietrich *et al.*,1990). In *Papaver*, there is some evidence that the SI response may involve these processes too. However, despite the identification of pollen "response" genes, their nature is not yet known. As yet, there has not been much clear evidence to form an opinion on this theory. Preliminary investigations of this idea by seeing whether stress response compounds are produced in pollen of *P. rhoeas* and, if they are, whether they can cause inhibition have been inconclusive, although they have established that stressed pollen produces cytotoxic molecules that inhibit pollen. Likewise, the phytoalexin sanguinarine, produced in the Papaveraceae, has been shown to inhibit *P. rhoeas* pollen tube growth *in vitro* (Atwal, 1993; Franklin *et al.*, 1994), which shows that such molecules are capable of the same sort of inhibitory activity against pollen tube growth that they exhibit in the host–pathogen interaction, although there is no evidence yet for the specific production of these molecules in the SI response.

b. Signal Transduction: Ca^{2+}-Mediated Signal Transduction for Self-Incompatibility The induction of events within the pollen that occur as the result of an external interaction between the stigmatic S glycoprotein and what is assumed to be some type of pollen receptor suggests the involvement of a signal transduction mechanism. The switching on or off of "response genes" within incompatible pollen of *Papaver rhoeas* as the result of an external signal provides further indirect evidence for the involvement of a "second messenger" signaling system in the SI response.

Using Ca^{2+}-selective dyes, the role of Ca^{2+} signaling during the self-incompatibility (SI) response in *P. rhoeas* has been investigated (Franklin-Tong *et al.*, 1993). Pollen tubes growing *in vitro* were microinjected with Calcium Green-1, which has high affinity and selectivity for free Ca^{2+}, and which undergoes a marked change in fluorescence on binding to Ca^{2+}, and cytosolic free Ca^{2+} ($[Ca^{2+}]_i$) in the pollen tubes was monitored during the SI response using laser scanning confocal microscopy (LSCM). This allowed detection and visualization of the intracellular localization of changes in $[Ca^{2+}]_i$ in the living cells during the response period, thus enabling any changes that occurred to be attributed to the signal–response coupling via Ca^{2+}. Addition of incompatible stigmatic S glycoproteins induced a transient increase in the level of $[Ca^{2+}]_i$ in pollen tubes. In contrast, no rise in $[Ca^{2+}]_i$ was detectable after addition of either compatible or heat-denatured incompatible stigmatic S glycoproteins. The elevation of $[Ca^{2+}]_i$ was followed by the specific inhibition of pollen tube growth in incompatible reactions. The photoactivation of caged Ca^{2+} to elevate $[Ca^{2+}]_i$ artifically resulted in the inhibition of pollen tube growth and thus mimicked the SI response. Taken together, these results provide strong evidence that $[Ca^{2+}]_i$ mediates the SI response in *P. rhoeas*, because a direct link between the transient rise in $[Ca^{2+}]_i$ and the biological phenomenon of inhibition of pollen tube growth has been shown. This demonstrates, for the first time, that the SI response in this species is mediated by $[Ca^{2+}]_i$.

As the rise in $[Ca^{2+}]_i$ appears to be generated intracellularly from a localized region of the cell, it suggests a role for the phosphoinositide pathway in the SI response, because it is generally believed that stimulus-evoked Ca^{2+} release from internal stores may be mediated through the release of inositol trisphosphate ($InsP_3$) (Berridge and Irvine, 1989; Drøbak, 1992). Inositol trisphosphate and other components of the phosphoinositide pathway are thought to be present in pollen tubes of *Papaver rhoeas* (B. K. Drøbak and V. E. Franklin-Tong, unpublished observations). However, at this stage the involvement of external Ca^{2+} cannot be ruled out, because external Ca^{2+} is necessary for pollen tube growth and Ca^{2+}-channel blockers are known to inhibit pollen tube growth. On the other hand, experiments have shown that release of caged IP_3 in pollen tubes results in a transient rise in $[Ca^{2+}]_i$ and in the inhibition of pollen tube growth (V. E. Franklin-Tong, unpublished data). This suggests that IP_3 receptors present in pollen tubes are capable of releasing Ca^{2+} from intracellular stores, and also that the phosphoinositide pathway could be involved in pollen tube inhibition. Investigations are currently underway to determine whether this pathway is involved in the SI response.

Further evidence of a signaling pathway operating in pollen of *Papaver* comes from preliminary experiments using the *in vitro* system. These have revealed rapid and transient phosphorylation of several pollen proteins, with differences between a challenge with an incompatible and a compatible stigma extract (Franklin *et al.*, 1992; Franklin-Tong *et al.*, 1992). Some proteins are newly phosphorylated within 1 min of the SI interaction, and others are dephosphorylated within 15 min. There is also evidence for phosphatase activity, because some phosphophoproteins are dephosphorylated during the SI response. These observations suggest a role for protein kinases and phosphatases in signal transduction in this SI response. An analogy may be drawn between these rapid and transient changes in phosphorylation in pollen and those observed in some host–pathogen interactions.

4. A Working Model for Self-Incompatibility in *Papaver rhoeas*

It is a little early to postulate how signal transduction might work in an SI system, but it is possible, nevertheless, to outline what is thought to occur in *Papaver rhoeas,* on the basis of experimental results (see Fig. 3).

The stigmatic glycoproteins act as signal molecules (the nature of allelic specificity is not yet known); these interact with receptors on the pollen grain or tube (the nature of these and how specificity is defined is unknown as yet). If a stigma signal molecule matches a pollen receptor molecule (i.e., in an incompatible pollination), it can bind; if it does not match (i.e., in a compatible interaction), it cannot bind. When a signal molecule binds the receptor, it triggers the cascade of signal transduction mechanisms inside the pollen, starting with the transient release of $[Ca^{2+}]_i$ in the pollen. Evidence to date suggests a Ca^{2+}-mediated signal transduction pathway. There is a possible role for IP_3 and also for the involvement of protein kinases. Following this message, the expression of "response" genes in the pollen will be altered; they might be switched on (or off), or the level

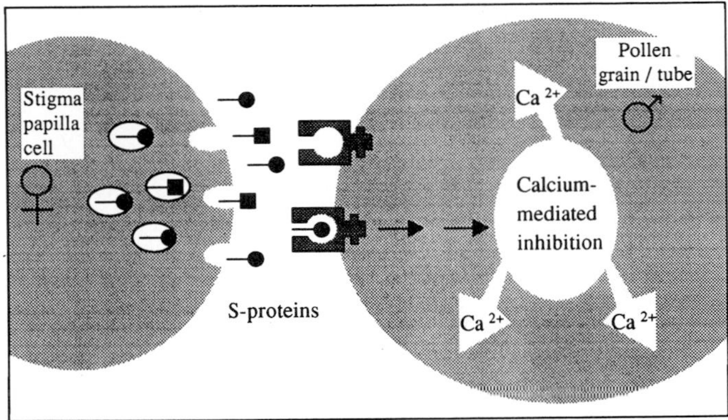

FIG. 3 Proposed mechanism for the self-incompatibility reaction in *Papaver rhoeas* (based on Dickinson, 1994; derived from Franklin-Tong and Franklin, 1993). Specific recognition of secreted stigmatic S protein by the pollen S gene product, thought to be a membrane-bound receptor, activates a Ca^{2+}-mediated signal tranduction pathway that leads to the inhibition of incompatible pollen. [Adapted with permission from Dickinson, H. G. (1994). *Nature* (*London*) **367,** 517–518. Copyright (1994) Macmillan Magazines Limited.]

of expression might be altered. We do not know the nature of these genes as yet, but pollen genes that have altered expression as a consequence of an incompatible response have been identified and isolated. One could speculate that these genes might be phytoalexins or other genes that are stress induced; however, it is possible that pollen-specific genes may be involved, especially from evidence obtained by cloning the pollen response genes. Transcription of these genes would result in products that would lead to a cellular response, involving inhibition of pollen germination or pollen tube growth. The attraction of a model like this is that one does not require active inhibitory molecules produced in the pistil or a specific uptake mechanism for these molecules. It also means that the inhibitory molecules produced in the pollen do not necessarily have to be encoded by the S gene. This provides a model and hypothesis to test experimentally and to investigate further the other components involved in the SI response.

D. Cell and Molecular Biology of Self-Incompatibility in the Poaceae (*Secale cereale* and *Phalaris coerulescens*)

As mentioned in Section I, in the grasses, SI is controlled by two multiallelic loci, S and Z (Lundqvist, 1956; Hayman, 1956). This makes the study of SI at the molecular level (as well as at the genetic level) potentially much more complex. Nevertheless, studies have been initiated in two grass species: *Phalaris coerulescens* and *Secale cereale*.

1. *Secale cereale* (Rye)

a. S-Linked Fragments with Homology to Brassica SLG Gene A method has been developed whereby S alleles in *Secale cereale* (rye) may be identified using molecular techniques. Wehling *et al.* (1994b) have developed denaturing gradient gel electrophoresis (DGGE) of PCR amplification products for *Secale,* using primers derived from conserved *Brassica* S sequences (SLG-13). This amplifies a large number of products from genomic DNA from rye. Although no S genotype-specific differences were detected using agarose gel electrophoresis, DGGE gels revealed complete S-linked cosegregation of a 280-bp fragment for S alleles S_3 and S_4. However, it has not yet been ascertained whether this is the S gene, or part of it, from rye. Fragments associated with the Z locus have not yet been identified. Nevertheless, this is a useful and important step in analyzing the S locus in a two-locus SI system.

The fact that conserved sequences from *Brassica* will amplify S-linked fragments in rye may be of great importance in our interpretation of the evolutionary relationships between different SI systems. It will be recalled that *Secale* has a gametophytic mode of control for SI and that it also has an additional Z locus, whereas *Brassica* has a single locus and a sporophytic mode of control. The implications of this are discussed in Section V. The detection of this polymorphism has the potential of great practical use in determining S genotypes of plants without the need for laborious pollinations, reducing the length of time of many experiments, because there would be no need to wait until flowering for S genotype identification. In addition, the identification of new S genes in other grass species may be possible.

b. Signal Transduction Associated with Self-Incompatibility Other studies on *Secale* has been directed at examining the possibility of the involvement of protein phosphorylation and Ca^{2+} as components of a signal transduction pathway in the SI response. Using [γ-^{33}P]ATP to label intact pollen grains, Wehling *et al.* (1994a) have found differences in the incorporation of ^{33}P, which is dependent on the type of stigmatic extract to which the pollen is challenged. Although a stimulation of phosphorylation in pollen preincubated with a compatible stigma extract was observed, the response was greatly increased in pollen challenged with an incompatible stigmatic extract. The timing of the response is rapid, with maximal levels of phosphorylation being observed within 90 sec; the difference between levels of phosphorylation in incompatible and compatible pollen stabilizes several minutes after the response, although these levels are still higher than those observed in control pollen where no reaction had been induced. These differences in phosphorylation levels are attributed solely to differences in stimulation of phosphorylation activity within the pollen grains. Interestingly, heat denaturation of incompatible stigma extracts had no effect on phosphorylation stimulation patterns induced. This contrasts with the observation in *Papaver* pollen that heat denaturation of incompatible S proteins resulted in no Ca^{2+} response, suggesting that no signal transduction ocurred if the S protein was denatured. Although no newly phosphorylated proteins were identified, four proteins (M_r 43K, 54K, 72K, and 81K) were found that showed quantitative differences in the degree of phosphorylation in the different SI reactions, with highest levels being found in incompatible pollen.

A comparison of phosphorylation patterns in pollen from self-incompatible and self-compatible mutants, in which the self-fertility (*sf*) trait is due to a pollen-part mutation, revealed that there were differences in the levels of ^{33}P incorporated into the pollen, with significantly less being incorporated into the self-compatible pollen. Of the four differentially phosphorylated proteins identified, three of these (43, 72, and 81 kDa) were not phosphorylated in the self-compatible mutants. These proteins could, therefore, be involved in the signal transduction pathway involved in pollination and the SI response.

These results suggest that, as in *Papaver,* the SI response in *Secale* involves protein phosphorylation, thereby implicating the involvement of protein kinases in the SI response. Experiments using protein kinase inhibitors and Ca^{2+}-channel blockers have complemented the phosphorylation studies, by providing further evidence for the involvement of protein kinases and Ca^{2+} in the mediation of the SI response in rye (Wehling *et al.*, 1994a). In *Secale,* it appears that a self-compatible interaction between pollen and S protein involves protein phosphorylation, presumably as a signal to activate germination. In an incompatible reaction, the same proteins are stimulated, but the level of phosphorylation is much higher. This response may possibly be comparable to that in *Papaver,* because in *Papaver* the proteins were analyzed on 2D gels, showing several different phosphorylation sites sequentially phosphorylated on an already phosphorylated phosphoprotein during an incompatible response, although the molecular masses of these protein were smaller (ca. 20 kDa).

c. A Model for the Mode of Action of Self-Incompatibility From studies with *sf* mutants, a model has been proposed for the initial steps of the SI response in *Secale* (Wehling *et al.,* 1994a). An SI-specific pollen-borne signal transduction pathway appears to operate, whereby self-compatible pollen grains carrying the *sf* allele cannot respond, owing to an interruption of the cascade of signals in the signal transduction pathway. Analogous situations have been observed in mammalian signaling pathways. If this is the case, any interruption of the signal transduction pathway may cause a breakdown in SI, thereby possibly explaining the fact that there are several *sf* mutants in rye that are not linked, or even on the same chromosome as either the S or Z genes; that is, these mutants that knock out a component of the signaling pathway involved in SI do not necessarily have to be linked to the self-incompatibility loci.

Thus, a model, based on the results described above, has been formulated (Wehling *et al.*, 1994a). It is proposed that there are a number of S- and Z-specific receptor molecules (possibly receptor protein kinases), found in the pollen grain, that have the capacity to bind stigmatic S- and Z-specific signal molecules that would act as ligands. The binding of these molecules to the receptors would excite an intracellular kinase domain, leading to the triggering of a cellular response in the pollen grain. There is some evidence that this response may be mediated by Ca^{2+}, where an influx of Ca^{2+} might trigger the pathway. Fundamental to this model is the idea that each S- and Z-specific receptor must be activated individually; the SI response would be the result of the additive effects of all the receptors in the pollen grain responding to the signals received from the stigmatic molecules. Thus, activation of only half of the receptors (e.g., in a situation in which a pollen grain carrying S_1Z_5 lands on a stigma carrying $S_1S_2Z_3Z_4$; only S_1 receptors would be activated) would not result in pollen tube inhibition; activation of both S_1 and Z_5 receptors would be required for an incompatible response (Wehling *et al.*, 1994a).

While admittedly still at a rather speculative stage, because the detailed elements of the signal transduction pathway have not yet been identified, this model is not inconsistent with any of the current concepts about the two-locus SI system. Thus, the model allows for the complementary action of the S and Z loci. Similarly, the concept that S and Z arose by gene duplication (Lundqvist, 1964) would fit this model admirably, because it is proposed that S and Z gene products function in the same manner. However, it is worth pointing out that this model, although consistent with the existing genetic evidence for the operation of SI in the grasses, is different from the model previously proposed (see Section I,A,4), in which it was envisaged that the S and Z components would act in concert by combining to form, perhaps, a dimer, which would act as a unique receptor for that S–Z combination of alleles. The advantage of having a signal transduction cascade system controlling whether or not pollen tubes are inhibited in the rapid SI response, as seen in the grasses, is obvious and credible. Thus, the rapidity of the inhibition of pollen tube emergence from the grain is dependent on the speed at which all the receptors are stimulated. A continuation of these studies will, no doubt, reveal in more detail how the SI system in *Secale* operates.

2. Cloning of a Pollen-Linked S Component from *Phalaris coerulescens*

Using differential screening with an *sf* pollen part mutant, a pollen-specific clone (Bm2) has been isolated, which when analyzed using RFLP analysis cosegregates perfectly with alleles of the S locus (Li *et al.*, submitted). Temporal expression is as expected for an S gene, with a dramatic increase at the mature pollen-shedding stage. This pollen-specific gene, which has been cloned, is 3 kb in length, with five introns. The deduced amino acid sequences for S_1, S_2, and part of the S_4 alleles exhibit a variable N terminus and a conserved C terminus. It is suggested that the 5′ variable region is responsible for encoding allelic specificity, whereas the conserved 3′ region could be involved in biological activity (Li *et al.*, submitted). The conserved region of this gene has 40% homology with a thioredoxin from tobacco and the catalytic site is conserved, suggesting that the protein encoded by this gene may actually be an active thioredoxin molecule. Because thioredoxin is usually thought to be responsible for reducing disulfide bridges, which is a farily nonspecific role, it is unusual to find it with this tissue specificity. A crucial question to ask is whether this thioredoxin merely overlaps the S gene coding sequence, or whether it actually plays a role in SI. Analysis of an S allele complete mutant, which is self-compatible, has revealed that there are major sequence alterations in this thioredoxin-like region, which strongly suggests that this part of the molecule is crucial for catalytic activity and that the thioredoxin does, in fact, play a role in the SI response, although how this functions in the SI response is far from clear. Further analysis of more mutants may clarify this. What remains unknown is what encodes the pollen Z alleles and how they function in concert with the S alleles. It is of great interest to note that no homologous sequences have been identified in pistil tissue. This, of course, points again to the idea that the pollen and pistil components are separate entities. The identity of the pistil component is currently unknown.

V. Summary

A. The Pistil S Genes: A Range of Different Self-Incompatibility Systems Emerging?

From what has been discussed in this article so far, it is clear that there are several different SI systems in the flowering plants. We mentioned earlier that SI systems are generally categorized according to the genetic control of SI, in the pollen which is either sporophytic (SSI) or gametophytic (GSI). Initially, there was the assumption that all SSI systems will be alike, and that all GSI systems would be alike; this does not seem to be true. When the two S genes from *Brassica* and *Nicotiana* were cloned, it was not thought surprising that they were different, because their genetic control is different (SSI and GSI, respectively). What was more surprising was that a different S gene was found in *Papaver rhoeas,* because this has GSI. Further, *Secale cereale* (which also has GSI) has an S-linked gene with homology to the SLG gene from *Brassica*. To complicate matters even further, in *Ipomoea* (SSI), although an SRK-like gene has been found, it is not linked to the S locus, raising the possibility that the S gene(s) from *Ipomoea* may be different from those in *Brassica*. Similarly, in *Pyrus serotina* (GSI), RNases have been identified, but whether they have a role in SI is unclear. The rather surprising findings open up the field for some exciting work to resolve how these very precisely controlled systems have evolved, how many types of different systems there may be, and how they operate. This must urge detailed molecular and biochemical studies on a variety of self-incompatible (and self-compatible) species in other genera.

B. Evolutionary Relationships of S Genes

The fact that there appear to be many different SI mechanisms for controlling self-fertilization has important implications in our interpretation of the evolutionary relationships between different SI systems, because it is clear that several unrelated S genes have been identified so far. This would strongly suggest that SI has arisen independently several times (how many times will not become clear for some time, because a wider range of species needs to be investigated).

With the intraspecific studies, relationships between the S genes are emerging in the Brassicaceae and Solanaceae, because these have been studied most intensively to date. In *Brassica* it is clear that the S locus is complex, with SLG, SLR, and SRK forming members of a multigene family. The three genes are clearly related because they have a number of features in common and have high homologies with each other. The SLR gene appears to be highly conserved, with an extremely low level of allelic variation, but no role in SI. The evidence so far suggests that the SLG and SLR loci are diverged products of gene duplication, sharing a common lineage; their products may have evolutionarily related functions. The high degree of homology between the SLG gene and the extracellular domain of the SRK gene suggests that these two genes were also generated as a result of an ancestral gene duplication. Because SRK-like genes have been found in several plant species so far, it would suggest that the SRK family is extremely ancient and implies that members of this family may have a diverse range of roles in plant signaling systems.

In the many solanaceous species studied, the story here is a little more simple than in *Brassica,* because the S RNases appear to be a single-copy gene. In contrast to *Brassica,* the S locus is highly polymorphic, with amino acid similarities within species as low as 40%. Thus, the S alleles appear to be highly divergent. There are also some interspecific similarities that are greater than intraspecific similarities; for example, some of the S alleles from *Nicotiana alata* exhibit a higher degree of sequence similarity to S alleles from *Petunia inflata* than to other *N. alata* alleles, indicating that the S polymorphism probably predates the divergence of these species. Because the divergence is thought to have occurred about 27 million years ago, SI must be considered as emerging even earlier than this.

C. How Is S-Allelic Specificity Encoded?

With a wealth of sequence data from Brassicaceae and the Solanaceae, we are beginning to get some idea of the sequences within these genes that could be involved in the generation of allelic variability. In *Brassica,* the SLG protein has two highly conserved and two highly polymorphic (44% homology) regions, which suggests that the latter are likely to have a key role in allelic specificity (Nasrallah *et al.,* 1987). In *Nicotiana,* in addition to five conserved regions that are important for function for the S RNase, there are several hypervariable regions that are candidates for determining allelic specificity (Ioerger *et al.,* 1991). In *Papaver,* although only two S genes have been sequenced so far, the amino acid sequences are only 55% homologous (Walker, 1994). Again, although there are conserved regions, several variable regions where allelic specificity could be encoded have also been identified.

D. The S Gene Pollen Components: Still Largely Unknown

It has become clear from this article that little still is known about the pollen component in any SI system. One question that has been asked for many years is whether the pistil and pollen components are identical or different, or whether they are encoded by separate, but tightly linked components of a supergene. We appear not to have progressed much further on this question.

In *Brassica,* transcripts with homology to SLG and SRK genes are expressed in the male reproductive tissues at very low levels (Nasrallah and Nasrallah, 1989; Guilluy *et al.,* 1991; Sato *et al.,* 1991). Similarly, in

Nicotiana, while a transcript with homology to the *S* RNase has been found in male reproductive tissues, this again occurs at low levels (Dodds *et al.,* 1993). Whether these transcripts play a role in SI is not known. Although these findings are not inconsistent with the dimer hypothesis of Lewis (1954), which predicts that the same S gene encodes both the stylar and the pollen component of the S gene, it would be prudent to reserve judgment on their significance until a more detailed analysis has taken place, because it is difficult to reconcile these concepts with the proposed models for the inhibition of incompatible pollen.

In *Brassica,* an alternative candidate for the male S determinant has emerged, because a pollen coat protein (PCP) that interacts with SLG has been identified (Doughty *et al.,* 1993). Similarly, in *Phalaris,* a pollen-specific gene linked to the S locus has been identified (Li *et al.,* submitted) that appears to have no homologous sequences in the pistil. These two findings are consistent with the idea that the pollen and pistil components are separate entities, which would argue for a "lock and key" type of model for the interaction of complementary S components.

E. Cell–Cell Signaling: Emerging Evidence for Signal Transduction in Self-Incompatibility

At the start of this article it was stated that not much was known about signal transduction systems in plants, let alone in SI. However, the concept that cell–cell communication and signaling plays an important role in SI has proved to be correct. Over the last few years real evidence has emerged that shows that signal transduction is involved in mediating SI in a range of species.

In *Brassica* the identification of the S receptor kinase (SRK) gene, which has homology with the gene encoding ZmPK1, a serine/threonine receptor kinase of unknown function from maize (Stein *et al.,* 1991; Walker and Zhang, 1990), has had a major impact on models of how SI operates in this species. The SRK gene encodes a functional serine/threonine protein kinase (Goring and Rothstein, 1992) that consists of a membrane-bound receptor kinase, with an extracellular domain homologous to SLG. Further studies will, no doubt, elucidate exactly what role SRK plays in the SI response in *Brassica.* Several other plant kinase genes with homology to the SRK gene have since been identified. In *Ipomoea* an SRK-like gene has been isolated (Y. Kowyama, personal communication), but this gene does not appear to be linked to the S locus. As previously mentioned, members of the SRK family may have a diverse range of roles in plant signaling systems, so it is possible that the gene from *Ipomoea* may play another role in pollination.

Good evidence for the involvement of signal transduction in the SI response in *Papaver* has emerged. Transient rises in cytosolic Ca^{2+}, specifically produced as a consequence of an incompatible reaction, provided strong evidence for a Ca^{2+}-dependent signal transduction pathway (Franklin-Tong *et al.*, 1993). Because the rise in $[Ca^{2+}]_i$ appears to be generated intracellularly, it suggests a role for the phosphoinositide pathway in the SI response, because it is generally believed that stimulus-evoked Ca^{2+} release from internal stores is mediated by $InsP_3$. However, at this stage the involvement of external Ca^{2+} cannot be ruled out. Further evidence for the operation of a signaling pathway in pollen of *Papaver* comes from the observation of rapid and transient phosphorylation of several pollen proteins, with differences between incompatible and compatible responses (Franklin-Tong *et al.*, 1992), which implicates the activation of protein kinases, and from the observation that expression of "response genes" within incompatible pollen of *Papaver rhoeas* is induced as the result of the SI interaction. In *Secale* there is further evidence of the involvement of protein kinase stimulation as a result of an SI response. Rapid increases in the levels of phosphorylation in pollen that has undergone either an incompatible or compatible SI response, with a differential between levels of phosphorylation in the two responses (Wehling *et al.*, 1994a), suggests that pollination may induce kinase activation, with SI operating over and above this. Inhibitor studies have provided further evidence for the involvement of protein kinases and Ca^{2+} in the mediation of the SI response in rye.

VI. Conclusions

There are still many questions remaining to be addressed with respect to the components involved in the determination and functioning of the SI response. Over the last few years real progress has been made, especially with respect to information on the S genes in the Brassicaceae and the Solanaceae. There is also firm evidence for the involvement of signaling pathways in mediating SI in a range of species. Many questions remain to be answered. For instance, how many SI systems are there and what mechanisms are involved? What is the distribution of SI systems and how have they evolved? Is self-compatibility ancestral to self-incompatibility? The question of the number of S alleles has been raised by population genetic studies. What is the molecular or biochemical basis underlying this? Now that signaling has been identified as being involved in SI, there is real scope for attempting to elucidate in detail the pathways involved, which could be a breakthrough for the study of signal transduction in plants, as there is a scarcity of model systems in which to study this. In addition, the pollen component remains, to a great extent, a mystery. Thus, there is much work still to be undertaken in this exciting field of research.

References

Ai, Y., Singh, A., Coleman, C. E., Ioerger, T. R., Kheyr-Pour, A., and Kao, T.-H. (1990). Self-incompatibility in *Petunia inflata:* Isolation and characterization of cDNAs encoding S-allele-associated proteins. *Sex. Plant Reprod.* **3,** 130–138.

Anderson, M. A., Cornish, E. C., Mau, S.-L., Williams, E. G., Hoggart, R., Atkinson, A., Bonig, I., Grego, B., Simpson, R., Roche, P. J., Haley, J. D., Penschow, J. D., Niall, H. D., Tregear, G. W., Cochlan, J. P., Crawford, R. J., and Clarke, A. E. (1986). Cloning of a cDNA for a stylar glycoprotein associated with expression of self-incompability in *Nicotiana alata. Nature* (*London*) **321,** 38–44.

Anderson, M. A., McFadden, G. I., Bernatzky, R., Atkinson, A., Orpin, T., Dedman, H., Tregear, G., Fernley, R., and Clarke, A. E. (1989). Sequence variability of three alleles of the self-incompatibility gene of *Nicotiana alata. Plant Cell* **1,** 483–491.

Annerstedt, I., and Lundqvist, A. (1967). Genetics of self-incompatibility in *Tradescantia paludosa* (Commelinaceae). *Hereditas* **58,** 13–30.

Arasu, N. T. (1968). Self-incompatibility in angiosperms: A review. *Genetica* **39,** 1–24.

Atwal, K. K. (1993). Studies on the mechanism of pollen inhibition during the self-incompatibility response of *Papaver rhoeas.* Ph.D. Thesis, University of Birmingham.

Atwood, S. S. (1944). Oppositional alleles in natural populations of *Trifolium repens. Genetics* **29,** 428–435.

Barlow, N. (1913). Preliminary note on heterostylism in *Oxalis* and *Lythrum. J. Genet.* **3,** 53–65.

Barlow, N. (1923). Inheritance of three forms in trimorphic species. *J. Genet.* **13,** 133–146.

Barrett, S. C. H., ed. (1992). "Evolution and Function of Heterostyly," Monogr. Theor. Appl. Genet., Vol. 15. Springer-Verlag, Berlin.

Bateman, A. J. (1955). Self-incompatibility in angiosperms. III. Cruciferae. *Heredity* **9,** 53–68.

Bateson, W., and Gregory, R. P. (1905). On the inheritance of heterostylism in *Primula. Proc. R. Soc. London, Ser. B* **76,** 581–586.

Bednarska, E. (1989). Localization of calcium on the stigma surface of *Ruscus asculatus* L. *Planta* **179,** 11–16.

Berridge, M. J., and Irvine, R. F. (1989). Inositol phosphates and cell signaling. *Nature* (*London*) **341,** 197–205.

Boyes, D. C., and Nasrallah, J. B. (1993). Physical linkage of the SLG and SRK genes at the self-incompatibility locus of *Brassica oleracea. Mol. Gen. Genet.* **236,** 369–373.

Boyes, D. C., Chen, C.-H., Tantikanjana, T., Esch, J. J., and Nasrallah, J. B. (1991). Isolation of a second S-locus-related cDNA from *Brassica oleracea:* Genetic relationships between the S-locus and two related loci. *Genetics* **127,** 221–228.

Bredemeijer, M. M., and Blaas, C. (1981). S-specific proteins in styles of self-incompatible *Nicotiana alata. Theor. Appl. Genet.* **59,** 185–190.

Brewbaker, J. L. (1959). Biology of the angiosperm pollen grain. *Indian J. Genet. Plant Breed.* **19,** 121–133.

Brewbaker, J. L., and Kwack, B. H. (1963). The essential role of calcium ions in pollen germination and pollen tube growth. *Am. J. Bot.* **50,** 859–863.

Brewbaker, J. L., and Majumder, S. D. (1961). Cultural studies of the pollen effect and the self-incompatibility inhibition. *Am. J. Bot.* **48,** 457–464.

Broothaerts, W. J., van Laere, A., Witters, R., Preaux, G., Decock, B., van Damme, J., and Vendrig, J. C. (1989). Purification and N-terminal sequencing of style glycoproteins associated with self-incompatibility in *Petunia hybrida. Plant Mol. Biol.* **14,** 93–102.

Bushnell, W. R. (1979). The nature of basic compatibility: Comparisons between pistil-pollen and host-parasite interaction. *In* "Recognition and Specificity in Plant Host-Parasite Interactions," (Daly, J. M., Uritani, I., eds.), pp. 211–227. University Park Press.

Campbell, J. M., and Lawrence, M. J. (1981a). The population genetics of the self-incompatibility polymorphism in *Papaver rhoeas*. I. The number of distribution of S-alleles in families from three localities. *Heredity* **46,** 69–79.

Campbell, J. M., and Lawrence, M. J. (1981b). The population genetics of the self-incompability polymorphism in *Papaver rhoeas*. II. The number and frequency of S-alleles a natural population. *Heredity* **46,** 81–90.

Chen, C-H., and Nasrallah, J. B. (1990). A new class of S sequences defined by a pollen reccesive self-incompability allele of *Brassica oleracea. Mol. Gen. Genet.* **222,** 241–248.

Clark, A. G., and Kao, T.-H. (1991). Excess nonsynonymous substitution at shared polymorphic sites among self-incompatibility alleles of Solanaceae. *Proc. Natl. Acad. Sci. U.S.A.* **88,** 9823–9827.

Clark, K. R., Okuley, J. J., Collins, P. D., and Sims, T. L. (1990). Sequence variability and developmental expression of S-alleles in self-compatible and pseudo-self-compatible *Petunia. Plant Cell* **2,** 815–826.

Clarke, A. E., Anderson, R., and Stone, B. A. (1979). Form and function of arabinogalactans and arabinogalactan-like proteins. *Phytochemistry* **18,** 521–540.

Clarke, A. E., Anderson, M. A., Bacic, T., Harris, P. J., and Mau, S.-L. (1985). Molecular basis of cell recognition during fertilization in higher plants. *J. Cell Sci., Suppl.* **2,** 261–285.

Collins, J. L. (1961). "The Pineapple: Botany, Cultivation and Utilization." Wiley (Interscience), New York.

Cope, F. W. (1962). The mechanism of pollen incompatibility in *Theobroma cacao* L. *Heredity* **17,** 157–182.

Cornish, E. C., Pettitt, J. M., Bonig, I., and Clarke, A. E. (1987). Developmentally controlled expression of a gene associated with self-incompatibility in *Nicotiana alata. Nature* (*London*) **326,** 99–101.

Cornish, E. C., Anderson, M. A., and Clarke, A. E. (1988). Molecular aspects of fertilization in flowering plants. *Annu. Rev. Cell Biol.* **4,** 209–228.

Correns, C. (1912). Selbststerilitat und individualstoffe. *Festschr. Med.-Naturwiss. Ges., Munster* **84,** 186–217.

Crowe, L. K. (1964). The evolution of outbreeding in plants. I. The angiosperms. *Heredity* **19,** 435–457.

Darlington, C. D., and Mather, K. (1949). "The Elements of Genetics." Allen & Unwin, London.

Darwin, C. (1876). "Effects of Cross and Self Fertilisation in the Vegetable Kingdom." John Murray, London.

de Nettancourt, D. (1977). "Incompatibility in Angiosperms." Springer Verlag, Berlin.

Dickinson, H. G. (1990). Self-incompatibility in flowering plants. *BioEssays* **12,** 155–161.

Dickinson, H. G. (1994). Self-pollination. Simply a social disease? *Nature* (*London*) **367,** 517–518.

Dickinson, H. G., Moriarty, J. F., and Lawson, J. R. (1982). Pollen-pistil interaction in *Lilium longiflorum:* The role of the pistil in controlling pollen tube growth following cross- and self-pollination. *Proc. R. Soc. London, Ser. B* **215,** 45–62.

Dickinson H. G., Crabbé, M. J. C., and Gaude, T. (1992). Sporophytic self-incompatibility systems: S-gene products. *Int. Rev. Cytol.* **140,** 525–561.

Dietrich, A., Mayer, J. E., and Hahlbrock, K. (1990). Fungal elicitors traigger rapid, transient and specific protein phosphorylation in parsley cell suspension cultures. *J. Biol. Chem.* **265,** 6360–6368.

Dodds, P. N., Bonig, I., Du, H., Rodin, J., McClure, B. A., Anderson, M. A., Newbiggin, E., and Clarke, A. E. (1993). The S-RNase gene of *Nicotiana alata* is expressed in developing pollen. *Plant Cell* **5,** 1771–1782.

Doughty, J., Hedderson, F., McCubbin, A., and Dickinson, H. G. (1993). Interaction between a coating-borne peptide of the *Brassica* pollen grain and stigmatic S (self-incompatibility)-locus-specific glycoproteins. *Proc. Natl. Acad. Sci. U.S.A.* **90,** 467–471.

Drøbak, B. K. (1992). The plant phosphoinositide system. *Biochem. J.* **288,** 697–712.

Dwyer, K. G., Balent, M. A., Nasrallah, J. B., and Nasrallah, M. E. (1991). DNA sequences of self-incompatibility genes from *Brassica campestris* and *B. oleracea:* Polymorphism predating speciation. *Plant Mol. Biol.* **16,** 481–486.

East, E. M. (1940). The distribution of self-sterility in flowering plants. *Proc. Am. Philos. Soc.* **82,** 449–518.

East, E. M., and Mangelsdorf, A. J. (1925). A new interpretation of the hereditary behavior of self-sterile plants. *Proc. Natl. Acad. Sci. U.S.A.* **11,** 166–183.

East, E. M., and Park, J. B. (1917). Studies on self-sterility. I. The behaviour of self-sterile plants. *Genetics* **2,** 505–609.

Ebert, P. R., Anderson, M. A., Bernatzky, R., Altschuler, M., and Clarke, A. E. (1989). Genetic polymorphism of self-incompatibility in flowering plants. *Cell (Cambridge, Mass.)* **56,** 255–262.

Emerson, S. (1939). A preliminary survey of the *Oenothera organensis* population. *Genetics* **24,** 524–537.

Emerson, S. (1940). Growth of incompatible pollen tubes in *Oenothera organensis. Bot. Gaz. (Chicago)* **101,** 890–911.

Emerson, S. (1941). Linkage relationships of two gametophytic characters in *Oenothera organensis. Genetics* **26,** 469–473.

Fearon, C. H., Hayward, M. D., and Lawrence, M. J. (1984a). Self-incompatibility in ryegrass. VII. The determination of incompatibility genotypes in autotetraploid families of *Lolium perenne* L. *Heredity* **53,** 403–413.

Fearon, C. H., Hayward, M. D., and Lawrence, M. J. (1984b). Self-incompatibility in ryegrass. VIII. The mode of action of S and Z alleles in the pollen of autotetraploids of *Lolium perenne* L. *Heredity* **53,** 415–422.

Fearon, C. H., Hayward, M. D., and Lawrence, M. J. (1984c). Self-incompatibility in ryegrass. IX. Cross-compatibility and seed-set in autotetraploid *Lolium perenne* L. *Heredity* **53,** 423–434.

Fearon, C. H., Cornish, M. A., Hayward, M. D., and Lawrence, M. J. (1994). Self-incompatibility in ryegrass. X. Number and frequency of alleles in a natural population of *Lolium perenne* L. *Heredity* **73,** 262–264.

Ferrari, T. E., and Wallace, D. H. (1975). Germination of *Brassica* pollen and expression of incompatibility *in vitro. Euphytica* **24,** 757–765.

Ferrari, T. E., Bruns, D., and Wallace, D. H. (1981). Isolation of a plant glycoprotein involved with control of intercellular recognition. *Plant Physiol.* **67,** 270–277.

Foote, H. C. C., Ride, J. P., Franklin-Tong, V. E., Walker, E. A., Lawrence, M. J., and Franklin, F. C. H. (1994). Cloning and expression of a distinctive class of self-incompatibility (S-) gene from *Papaver rhoeas* L. *Proc. Natl. Acad. Sci. U.S.A.* **91,** 2265–2269.

Ford, M. A., and Kay, Q. O. N. (1985). The genetics of incompatibility in *Sinapis arvensis* L. *Heredity* **54,** 99–102.

Franklin, F. C. H., Franklin-Tong, V. E., Thorlby, G. J., Atwal, K. K., and Lawrence,

M. J. (1991). Molecular basis of the self-incompatibility mechanism in *Papaver rhoeas* L. *Plant Growth Regul.* **11,** 5–12.

Franklin, F. C. H., Hackett, R. M., and Franklin-Tong, V. E. (1992). Molecular studies of the pollen-stigma interaction. *In* "Perspectives in Plant Cell Recognition" (J. A. Callow and J. G. Green, eds.), SEB Seminar. Ser., pp. 79–103. Cambridge Univ. Press, Cambridge, UK.

Franklin, F. C. H., Atwal, K. K., Ride, J. P., and Franklin-Tong, V. E. (1994). Towards an elucidation of the mechanisms of pollen tube inhibition during the self-incompatibility response in *Papaver rhoeas. In* "Molecular and Cellular Aspects of Plant Reproduction" (R. Scott and A. J. Stead, eds.), pp. 173–190. Cambridge Univ. Press, Cambridge, UK.

Franklin-Tong, V. E., and Franklin, F. C. H. (1992). Gametophytic self-incompatibility in *Papaver rhoeas* L. *Sex. Plant Reprod.* **5,** 1–7.

Franklin-Tong, V. E., Lawrence, M. J., and Franklin, F. C. H. (1988). An *in vitro* bioassay for the stigmatic product of the self-incompatiblity gene in *Papaver rhoeas* L. *New Phytol.* **110,** 109–118.

Franklin-Tong, V. E., Ruuth, E., Marmey, P., Lawrence, M. J., and Franklin, F. C. H. (1989). Characterization of a stigmatic component from *Papaver rhoeas* L. which exhibits the specific activity of a self-incompatibility (S-) gene product. *New Phytol.* **112,** 307–315.

Franklin-Tong, V. E., Lawrence, M. J., and Franklin, F. C. H. (1990). Self-incompatibility in *Papaver rhoeas* L: Inhibition of incompatible pollen tube growth is dependent on pollen gene expression. *New Phytol.* **116,** 319–324.

Franklin-Tong, V. E., Atwal, K. K., Howell, E. C., Lawrence, M. J., and Franklin, F. C. H. (1991). Self-incompatibility in *Papaver rhoeas* L.: There is no evidence for the involvement of stigmatic ribonuclease activity. *Plant, Cell. Environ.* **14,** 423–429.

Franklin-Tong, V. E., Lawrence, M. J., Thorlby, G. J., and Franklin, F. C. H. (1992). Recognition, signals and pollen responses in the incompatibility reaction in *Papaver rhoeas. In* "Angiosperm Pollen and Ovules" (E. Ottaviano, D. L. Mulcahy, G. Sari-Gorla, and G. Mulcahy-Bergamini, eds.), pp. 84–93. Springer-Verlag, New York.

Franklin-Tong, V. E., Ride, J. P., Read, N. D., Trewavas, A. J., and Franklin, F. C. H. (1993). The self-incompatibility response in *Papaver rhoeas* is mediated by cytosolic free calcium. *Plant J.* **4,** 163–177.

Ganders, F. R. (1979). The biology of heterostyly. *N. Z. J. Bot.* **17,** 607–635.

Germain, E., Leglise, P., and Delorty, F. (1981). Analyse du système d'incompatibilite pollinique observé chez le noisetier *Corylus avellana* L. *Colloq. Rech. Fruitieres, Bordeaux, 1st,* pp. 197–216.

Gerstel, D. U. (1950). Self-incompatibility studies in Guayule. II. Inheritance. *Genetics* **35,** 482–506.

Goring, D. R., and Rothstein, S. J. (1992). The *Brassica* receptor kinase gene in a self-incompatible *Brassica napus* line encodes a functional serine/threonine kinase. *Plant Cell* **4,** 1273–1281.

Goring, D. R., Glavin, T. L., Schafer, U., and Rothstein, S. J. (1993). An S-receptor kinase gene in self-compatible *Brassica napus* has a 1-bp deletion. *Plant Cell* **5,** 531–539.

Gray, J. E., McClure, B. A., Bonig, I., Anderson, M. A., and Clarke, A. E. (1991). Action of the style product of the self-incompatibility gene of *Nicotiana alata* (S-RNase) on *in vitro*-grown pollen tubes. *Plant Cell* **3,** 271–283.

Guilluy, C-M., Trick, M., Heizmann, P., and Dumas, C. (1991). PCR detection of transcripts homologous to the self-incompatibility gene in anthers of *Brassica. Theor. Appl. Genet.* **82,** 466–472.

Haring, V., Gray, J. E., McClure, B. A., Anderson, M. A., and Clarke, A. E. (1990). Self-incompatibility. A self-recognition system in plants. *Science* **250,** 937–941.

Harris, P. J., Freed, K., Anderson, M. A., Weinhandl, J. A., and Clarke, A. E. (1987). An enzyme-linked immunoabsorbent assay (ELISA) for *in vitro* pollen growth based on binding of a monoclonal antibody to the pollen tube surface. *Plant Physiol.* **84,** 851–855.

Harris, P. J., Weinhandl, J. A., and Clarke, A. E. (1989). Effect of in vitro pollen growth of an isolated style glycoprotein associated with self-incompatibility in *Nicotiana alata. Plant Physiol.* **89,** 360–367.

Hayman, D. L. (1956). The genetical control of incompatibility in *Phalaris coerulescens* Desf. *Aust. J. Biol. Sci.* **9,** 321–331.

Hayman, D. L., and Richter, J. (1992). Mutations affecting self-incompatibility in *Phalaris coerulescens* Desf. (Poaceae). *Heredity* **68,** 495–503.

Hepler, P. K., and Wayne, R. O. (1985). Calcium and plant development. *Annu. Rev. Plant Physiol.* **36,** 397–439.

Heslop-Harrison, J. (1975). Incompatibility and the pollen-stigma interaction. *Annu. Rev. Plant Physiol.* **26,** 403–425.

Heslop-Harrison, J. (1983). Self-incompatibility: Phenomenology and physiology. *Proc. R. Soc. London, B Ser.* **218,** 371–395.

Heslop-Harrison, Y., and Shivanna, K. R. (1977). The receptive surface of the angiosperm stigma. *Ann. Bot. (London)* [N.S.] **41,** 1233–1258.

Hinata, K., Watanabe, M., Toriyama, K., and Isogai, A. (1993). A review of recent studies on homomorphic self-incompatibility. *Int. Rev. Cytol.* **143,** 257–296.

Hiratsuka, S. (1992a). Detection and inheritance of a stylar protein associated with a self-incompatibility genotype of Japanese pear. *Euphytica* **61,** 55–59.

Hiratsuka, S. (1992b). Characterization of an S-allele-associated protein in Japanese pear. *Euphytica* **62,** 103–110.

Hughes, M. B., and Babcock, E. B. (1950). Self-incompatibility in *Crepis foetida* (L.) subsp. *rhoeadifolia* (Bieb.) Schinz et Keller. *Genetics* **35,** 570–588.

Ioerger, T. R., Clark, A. G., and Kao, T.-H. (1990). Polymorphism at the self-incompatibility locus in Solanaceae predates speciation. *Proc. Natl. Acad. Sci. U.S.A.* **87,** 9732–9735.

Ioerger, T. R., Golke, J. R., Xu, B., and Kao, T.-H. (1991). Primary structural features of the self-incompatibility protein in the Solanaceae. *Sex. Plant Reprod.* **4,** 81–87.

Isogai, A., Takayama, S., Tsukomoto, C., Ueda, Y., Shiozawa, H., Hinata, K., Okazaki, K., and Suzuki, A. (1987). S-locus-specific glycoproteins associated with self-incompatibility in *Brassica campestris. Plant Cell Physiol.* **28,** 1279–1291.

Jacob, V. J. (1980). Pollination, fruit setting and incompatibility in *Cola nitida. Incompatibility Newsl.* **12,** 50–56.

Jaffe, L., Weisenseel, M. H., and Jaffe, L. F. (1975). Calcium accumulation within the growing tips of pollen tubes. *J. Cell Biol.* **67,** 488–492.

Jahnen, W., Lush, W. M., and Clarke, A. E. (1989). Inhibition of in-vitro pollen tube growth by isolated S-glycoproteins of *Nicotiana alata. Plant Cell* **1,** 501–510.

Kamboj, R., and Jackson, J. F. (1986). Self-incompatibility alleles control low molecular weight, basic proteins in pistils of *Petunia hybrida. Theor. Appl. Genet.* **71,** 815–819.

Kandasamy, M. K., Paolillo, D. J., Nasrallah, J. B., and Nasrallah, M. E. (1989). The S-locus specific glycoproteins of *Brassica* accumulate in the cell wall of developing stigma papillae. *Dev. Biol.* **134,** 462–472.

Kaufmann, H., Salamini, F., and Thompson, R. D. (1991). Sequence variability and gene structure at the self-incompatibility locus of *Solanum tuberosum. Mol. Gen. Genet.* **226,** 457–466.

Kheyr-Pour, A. Bintrim, S. B., Ioerger, T. R., Remy, R., Hammond, S. A., and Kao, T.-H. (1990). Sequence diversity of pistil S-proteins associated with gametophytic self-incompatibility in *Nicotiana alata. Sex. Plant Reprod.* **3,** 88–97.

Kirch, H. H., Uhrig, H., Lottspeich, F., Salamini, F., and Thompson, R. D. (1989). Characterization of proteins associated with self-incompatibility in *Solanum tuberosum. Theor. Appl. Genet.* **78,** 581–588.

Knight, R., and Rogers, H. H. (1953). Sterility in *Theobroma cacao* L. *Nature (London)* **72,** 164.

Knight, R., and Rogers, H. H. (1955). Incompatibility in *Theobroma cacao*. *Heredity* **9,** 69–77.

Kowyama, Y., Shimano, N., and Kawase, T. (1980). Genetic analysis of incompatibility in the diploid *Ipomoea* species closely related to the sweet potato. *Theor. Appl. Genet.* **58,** 149–155.

Kowyama, Y., Takahasi, H., Muraoka, K., Tani, T., Hara, K., and Shiotani, I. (1994). The number, frequency and dominance relationships of S-alleles in diploid *Ipomoea trifida*. *Heredity* (in press).

Lalonde, B. A., Nasrallah, M. E., Dwyer, K. G., Chen, C.-H., Barlow, B., and Nasrallah, J. B. (1989). A highly conserved *Brassica* gene with homology to the S-locus-specific glycoprotein structural gene. *Plant Cell* **1,** 249–258.

Lane, M. D., and Lawrence, M. J. (1993). The population genetics of the self-incompatibility polymorphism in *Papaver rhoeas*. VII. The number of S-alleles in the species. *Heredity* **71,** 596–602.

Lane, M. D., and Lawrence, M. J. (1995). The population genetics of the self-incompatibility polymorphism in *Papaver rhoeas*. X. An association between incompatibility genotype and seed dormancy. *Heredity* (in press).

Larsen, K. (1977). Self-incompatibility in *Beta vulgaris* L. I. Four gametophytic, complimentary S-loci in sugar beet. *Hereditas* **85,** 227–248.

Lawrence, M. J., and Franklin-Tong, V. E. (1994). The population genetics of the self-incompatibility polymorphism in *Papaver rhoeas*. IX. Evidence of an extra effect of selection acting on the S-locus. *Heredity* **72,** 353–364.

Lawrence, M. J., and O'Donnell, S. (1981). The population genetics of the self incompatibility polymorphism in *Papaver rhoeas*. III. The number and frequency of S-alleles in two further natural populations (R102 and R104). *Heredity* **47,** 53–61.

Lawrence, M. J., Lane, M. D., O'Donnell, S., and Franklin-Tong, V. E. (1993). The population genetics of the self-incompatibility polymorphism in *Papaver rhoeas*. V. Cross-classification of the S-alleles of samples from three natural populations. *Heredity* **71,** 581–590.

Lawrence, M. J., O'Donnell, S., Lane, M. D., and Marshall, D. F. (1994). The population genetics of the self-incompatibility polymorphism in *Papaver rhoeas*. VIII. Sampling effects as a possible cause of unequal allele frequencies. *Heredity* **72,** 345–352.

Lee, H. S., Huang, S., and Kao, T.-H. (1994). S-proteins control rejection of incompatible pollen in *Petunia inflata*. *Nature (London)* **367,** 560–563.

Lewis, D. (1947). Competition and dominance of incompatibility alleles in diploid pollen. *Heredity* **1,** 85–108.

Lewis, D. (1948). Structure of the incompatibility gene. I. Spontaneous mutation rate. *Heredity* **2,** 219–236.

Lewis, D. (1949a). Incompatibility in flowering plants. *Biol. Rev. Cambridge Philos. Soc.* **24,** 472–496.

Lewis, D. (1949b). Structure of the incompatibility gene. II. Induced mutation rate. *Heredity* **3,** 339–355.

Lewis, D. (1951). Structure of the incompatibility gene. III. Types of spontaneous and induced mutation. *Heredity* **5,** 399–414.

Lewis, D. (1954). Comparative incompatibility in angiosperms and fungi. *Adv. Genet.* **6,** 235–285.

Lewis, D. (1960). Genetic control of specificity and activity of the S antigen in plants. *Proc. R. Soc. London, Ser. B* **151,** 468–477.

Lewis, D. (1976). Incompatibility in flowering plants. *Recept. Recognition, Ser. A* **2,** 167–198.

Lewis, D., and Crowe, L. K. (1954). Structure of the incompatibility gene. IV. Types of mutation in *Prunus avium* L. *Heredity* **8,** 357–363.

Lewis, D., Verma, S. C., and Zuberi, M. I. (1988). Gametophytic-sporophytic incompatibility in the Cruciferae—*Raphanus sativus*. *Heredity* **61,** 355–366.

Lewis, D. H. (1980). Are there inter-relations between the metabolic role of boron, synthesis of phenolic phytoalexins and the germination of pollen? *New Phytol.* **84,** 261–270.

Li, X., Nield, J., Hayman, D., and Langridge, P. Cloning the self-incompatibility gene from pollen of the grass *Phalaris coerulescens* Desf. *Plant Cell* (submitted for publication).

Lundqvist, A. (1956). Self-incompatibility in rye. I. Genetic control in the diploid. *Hereditas* **42,** 293–348.

Lundqvist, A. (1964). The genetics of incompatibility. *Genet. Today, Proc. Int. Congr., 11th, 1963,* Vol. 1, pp. 637–647.

Lundqvist, A. (1979). One-locus sporophytic self-incompatibility in the carnation family, Caryophyllaceae. *Hereditas* **91,** 307.

Lundqvist, A. (1990a). Variability within and among populations in the 4-gene system for control of self-incompatibility in *Ranunculus polyanthemos*. *Hereditas* **113,** 47–61.

Lundqvist, A. (1990b). One-locus sporophytic S-gene system with traces of gametophytic pollen control in *Cerastium arvense* ssp. *strictum* (Caryophyllaceae). *Hereditas* **113,** 203–215.

Lundqvist, A. (1991). Four-locus S-gene control of self-incompatibility made probable in *Lilium martagon* (Liliaceae). *Hereditas* **114,** 57–63.

Lundqvist, A., Osterbye, U., Larsen, K., and Linde-Laursen, I. (1973). Complex self-incompatibility systems in *Ranunculus acris* L. and *Beta vulgaris* L. *Hereditas* **74,** 161–168.

Martin, F. W. (1968). The system of self-incompatibility in *Ipomoea*. *J. Hered.* **59,** 263–267.

Mascarenhas, J. P., and Lafountain, J. (1972). Protoplasmic streaming, cytochalasin B and growth of pollen tubes. *Tissue Cell* **4,** 11–14.

Mau, S.-L., Raff, J. W., and Clarke, A. E. (1982). Isolation and partial characterization of components of *Prunus avium* L. styles, including an antigenic glycoprotein associated with a self-incompatibility genotype. *Planta* **156,** 505–516.

Mau, S.-L., Williams, E. G., Atkinson, A., Cornish, E. C., Grego, B., Simpson, R., Kheyr-Pour, A., and Clarke, A. E. (1986). Style proteins of a wild tomato (*Lycopersicon peruvianum*) associated with expression of self-incompatibility. *Planta* **169,** 184–191.

McClure, B. A., Haring, V., Ebert, P. R., Anderson, M. A., Simpson, R. J., Sakiyama, F., and Clarke, A. E. (1989). Style self-incompatibility gene products of *Nicotiana alata* are ribonucleases. *Nature (London)* **342,** 955–957.

McClure, B. A., Gray, J. E., Anderson, M. A., and Clarke, A. E. (1990). Self-incompatibility in *Nicotiana alata* involves degradation of pollen rRNA. *Nature (London)* **347,** 757–760.

Me, G., and Radicati, L. (1983). Studies of pollen incompatibility in some filbert (*Corylus avellana* L.) cvs. and selections. *In* "Pollen: Biology and Implications for Plant Breeding" (D. L. Mulcahy and E. Ottaviano, eds.), pp. 237–242. Elsevier, New York.

Michell, R. H. (1975). Inositol phospholipids and cell surface receptor function. *Biochim. Biophys. Acta* **415,** 81–147.

Miller, D. D., Callaham, D. A., Gross, D. J., and Hepler, P. K. (1992). Free Ca^{2+} gradient in growing pollen tubes of *Lilium*. *J. Cell Sci.* **101,** 7–12.

Murfett, J., Atherton, T., Mou, B., Gasser, C., and McClure, B. A. (1994). S-RNase expressed in transgenic *Nicotiana* causes S-allele-specific pollen rejection. *Nature (London)* **367,** 563–566.

Nasrallah, J. B., and Nasrallah, M. E. (1989). The molecular genetics of self-incompatibility in *Brassica*. *Annu. Rev. Genet.* **23,** 121–139.

Nasrallah, J. B., Kao, T.-H., Goldberg, M. L., and Nasrallah, M. E. (1985). A cDNA clone encoding an S-specific glycoprotein from *Brassica oleracea*. *Nature (London)* **318,** 263–267.

Nasrallah, J. B., Kao, T.-H., Chen, C.-H., Goldberg, M. L., and Nasrallah, M. E. (1987).

Amino acid sequence of glycoproteins encoded by three alleles of the S-locus of *Brassica oleracea*. *Nature (London)* **326,** 617–619.

Nasrallah, J. B., Yu, S.-M., and Nasrallah, M. E. (1988). Self-incompatibility genes of *Brassica oleracea:* Expression, isolation and structure. *Proc. Natl. Acad. Sci. U.S.A.* **85,** 5551–5555.

Nasrallah, J. B., Rundle, S. J., and Nasrallah, M. E. (1994). Genetic evidence for the requirement of the Brassica S-locus receptor kinase gene in the self-incompatibility response. *Plant J.* **5,** 373–384.

Nasrallah, M. E., and Wallace, D. H. (1967a). Immunogenetics of self-incompatibility in *Brassica oleracea* L. *Heredity* **22,** 519–527.

Nasrallah, M. E., and Wallace, D. H. (1967b). Immunochemical detection of antigens in self-incompatibility genotypes of cabbage. *Nature (London)* **213,** 700–701.

Nasrallah, M. E., Barber, J., and Wallace, D. H. (1970). Self-incompatibility proteins in plants: Detection, genetics and possible mode of action. *Heredity* **25,** 23–27.

Nasrallah, M. E., Wallace, D. H., and Savo, R. M. (1972). Genotype, protein and phenotype relationships in self-incompatibility of *Brassica*. *Genet. Res.* **20,** 151–160.

Nasrallah, M. E., Kandasamy, M. K., and Nasrallah, J. B. (1992). A genetically defined trans-acting locus regulates S-locus function in *Brassica*. *Plant. J.* **2,** 497–506.

Nishio, T., and Hinata, K. (1977a). Analysis of S-specific proteins in stigmas of *Brassica oleracea* L. by isoelectric focussing. *Heredity* **38,** 391–396.

Nishio, T., and Hinata, K. (1977b). Positive PAS reaction of S-specific proteins in stigmas of *Brassica oleracea* L. *Incompatibility Newsl.* **8,** 31–33.

Nishio, T., and Hinata, K. (1978a). Stigma proteins in self-incompatible *Brassica campestris* L. and self-compatible relatives, with special reference to S-allele specificity. *Jpn. J. Genet.* **53,** 27–33.

Nishio, T., and Hinata, K. (1978b). S-allele specificity of stigma proteins in *Brassica oleracea* and *B. campestris*. *Heredity* **41,** 93–100.

Nishio, T., and Hinata, K. (1979). Purification of an S-specific glycoprotein in *Brassica campestris* L. and self-incompatible relatives, with special reference to S-allele specificity. *Jpn. J. Genet.* **53,** 27–33.

Nishio, T., and Hinata, K. (1980). Rapid detection of S-glycoproteins of self-incompatible crucifers using the conA reaction. *Euphytica* **29,** 217–221.

Nishio, T., and Hinata, K. (1982). Comparative studies on S-glycoproteins purified from different S-genotypes in self-incompatible *Brassica* species. I. Purification and chemical properties. *Genetics* **100,** 641–647.

Obermeyer, G., and Weisenseel, M. H. (1991). Calcium channel blocker and calmodulin antagonists affect the gradient of free calcium ions in lily pollen tubes. *Eur. J. Cell Biol.* **56,** 319–327.

Ockendon, D. J. (1974). Distribution of self-incompatibility alleles and breeding structure of open-pollinated cultivars of Brussels sprouts. *Heredity* **33,** 159–171.

Ockendon, D. J. (1985). Genetics and physiology of self-incompatibility in *Brassica*. *In* "Current Communications in Molecular Biology," (I. Sussex, ed.), pp. 1–6. Cold Spring Harbor Lab., Cold Spring Harbor, NY.

O'Donnell, S., and Lawrence, M. J. (1984). The population genetics of the self-incompatibility polymorphism in *Papaver rhoeas*. IV. The estimation of the number of alleles in a population. *Heredity* **53,** 495–507.

O'Donnell, S., Lane, M. D., and Lawrence, M. J. (1993). The population genetics of the self-incompatibility polymorphism in *Papaver rhoeas*. VI. Estimation of the overlap between the allelic complements of a pair of populations *Heredity* **71,** 591–595.

Østerbye, U. (1975). Self-incompatibility in *Ranunculus acris* L. I. Genetic interpretation and evolutionary aspects. *Hereditas* **80,** 91–112.

Østerbye, U. (1977). Self-incompatibility in *Ranunculus acris* L. II. Four S-loci in a german population. *Hereditas* **87,** 173–178.

Picton, J. M., and Steer, M. W. (1983). Evidence for the role of calcium ions in extension in pollen tubes. *Protoplasma* **115,** 11–17.

Picton, J. M., and Steer, M. W. (1985). The effects of ruthenium red, lanthanum, isothiocyanate and trifluoperazine on vesicle transport, vesicle fusion and tip extension in pollen tubes. *Planta* **163,** 20–26.

Polya, G. M., Micucci, V., Rae, A. L., Harris, P. J., and Clarke, A. E. (1986). Ca^{2+}-dependent protein phosphorylation in germinated pollen of *Nicotiana alata,* an ornamental tobacco. *Physiol. Plant.* **67,** 151–157.

Raff, J. W., Knox, R. B., and Clarke, A. E. (1981). Style antigens of *Prunus avium. Planta* **153,** 125–129.

Rathore, K. S., Cork, R. J., and Robinson, K. R. (1991). A cytoplasmic gradient of Ca^{2+} is correlated with the growth of lily pollen tubes. *Dev. Biol.* **148,** 612–619.

Reiss, H.-D., and Herth, W. (1978). Visualization of the Ca^{2+} gradient in growing pollen tubes of *Lilium longiflorum* with chlortetracycline fluorescence. *Protoplasma* **97,** 373–377.

Reiss, H.-D., and Herth, W. (1985). Nifedipine-sensitive calcium channels are involved in polar growth of *Lilium* pollen tubes. *J. Cell Sci.* **76,** 247–254.

Richards, A. J. (1986). "Plant Breeding Systems." Allen & Unwin, London.

Riley, H. P. (1932). Self-sterility in sheperd's purse. *Genetics* **17,** 231–295.

Roberts, I. N., Stead, A. D., Ockendon, D. J., and Dickinson, H. G. (1979). A glycoprotein associated with the acquisition of the self-incompatibility system by maturing stigmas of *Brassica oleracea. Planta* **146,** 179–183.

Roberts, I. N., Gaude, T. C., Harrod, G., and Dickinson, H. G. (1983). Pollen-stigma interactions in *Brassica oleracea:* A new pollen germination medium and its uses in elucidating the mechanism of self-incompatibility. *Theor. Appl. Genet.* **65,** 231–238.

Rundle, S. J., Nasrallah, M. E., and Nasrallah, J. B. (1993). Effects of inhibitors of proteinserine/threonine phosphatases on pollination in *Brassica. Plant Physiol.* **103,** 1165–1171.

Sampson, D. R. (1957). The genetics of self- and cross-incompatibility in *Brassica oleracea. Genetics* **42,** 253–263.

Sarker, R. H., Elleman, C. J., and Dickinson, H. G. (1988). The control of pollen hydration in *Brassica* requires continued protein synthesis whilst glycosylation is necessary for intra-specific incompatibility. *Proc. Natl. Acad. Sci. U.S.A.* **85,** 4340–4344.

Sato, T., Thorsness, M. K., Kandasay, M. K., Nishio, T., Hirai, M., Nasrallah, J. B., and Nasrallah, M. E. (1991). Activity of an S-locus gene promoter in pistils and anthers of transgeneic *Brassica. Plant Cell* **3,** 867–876.

Scutt, C. P., Gates, P. J., Gatehouse, J. A., Boulter, D., and Croy, R. R. D. (1990). A cDNA-encoding an S-locus specific glycoprotein from *Brassica oleracea* plants containing the S5 self-incompatibility allele. *Mol. Gen. Genet.* **220,** 409–413.

Scutt, C. P., Fordham-Skelton, A. P., and Croy, R. R. D. (1993). Ocadaic acid causes breakdown of self-incompatibility in *Brassica oleracea:* Evidence for the involvement of proetin phosphatases in the incompatiblity response. *Sex. Plant Reprod.* **6,** 282–285.

Sedgley, M. (1974a). Assessment of serological techniques for S-allele identification in *Brassica oleracea. Euphytica* **23,** 543–551.

Sedgley, M. (1974b). The concentration of S-protein in stigmas of *Brassica oleracea* plants homozygous and heterozygous for a given S-allele. *Heredity* **33,** 412–416.

Sharma, N., and Shivanna, K. R. (1983). Lectin-like components of pollen and complementary saccharide moiety of the pistil are involved in self-incompatibility recognition. *Curr. Sci.* **52,** 913–916.

Sharma, N., and Shivanna, K. R. (1986). Self-incompatibility, recognition and inhibition in *Nicotiana alata. In* "Biotechnology and Ecology of Pollen" (D. L. Mulcahy, G. B. Mulcahy, and E. Ottaviano, eds.), pp. 179–184. Springer-Verlag, Berlin.

Singh, A., and Kao, T.-H. (1992). Gametophytic self-incompatibility: Biochemical, molecular genetic and evolutionary aspects. *Int. Rev. Cytol.* **140,** 449–483.

Singh, A., and Paolillo, D. J. (1989). Towards an *in vitro* bioassay for the self-incompatibility response in *Brassica oleracea. Sex. Plant Reprod.* **2,** 277–280.

Singh, A., Perdue, T. D., and Pailillo, D. J. (1989). Pollen-pistil interactions in *Brassica oleracea:* Cell Calcium in self and cross pollen grains. *Protoplasma* **151,** 57–61.

Singh, A., Ai, Y., and Kao, T.-H. (1991). Characterization of ribonuclease activity of three S-allele-associated proteins of *Petunia inflata. Plant. Physiol.* **96,** 61–68.

Stein, J. C., Howlett, B., Boyes, D. C., Nasrallah, M. E., and Nasrallah, J. B. (1991). Molecular cloning of a putative receptor protein kinase gene encoded at the self-incompatibility locus of *Brassica oleracea. Proc. Natl. Acad. Sci. U.S.A.* **88,** 8816–8820.

Stevens, J. P., and Kay, Q. O. N. (1989). The number, dominance relationships and frequencies of self-incompatibility alleles in a natural population of *Sinapis arvensis* L. in South Wales. *Heredity* **62,** 199–205.

Stout, A. B. (1916). Self- and cross-pollination in *Cichorium intybus* with reference to sterility. *Mem. N.Y. Bot. Gard.* **6,** 333–454.

Takayama, S., Isogai, A., Tsukomoto, C., Ueda, Y., Hinata, K., Okazaki, K., and Suzuki, A. (1986a). Isolation and some characterization of S-locus-specific glycoproteins associated with self-incompatibility in *Brassica campestris. Agric. Biol. Chem.* **50,** 1365–1367.

Takayama, S., Isogai, A., Tsukomoto, C., Ueda, Y., Hinata, K., Okazaki, K., Koseki, K., and Suzuki, A. (1986b). Structure of carbohydrate chains of S-glycoproteins in *Brassica campestris* associated with self-incompatibility. *Agric. Biol. Chem.* **50,** 1673–1676.

Takayama, S., Isogai, A., Tsukomoto, C., Ueda, Y., Hinata, K., Okazaki, K., and Suzuki, A. (1987). Sequences of S-glycoproteins, products of the *Brassica campestris* self-incompatibility locus. *Nature (London)* **326,** 102–105.

Tantikanjana, T., Nasrallah, M. E., Stein, J. C., Chen, C.-H., and Nasrallah, J. B. (1993). An alternative transcript of the S locus glycoprotein gene in a class II pollen-recessive self-incompatibility haplotype of *Brassica oleracea* encodes a membrane-anchored protein. *Plant Cell* **5,** 657–666.

Teasdale, J., Daniels, D., Davis, W. C., Eddy, R., and Hadwiger, L. A. (1974). Physiological and cytological similarities between disease resistance and cellular incompatibility responses. *Plant Physiol.* **54,** 690–695.

Thompson, K. F. (157). Self-incompatibility in marrow-stem kale. *Brassica oleracea* var. *acephala.* I. Demonstration of a sporophytic system. *J. Genet.* **55,** 45–60.

Thompson, K. F., and Taylor, J. P. (1966). The breakdown of self-incompatibility in cultivars of *Brassica oleracea. Heredity* **21,** 637–648.

Thompson, M. M. (1979). Genetics of incompatibility in *Corylus avellana* L. *Theor. Appl. Genet.* **54,** 113–116.

Thompson, R. D., and Kirch, H.-H. (1992). The S-locus of flowering plants: When self-rejection is self interest. *Trends Genet.* **8,** 381–387.

Tobias, C. M., Howlett, B., and Nasrallah, J. B. (1992). An *Arabidopsis thaliana* gene with sequence similarity to the S-locus receptor kinase of *Brassica oleracea. Plant Physiol.* **99,** 284–290.

Toriyama, K., Thorsness, M. K., Nasrallah, J. B., and Nasrallah, M. E. (1991). A *Brassica* S-locus gene promoter directs sporophytic expression in the anther tapetum of transgenic *Arabidopsis. Dev. Biol.* **143,** 427–431.

Thewavas, A. J., and Gilroy, S. (1991). Signal transduction in plant cells. *Trends Genet.* **7,** 356–361.

Trick, M. (1990). Genomic sequence of a *Brassica* S-locus-related gene. *Plant Mol. Biol.* **15,** 203–205.

Trick, M., and Flavell, R. B. (1989). A homozygous S-genotype of *Brassica oleracea* expresses two S-like genes. *Mol. Gen. Genet.* **218,** 112–117.

Trick, M., and Heizmann, P. (1992). Sporophytic self-incompatibility systems: *Brassica* S gene family. *Int. Rev. Cytol.* **140,** 485–524.

Umbach, A. L., Lalonde, B. A., Kamdaswamy, M. K., Nasrallah, J. B., and Nasrallah, M. E. (1990). Immunodetection of protein glycoforms encoded by two independent genes of the self-incompatibility multigene family of *Brassica. Plant Physiol.* **93,** 739–747.

van der Donk, J. A. W. M. (1974). Differential synthesis of RNA in self- and cross-pollinated styles of *Petunia hybrida.* L. *Mol. Gen. Genet.* **131,** 1–8.

Verma, S. C., Malik, R., and Dhir, I. (1977). Genetics of the incompatibility system in the crucifer *Eruca sativa* L. *Proc. R. Soc. London, Ser. B* **196,** 131–159.

Vuilleumier, B. (1967). The origin and evolutionary development of heterostyly in the angiosperms. *Evolution* (*Lawrence, Kans.*) **21,** 210–226.

Walker, E. A. (1994). Cloning, characterizationa and expression in *E. coli* of S-(self-incompatibility) alleles from *Papaver rhoeas.* Ph.D. Thesis, University of Birmingham.

Walker, J. C., and Zhang, R. (1990). Relationship of a putative receptor from maize to the S-locus glycoprotein of *Brassica. Nature* (*London*) **345,** 743–746.

Wehling, P., Hackauf, B., and Wricke, G. (1994a). Phosphorylation of pollen proteins in relation to self-incompatibility in rye (*Secale cereale* L). *Sex. Plant Reprod.* **7,** 1–11.

Wehling, P., Hackauf, B., and Wricke, G. (1994b). Identification of self-incompatibility related PCR fragments in rye (*Secale cereale* L) by DGGE. *Plant J.* (submitted for publication).

Whitehouse, H. L. K. (1950) Multiple allelomorph incompatibility of pollen and style in the evolution of the angiosperms. *Ann. Bot* (*London*) [N.S.] **14,** 199–216.

Williams, E. G., Ramm-Anderson, S., Dumas, C., Mau, S.-L., and Clarke, A. E. (1982). The effect of isolated components of *Prunus avium* L. styles on in-vitro growth of pollen tubes. *Planta* **156,** 517–519.

Williams, R. D., and Williams, W. (1947). Genetics of red clover (*Trifolium pratense* L.) compatibility. III. The frequency of incompatibility S-alleles in two non-pedigree populations of red clover. *J. Genet.* **48,** 67–79.

Woodward, J. R., Bacic, A., Jahnen, W., and Clarke, A. E. (1989). N-linked glycan chains on S-allele-associated glycoproteins from *Nicotiana alata. Plant Cell* **1,** 511–514.

Wright, S. (1939). The distribution of self-sterility alleles in populations. *Genetics* **24,** 538–552.

Xu, B., Grun, P., Kheyr-Pour, A., and Kao, T.-H. (1990a). Identification of pistil-specific proteins associated with three self-incompatibility alleles in *Solanum chacoense. Sex. Plant Reprod.* **3,** 54–60.

Xu, B., Mu, J., Nevins, D., Grun, P., and Kao, T.-H. (1990b). Cloning and sequencing of cDNAs encoding two self-incompatibility associated proteins in *Solanum chacoense. Mol. Gen. Genet.* **224,** 341–346.

Zuberi, M. I., and Lewis, D. (1988). Gametophytic-sporophytic incompatibility in the Cruciferae—*Brassica campestris. Heredity* **61,** 367–377.

Note Added in Proof. Two papers have been recently published which demonstrate that the RNase activity of the S glycoproteins from *Petunia* and *Lycopersicon* is essential for rejection of incompatible pollen: Huang, S., Lee, H-S., Karunanandaa, B., and Kao, T-H. (1994). Ribonuclease activity of *Petunia inflata* S proteins is essential for rejection of self-pollen. *Plant Cell,* **6,** 1021–1028; Royo, J., Kunz, C., Kowyama, Y., Anderson, M., Clarke, A. E., and Newbigin, E. (1994). Loss of a histidine residue at the active site of *S*-locus ribonuclease is associated with self-compatibility in *Lycopersicon peruvianum. Proc. Natl. Acad. Sci. U.S.A.* **91,** 6511–6514.

Mitosis and Meiosis in Cultured Plant Cells and Their Relationship to Variant Cell Types Arising in Culture

Vittoria Nuti Ronchi
Institute of Mutagenesis and Differentiation, CNR, 56124 Pisa, Italy

Totipotency is an unique property of plant cells. It is the capacity to repeat, starting from a single somatic cell, the developmental pathway normally followed by a fertilized egg in the ovary. This remarkable capacity, until now a puzzling phenomenon, is fully expressed when plant cells are cultured *in vitro,* allowing the regeneration of new plants completely modeled on their mother plant. The finding that transition from mitosis to meiosis may occur in cultured plant cells, and therefore possibly an alternative pathway of the mitotic process under stressing conditions, may offer a possible solution for the totipotency enigma. These data and the unexpected finding that plants regenerated from cultured plant cells have often deviated from the norm, have promoted the study of plant cell behavior in culture. This article gives details about the different deregulations of the cell cycle occurring in culture, in comparison with similar phenomena described in other organisms. The meiotic process, considered an alternative of the mitotic process when it occurs in somatic cultured plant cells, is discussed in detail, taking into account the importance, for evolutionary purposes, of this source of genetic variation. A new concept of cell totipotency is formulated, as expressed in plant cells when, by means of a meiotic-like process, they acquire a gametic condition.

KEY WORDS: Totipotency, Mitosis, Meiosis, Cell cycle, Somatic embryogenesis, Plant regeneration, Polyploidy, Somaclonal variation.

I. Introduction

All living somatic plant cells have been endowed by nature with a remarkable plasticity that has been cleverly exploited by humans, in the course of time, to improve the quality and quantity of crop plants; this plasticity also means rapid acquisition of fitness to the most unlikely environments, and an aptitude to respond to different stimuli and to various manipulations. When plant tissues and/or cells are cultured *in vitro,* they may express the ability to generate new plants completely modeled on the mother plant. Some plant cells, therefore, possess a property known as totipotency, namely the capacity to repeat, starting from a single somatic cell, the developmental pathway normally followed by a fertilized egg in the ovary. Although it is possible, at present, to handle skillfully, *in vitro,* most of the plants of the world, totipotency is still a puzzling phenomenon. Still mysterious to us is what triggers a somatic differentiated plant cell to reenter mitosis, and even more unexplained is what distinguishes the cells destined to develop as an embryo from those allowed only to proliferate as undifferentiated callus. Moreover, the manipulation of organs, tissues, and cells removed from the plant and cultured on completely synthetic media has provided the unexpected finding that plants regenerated *in vitro* are often strikingly deviating from the norm and from the mother plant. The phenomenon, known as *somaclonal variation* (Larkin and Scowcroft, 1981), has been described in different systems; variations in both qualitative and quantitative genetic traits have also been observed, besides in regenerates, in regenerate-derived families. The mutant characters are, in most cases, similar to the ones induced by means of physical or chemical agents; but the observed frequencies are far in excess of those reported for classic mutagenesis. Up to now, several hypotheses have been put forward to explain the phenomenon; it is commonly thought that events at the molecular and chromosomal level contribute to induce tissue culture genetic changes, all tracing back to DNA modifications. Somaclonal variation has in fact been ascribed to single-gene variation, transposable element activation, quantitative trait variation, somatic recombination, and changes in chromosome structure and number. In an attempt to identify a single cause of the tissue culture effects, some authors (Phillips *et al.,* 1990) have suggested that processes of hypo/hypermethylation could be directly or indirectly connected to mutational events. More data are being produced by means of analysis of the offspring of regenerated plants, and more evidence is accumulating about unexpected results that are difficult to explain by any of the above-mentioned hypotheses. Particularly enigmatic are the results showing the exceedingly frequent induction, via tissue culture, of single recessive mutations, breeding true in subsequent generations; or the heterosis effect shown in *Zea mays* by culture-derived regenerated plants when crossed with the donor inbred plant. This heterosis effect, and the fact that many quantitative traits could be modified in a single experiment, are difficult to explain by simple mutation events (Phillips *et al.,* 1990; Lee *et al.,* 1988). Somatic recombination, as suggested by Evans and Sharp (1983) for tomato regenerated mutants, could better define these results.

Other impressive effects induced by tissue culture concern changes both at the heterochromatin and chromosomal levels. The extension of the phenomenon may be conspicuous in some species (such as *Z. mays*), it may be inherited in subsequent generations of the regenerated plant, and expressed both in mitosis, all through the development, and/or in the course of meiosis. Two important concepts emerge from an analysis of the more recent data on the frequency of chromosomal aberrations in regenerated plants: most authors ascribe the occurrence of the chromosomal aberrations to time early in *in vitro* culture, and the frequent incidence of sectors or mosaics for different karyotypes suggests a multicellular origin of the regenerated plant (Benzion and Phillips, 1988; Rhodes *et al.*, 1986). Because there is also the suggestion that initial conditions for chromosome damage exist in the explant tissue (Barbier and Dulieu, 1980), to understand the range and the inference of the phenomenon it is necessary to go back to the early events occurring in the explant and in the culture to trace the origin of the cytological aberrations affecting the genetic background of the regenerated plants. The knowledge of these mechanisms of variation will also allow, by determining the deregulated sequential steps concerning cell division, investigators to reveal some of the genetic programs controlling the decision of somatic cells to change their developmental destiny.

This article discusses the different deregulations of the cell cycle affecting a series of sequential events during mitosis in cultured plant cells, deregulation that may be so extreme as to allow cells to adopt the meiotic devices that trigger passage from mitosis to meiosis. The surprising inference is that some of these different manners of division displayed by cultured plant cells recall ancestral behavior.

The striking analogies of cell cycle-related proteins in different biological systems (mammals, yeast, frogs, and sea urchins), owing to the fact that, in nature, these genes are highly conserved, may facilitate a better understanding of the comparable processes involved.

II. Evolution of Plant Somatic Cell Suspension Cultures

A. Establishment of a Cell Culture

The establishment of a somatic cell suspension culture, starting from living plant tissue, involves a series of specific choices with regard to plant genotype, explant tissue, medium components, and so on, all factors that have a deep influence on the proliferative and regenerative capacity of the specific cell line. The reports on this subject are numerous, dedicated to most worldwide plants, and surveyed by exhaustive reviews focused on various aspects, problems, and possibilities of plant cell culture. Leaving out, therefore, all details concerned with the above aspects, this article examines the evolution of somatic cell suspension cultures from the moment an explant is removed from the plant up to when small clumps of cells are suspended in a liquid medium and submitted to a recurrent medium change. Whatever the type of explant (hypocotyls, leaves, roots, stems) and medium and growth factors used, the sequence of events follows a precise pattern in relation to cell division; whereas most of the dividing cells appear, after DNA synthesis by G_1 cells, about 48 hr after excision, earlier cytological inspection reveals that a cell population located close to the xylem (wood parenchyma) begins dividing as soon as 3 hr after excision, with a peak, at least in tomato, at 18–20 hr (G. Martini, unpublished observations; Lipucci di Paola *et al.*, 1987). Work has revealed that most of these cells show a deviant pattern of division, leading to a chromosome segregation event. It is worth noting the urgency of the response to the condition of stress due to the excision and to the culture medium submission, the earliest sign of which could be abnormal cell division. Because it is conceivable that these divisions could start a clonal series of cells with similar characteristics, an early aberration occurring at this stage may be recovered later on, even in regenerated plants (Benzion and Phillips, 1988). On the other hand, it must be noted that the origin of an aberrant chromosome constitution may date back to the explant source tissue, to the age of the culture (Nuti Ronchi *et al.*, 1981), or to the process of plant regeneration.

We should therefore study these parameters to decide when the deviant behavior of a dividing cell is initiated.

1. Explant Source

Changes in chromosome number complements are the most common anomalies present in plant tissue used as explant source, and these abnormalities are often accompanied by structural chromosome aberrations; both types of variations may be transferred to the cells growing in culture. This may be particularly true when the plant tissue includes endopolyploid cells, which often show, on reentering mitosis in culture, a high incidence of chromosome breaks and reunion, as evidenced in metaphase and anaphase by fragments and bridges (D'Amato and Avanzi, 1948). But the more frequent phenomenon is the transfer, to the cultured cell, of the ploidy of the explant tissue, ploidy that is likely enough to be different, or at least varying compared to the normal diploid value, because differentiated tissues of most plant species undergo spontaneous chromosomal polyploidization. Even if this aspect is not always antithetical to plant regeneration, a high or mixed ploidy is not the best attribute of a cell suspension culture.

A significant difference, in terms of the response of plant tissue to excision and culture conditions is made by the presence or absence, in the explant, of vascular tissue. Studies with the *Nicotiana glauca* pith tissue system have revealed that a specific amplification of some DNA sequences and a process of endoreduplication were the prerequisites leading to cell proliferation which occurred by means of an amitotic process (Nuti Ronchi *et al.*, 1973; Parenti *et al.*, 1973; Martini and Nuti Ronchi, 1974). On the other hand, when the tissue to be cultured consists of hypocotyls of carrot or tomato, which are provided with a vascular cylinder, initial cell proliferation occurs mainly by means of processes of chromosomal segregation. It is therefore conceivable that processes such as amitosis (nuclear fragmentation), somatic meiosis, or prophase segregation, not being exact processes like mitosis, may be more liable to error during chromosome division and distribution. In both cases, that is, in the presence or absence of vascular tissue, cytological events deviating from the norm may influence, at the beginning, and in cells still belonging to the explant tissue, the future cell composition of the cultures, and originate a clonal population of affected cells (if abnormal) or committed cells (if able to regenerate).

2. Age of the Cell Culture

The age of the cell culture is important, because variations in chromosome number and structure are known to increase with the age of the cell culture, even if this aspect has not been frequently monitored, and to be connected mostly to the loss of regenerative capacity of the cell lines. Data confirm that, even when cell cultures are initiated from a stable diploid tissue, chromosome stability is not favored, and the rearrangement of chromosomal numbers and structures increases with the age of the culture. For example, two 4-month-old suspension cultures of *Nicotiana* species showed chromosome numbers to be variable around the diploid or tetraploid values up to more than 100 chromosomes, but exhibiting a good regenerative capacity; the same culture, studied 2 years later, presented a chromosome number stabilized around the hexaploid value (mode at 72), but with a loss of organogenic capacity (Nuti Ronchi *et al.*, 1981). In *Hordeum* cultures, Orton (1980) found stable variants shortly after callus induction. Although this point is important, and a large amount of information is available from the published data (Armstrong and Phillips, 1988; Lee and Phillips, 1987), little work has been dedicated to determine the cause of the abnormal cytological divisions, and the existence of any relationship with cell cycle variants studied in other organisms. For most authors, all of the alterations appear to be induced by the tissue culture process, namely by the medium components, and more specifically by the growth factors, which indeed may, under particular conditions, operate in this manner (Nuti Ronchi *et al.*, 1976; Singh *et al.*, 1972).

However, the exceedingly high frequencies of the "mutational" events occurring in plant tissue culture point to mechanisms differing from the ones operating after treatment with chemical or physical mutagens.

B. Embryogenic and Organogenic Cell Suspension Cultures

Plant regeneration *in vitro* may occur either via embryogenesis or organogenesis. The cells may be cultured in suspension in liquid medium, or on solid agarose medium. Because carrot is considered one of the best model systems *in vitro,* presenting several optimal features both for ease of culturing and for regenerating somatic embryos from a single somatic cell (regenerated embryos being therefore considered clones), most of the data discussed in this article concern this plant species, cultured in liquid medium. Cell suspension cultures present several advantages compared to the ones in solid medium, especially because cells are continuously bathed in the medium, allowing an even distribution of nourishment (or of any other specific molecules added for experimental reasons); the cellular density can be easily measured, and the growth rate can be monitored as well.

The most common way in which to start the process of somatic embryogenesis is to remove the growth factors from the medium. In carrot, this procedure must be accompanied by a proper dilution of cell density. Moreover, the change to hormone-free medium must be performed after 7 days of subculture, an amount of time allowing the generation, later on, of higher numbers of somatic embryos. For some species, such as *Medicago sativa,* even if the cells are cultured in liquid medium, the regeneration of the embryos occurs only when cells are anchored to a solid substrate.

Carrot somatic embryos follow almost the same pattern of development as zygotic embryos because, in culture, the regenerated embryos do not enter a resting dry period, as occurs in plant seeds, but continue to grow directly to the plantlet stage.

Organogenesis may occur both in liquid and solid medium, but for the advantages already discussed only liquid culture is considered here. A complete cytohistological study has been carried out in *N. glauca* (Nuti Ronchi, 1981), showing that the making of the promeristem (formed on the cultured cell clumps), from which the buds develop is possible only after the formation of a vascular center encircled by a cell layer similar in appearance and function to the endodermis; this structure, allowing the solutes normally distributed by diffusion in undifferentiated callus to be canalized instead (as in the root), seems to provide the basis for the subsequent growth of the shoot meristem.

An interesting hypothesis has been formulated on the basis of a comparison of carrot and *M. sativa* somatic embryogenesis processes, which showed that the differences between organogenesis and embryogenesis could be due solely to changing environmental conditions. In fact, carrot somatic embryos may be replaced by a shooty monopolar structure when cells are treated with proline (Caligo *et al.*, 1985); also, *M. sativa* may show the passage, in the same petri dish transferred to a different light intensity, from bipolar embryos to monopolar rootless shoots (roots needing another hormone treatment to develop) (Nuti Ronchi *et al.*, 1986c).

These observations suggest that the main potentialities of cultured cells are directed toward the development of a complete organism, provided with roots and vegetative apex, similar to the zygotic embryo that grows out of a fertilization process; only our ignorance of the true nature of totipotency and of the right conditions required to induce its full expression prevents our creation of somatic embryos from all plant species.

III. Deregulation of the Cell Cycle: Mitosis

The available data suggest that most of the variability at the chromosome and gene level detected *in vitro* is due to unknown factors acting on the explants at the earliest time of culture, or later on the derived cell suspensions, mutant phenotypes being induced in cells or in regenerated plants. Therefore, to illustrate the role of variants of the cell division processes in originating "mutations" in cell cultures, and to allow a better understanding of their role in somaclonal variation, it is convenient to give details about the different deregulation of the cell cycle affecting mitotic events in cultured cells. Avoiding at present a description of chromosomal aberrations frequently shown to occur in culture (breaks, transposition, deletion, etc.), attention here is given to anomalies of different phases of the cell cycle, in comparison also with those occurring in other eukaryotes.

The study of the regulation of the cell cycle, and especially of the mitotic process (which engages a large part of the entire cycle), has allowed dramatic progress due particularly to the ascertainment of the evolutionary conservation of the cell cycle, the basic functions being apparently conserved from higher to lower eukaryotes. This fact has allowed similar experimental approaches to be used in different organisms, in a search for homologies among related proteins and in comparing the more conserved and most important parts of a sequence or process (Kirschner, 1992). Different organisms, ranging from fission yeast and fungi such as *Aspergillus* to more complex organisms such as frogs, sea urchins, and mammals, were found to share most genes related to the cell cycle; the genes involved in the physical events of the mitotic process (spindle and chromosome condensation), the key molecules (tubulin and histones), and the general organization of the nucleus are particularly highly conserved. Notwithstanding the wide range of variation at the structural and molecular levels among the different organisms, at the moment it is possible to trace a general model of the regulation of cell growth and division that is sufficiently substantiated in a unicellular organism such as yeast; in a multicellular organism, in which cell division must be coordinated during development for organogenesis, more specialized proteins need to be integrated into the model.

A key strategy in the investigation of the biochemistry and physical events of the cell cycle and mitosis has been offered by the study of cell cycle mutants and temperature-sensitive (*ts*) mutants. This last type of mutation is given by temperature-sensitive cells that traverse the cell cycle undisturbed at their permissive temperature, but become arrested at a specific step when transferred to the nonpermissive (restrictive) temperature. This article refers only to the phases, during the cell cycle and mitosis, that appear most affected in plant cell cultures and more suitable in a comparative effort. Because few cell cycle-related genes have been studied in plants, the references would mainly concern the mammalian and yeast genes in the attempt to explain the mutant phenotypes arising in cultured plant cells.

A. Failure of the Breakdown of the Nuclear Envelope

The mitotic process is generally considered to be composed of different subsequent events, each of them dissociable from the others and inducible, and each governed by independent factors and not causally coupled (Ghosh and Paweletz, 1993).

In plant cells, the process may briefly be summarized as composed of four phases: the gap before DNA synthesis (G_1), the DNA synthetic phase (S), the gap after DNA replication (G_2), and the mitotic phase, which ends with cell division (M). Even if this devising has been helpful up to now to facilitate research approaches, the last decade of research at biochemical, genetic, and cytological levels has largely increased our information in this field; the cell cycle is now understood to be composed of complex manifold processes involving chromosomal organization and movement, spindle structure and function, nuclear envelope formation and breakdown, and establishment of a new cell wall. All of these functions (and more besides) are accomplished by means of a great number of macromolecular components that are formed, activated, and moved. A remarkable aspect is given by the extent of conservation, of the proteins involved, between species that are phylogenically distant. In higher plants the research on this aspect has started only recently, but has already confirmed, for some species *(Arabidopsis,* oats, pea, *Z. mays),* a strong homology to p34 cdc2 protein (John *et al.*, 1989; Feiler and Jacobs, 1990; Colasanti *et al.*, 1991).

The first signal that a cell is undergoing mitosis, and committed to the cell cycle, is connected with its progression through G_1 and the onset of DNA synthesis. These two main events are followed by the mitotic chromosome condensation and breakdown of the nuclear envelope. Microtubule associate protein (MAP) kinases are a family of protein kinases involved in the reentry into the cell cycle at G_0/G_1 and in entry to mitosis. Members of this family have been shown to be present in mammals and yeast; the *M. sativa* cDNA clone MsK7, which shows 52% identity to animal MAP kinases, has been isolated and its expression detected during the G_1-toS and G_2 phases of the alfalfa cell cycle (Jonak *et al.*, 1993). It will be the object of further research to find out which role MAP kinases may have in mediating responses to hormones and to ascertain a possible tissue- and stage-specific developmental expression. Like *M. sativa,* alfalfa MAP kinases are homologous to animal and yeast MAP kinases, reinforcing the supposition that plant tissue culture mitotic variations may be compared and identified with analogous variations occurring in higher eukaryotes, for which variations have already been detected and studied, often also at the molecular level. This article does not focus on the wide range of variations of the cell cycle described to occur in organisms, but dwells only on those related to the ones discovered in cultured plant cells.

1. Endopolyploidy

Failure of the breakdown of the nuclear envelope occurs frequently in cultured plant cells, which thus have a remarkable capacity to fail to proceed through mitosis, producing instead a restitution nucleus with double the number of chromosomes. The mechanisms adopted to produce a polyploid nucleus may be of differing origins, depending on the absence or presence of some mitotic stages within the nuclear envelope. In the first case, polyploid cells arise from endoreduplication cycles in which two or more rounds of DNA replication occur without chromosome segregation and breakdown of the nuclear envelope. The second case consists of an aberrant mitotic process inside a persistent nuclear envelope, chromosomes being arrested in metaphase but slightly separating after the splitting of the centromeres; the full process, occurring within the nuclear envelope, ends with the formation of a restitution polyploid nucleus. A frequent variant of this last process is given by an arrested prometaphase, with chromosomes arranged in a circle along the nuclear envelope, which persists until the formation of the restitution nucleus. It is important to note that it is possible to discern between the two methods of generating polyploidy in the following mitosis, because the chromosomes derived by endoreduplication appear tied, in prophase, by the relational coiling of the two pairs of chromatids, which also remain connected, up to metaphase, by the centromeres (diplochromosomes). When, instead, chromosomes attain within the nuclear envelope some endomitotic stages with contracted chromosomes, thus releasing the relational coiling, the next mitosis will be indistinct from a polyploid mitosis derived from other anomalies. Quadruplochromosomes may also occur when the process is repeated (Nuti Ronchi *et al.*, 1965; Rizzoni and Palitti, 1973). Higher endoreduplication cycles result in a structure similar to polytene chromosomes, as has been reported to occur in the tissue culture of barley (Gaponenko *et al.*, 1988) and often observed in tomato cultures (V. Nuti Ronchi, unpublished observation).

It is interesting to make some pertinent remarks on this subject, particularly because there have been several reports concerning the control of early cell cycle events. The first important point is that multiple forms of protein kinases appear responsible for mitotic induction and initiation (Broek *et al.*, 1991; Dutta and Stillman, 1992; Hamaguchi *et al.*, 1992). A second point concerns the controls regulating the onset of S phase and mitosis, on the supposition that G_2 cells are forbidden to undergo S phase if they have not completed mitosis. On this last point important progress has been made by Moreno and Nurse (1994), who have identified the fission yeast gene called *rum1+* (for replication uncoupled from mitosis) as a regulator of DNA replication and progression through the G_1 phase. A detailed account of this work is not provided in this article, but the model proposed to explain the results is exciting because, as the authors state, "given the conservation of cell cycle controls, similar gene functions may have related functions in other eukaryotic cells." Mutants of fission yeast that overexpress *rum1+* are allowed to undergo two successive rounds of DNA synthesis without mitosis, whereas deletion of the gene allows cells to undergo two successive mitoses without an intervening S phase. Transposing these results to the variant plant mitoses just discussed, it is possible to recognize the likeness to the endopolyploidizing events, acknowledging the fact that specific proteins play a central role in regulating these events connected with the cell cycle: any perturbation in the form of temperature shift or other situation of stress may affect their function, changing it from mitosis to meiosis (two successive mitoses without an S phase). But, similarly, it may be sufficient for a single gene to change function to divert cells to endopolyploidy (two successive rounds of DNA synthesis without mitosis).

The different potentialities of the early stages of the plant cell cycle to progress either toward polyploidy (strictly connected, in plant cells, with developmental decisions), or to segregational chromosome reduction (normally considered only particular to reproductive cells), appear properly represented by the Moreno and Nurse (1994) model.

It is also pertinent to recall that endoreduplication not only occurs spontaneously in different cell types, but may also be induced by a number of chemical and physical agents. The case described in pea root tips (Nuti Ronchi *et al.*, 1965) seems to correspond to the Moreno and Nurse (1994) finding occurring in cell cycles, in which DNA replication and chromosome segregation do not alternate with each other. In pea root tips, treatments with 8-azaguanine and labeling experiments with [^{3}H]thymidine have shown that the drug arrests meristematic cells, accumulating them in G_2; although allowing, during treatment, the completion of the DNA synthesis already started, the drug inhibits a new round. After recovery in the presence of guanine, G_2 cells start anew a second round of DNA synthesis, followed again by the G_2 phase. A synchronous wave of endopolyploid prophases with diplochromosomes appears after about 18 hr of recovery in water. A treatment has also been found that induces quadruplochromosomes (eight-chromatid chromosomes). Unfortunately, only negligible frequencies of diplochromosome mitoses have been found after treating root tips of *Vicia faba, Bellevalia romana,* and *Allium sativum* with 8-azaguanine. A possible deamination of 8-azaguanine to azaxanthine, a biologically relatively inert compound (Handschumacher and Welch, 1960), has been suggested as the cause of the lack of effect in species other than *Pisum*.

A dramatic increase in both the number of polyploid cells and the maximum degree of polyploidy (up to 2048 C) has been obtained by treating suspension cultures of *Phaseolus coccineus* with the staurosporine analog K-252, an inhibitor of protein kinase activity. It is suggested that phosphorylation of a protein kinase, probably of the cell cycle-linked p34cdc2 type, is involved in the control of endoreduplication (Nagl, 1993).

This finding seems to confirm, for plant species, the already discussed proposal that entry into S phase and mitosis is determined by the state of p34cdc2 (Broek *et al.*, 1991; Moreno and Nurse, 1994).

The use, for experimental purposes, of suspension cell cultures derived from plants already known to undergo, *in vivo,* various cycles of endoreduplication, like the embryo suspensor of *Phaseolus* (Nagl, 1974), is an example of how cell cultures may be suitable for the study of cell cycle regulation *in vitro*.

In plant cell cultures the switch from the mitotic to the endomitotic cycle has been studied and appears to be mediated mainly by plant growth regulators (Libbenga and Torrey, 1973; Nagl, 1971).

2. Amitosis (Nuclear Fragmentation)

Two other variants of the mitotic process, that is, amitosis and prophase chromosome reduction, which cause the opposite of the polyploidization event (i.e., ploidy reduction), are also to be included among those occurring within the nuclear envelope; for both, ultrastructural evidence on this point is still lacking. Nuclear fragmentation (amitosis) has been shown to be the main mode of division of *N. glauca* pith explant cell nuclei undergoing endoreduplication during the first days of culture (Nuti Ronchi *et al.*, 1973; Martini and Nuti Ronchi, 1974); Fasseas and Bowes (1980) later confirmed the phenomenon in proliferating storage cells in mature cotyledons of *Phaseolus vulgaris*, showing, by electron microscope analysis, that microtubules were not evident at the margin of the freely forming walls. Amitosis is considered, as are many other variants of the mitotic process, a primitive karyokinetic system still occurring in some plants *in vivo*. *Tradescantia* stem cells divide by this mechanism when, in spring time, a fast-growing cell population is required to produce more cells in the shortest time (Conard, 1928); similarly, amitosis occurs in *Malus silvestris* cork spot necrosis (Miller, 1980), pea (*Pisum sativum* L. cv. Kiir), tobacco (*Nicotiana tabacum* L.), and *Haplopappus gracilis* (Nutt.) (Kallak and Yarvekylg, 1977). Amitosis has also been induced in *Allium cepa* L. root tips by a 6-hr treatment with ethidium bromide (de la Torre and Gimenez-Martin, 1975), and it has been found in mammals (Schröder and Kurth, 1972) and in amebas (Band and Mohrlok, 1973).

The occurrence of amitosis, or nuclear fragmentation, in cultured cells may be considered as part of the manifold devices required to ensure the amount of variability needed for selection to operate for the fittest karyotype. Even if probably of archaic origin, amitosis is nevertheless a fairly precise mechanism, because the division of the large lobed polyploid nucleus in more nuclei, including one or more sets of the haploid complement, is accomplished by means of ingrowth of the mother cell wall and of freely forming walls. These newly formed walls, growing through the cell, separate the reduced nuclei, thus completing the cellularization of the cell (Nuti Ronchi and Terzi, 1988); similar processes of new cell wall formation, performed in addition to the one directed by the fragmoplast, have also been shown to be adopted during the cellularization process of syncytial endosperm (Newcombe and Fowke, 1973).

A process similar to amitosis has been described to occur in large amebas (Afon'kin, 1989). The processes of induced polyploidization and depolyploidization in amebas is described and discussed as a sort of parasexual cycle that serves as a source of genetic variation and as a mechanism that provides segregation in agamously reproducing cells.

Also found in agamously reproducing protozoans are similar examples of "noncanonical" mechanisms of variability that allow the genetic variation of these organisms.

Amitosis has been the first process (other than mitosis) found, in plant suspension cultures, to attain the double goal of a fast proliferation rate and the creation of different chromosome and nuclear constitutions (Nuti Ronchi *et al.*, 1973; Martini and Nuti Ronchi, 1974, 1977; Nuti Ronchi and Terzi, 1988).

3. Prophase Chromosome Reduction

The other process, which occurs in most plant species cultured *in vitro*, has been named *prophase chromosome reduction* (Nuti Ronchi, 1990; Nuti Ronchi *et al.*, 1992a) because it occurs in nuclei that appear to divide in prophases, thus causing ploidy reduction and chromosomal segregation. Most typically, nuclear division performed by means of this mechanism results in the separation of chromosomes into two or more groups, according to ploidy, during different prophase stages. The unthreaded groups of chromatids may proceed directly into interphase, reforming two or more reduced nuclei included later in newly separated cells after cell wall formation. At the moment there is a lack of ultrastructural evidence concerning the persistence of the nuclear envelope during the whole process; this mechanism has been placed among this group of variant divisions because it apparently occurs in a limited, enclosed space inside the cell. An alternative process (see also Section III,B,2) consists of the progression of two or more reduced prophases to two distinct metaphases (which may sometimes appear arrested as in colchicine-treated cells), which then proceed directly to interphase after cytokinesis, omitting entirely the anaphase and telophase stages.

The first manner of division is widespread in cultured cells of different plant species, in which an actively proliferating callus often appears almost devoid of mitotic figures but remains as in a permanent prophase state. Carrot suspension lines T2 and A+, which divide only according to these mechanisms (Nuti Ronchi *et al.*, 1992a), have allowed an accurate study of the phenomenon. It is interesting to note that, whereas the DNA endoreduplication and the intraenvelope endomitosis processes may induce a high level of polyploidy, depolyploidization events may occur in the same culture, thus restoring the diploid number of the species, or other chromosome numbers more fitting for the present environment, or even the haploid complement, providing therefore a mechanism of chromosome segregation normally offered by the sexual cycle. The flow of variability from which to draw for possible developmental choices appears tremendous.

4. Indefinite Growth

Another interesting variant of the mitotic cycle that is possibly related to the end of S phase and progression through the cell cycle is offered by the carrot cell line E^2A^1C6 (Nuti Ronchi *et al.*, 1992a), which showed a higher than normal mitotic index (over 7%) at all time points on the growth curve, indicating a capacity for indefinite growth that recalls a tumorous phenotype; the control line A+ instead showed the usual mitotic peak on the third day after subculturing (Fig. 1).

The irregularity of this carrot line may be related to the mechanism of some regulators of DNA synthesis and mitosis, like some factors whose absence allows mitosis to initiate before DNA synthesis is finished. The ability to identify homologous genes from different organisms by function,

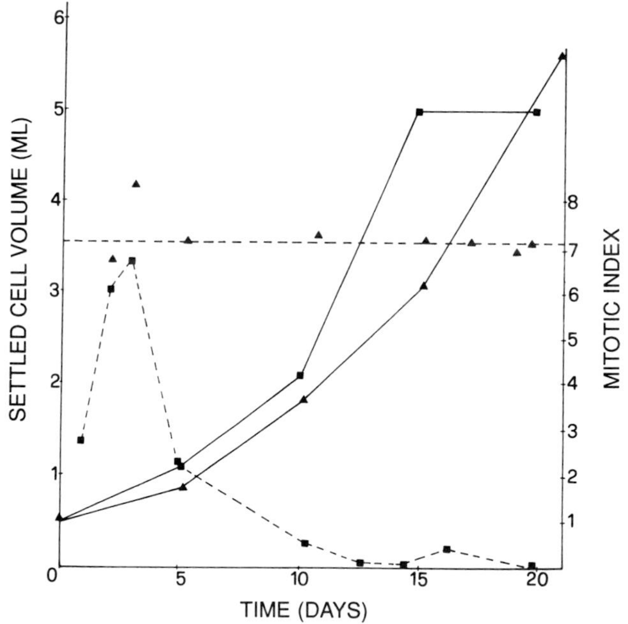

FIG. 1 Growth curves (—) and mitotic index (---) of carrot cell lines A+ (■) and E^2A^1C6 (▲). [From Nuti Ronchi *et al.* (1992a). Reprinted by permission of Kluwer Academic Publisher.]

using simple eukaryotes such as yeast, has already provided new strategies for the study of the cell cycle in mammals. These approaches can be used for the study of gene function in any other species. From these data it is clear that plant cell cultures that express a mutated cell cycle-related phenotype may offer a good system for studies in this field. Efforts to isolate mutants of the conditional type [e.g., temperature-sensitive (*ts*) embryogenic cell lines] have already provided interesting results in carrot, concerning mutants impaired in development (Breton and Sung, 1982; Terzi *et al.*, 1982; Giuliano *et al.*, 1984; Lo Schiavo *et al.*, 1988, 1990). So far, the instability that may often be found in the cultures has limited the use of this approach in the study of cell cycle variants. It is worth noting that the study of mutants induced in cell cultures has revealed the unexpected fact that recessive mutants (as are most of the *ts* mutants) are found in diploid cell lines at frequencies similar to those obtained in haploid cells. Until now inexplicable, these data presently lend strong support to the occurrence of mechanisms of chromosome segregation or parasexual cycle in plant cell cultures (Nuti Ronchi *et al.*, 1992b; Giorgetti *et al.*, in press).

B. Failure of Spindle Assembly

Mitotic phase-specific events succeed the early events discussed in the previous section, and include chromosome condensation, spindle assembly, and breakdown of the nuclear envelope. These phases, which are strictly related in plant cells, also appear sensitive to any sort of environmental changes, which therefore induce the variants of normal behavior often found in plant cell cultures. The passage from prophase to metaphase is mediated by a prometaphase stage in which the chromosomes, almost fully condensed, reach slowly to the cell equator. From a structural point of view, spindle assembly culminates with the formation of the mitotic apparatus (MA), which consists of the assemblage of microtubules and associated proteins responsible for separating the duplicate chromosomes (Palevitz, 1993).

1. C Metaphases

One of the most frequent anomalies of mitosis is due to the failure of the separated chromatids to move to the pole, leading to the formation of a restitution nucleus (i.e., one with double the number of chromosomes). This polyploidizing variant, which operates in a different way than the ones occurring inside the nuclear envelope, presents metaphase chromosomes that remain dispersed and motionless in the cell for a relatively long time (hence the name "C-metaphase" or "colchicine-metaphase," for its likeness to the colchicine-induced mitotic block); anaphase does not follow, cytokinesis is inhibited, and a restitution polyploid nucleus is formed. It is worth noting that the C-metaphase feature is the most common variant induced in the plant meristems by even a short-term physical or chemical treatment. Because colchicine is the most specific poison acting on the microtubules (combining with tubulin and inhibiting its polymerization), it is conceivable that various conditions of stress may similarly affect the mitotic apparatus and chromosome segregation in anaphase as well. In proliferating callus, C-metaphases are the most frequent cause of the shift of the cells to higher ploidies, randomly affecting a low percentage of the cell population. Various reports have thoroughly discussed this point, particularly the role in the occurrence of the phenomenon of growth factors and other medium constituents, culture age, and the possible influence of specific microenvironments, often induced by long subculture times (Nuti Ronchi *et al.*, 1981; Nuti Ronchi, 1990; Bayliss, 1974, 1980; Singh *et al.*, 1972); frequent medium changes appear to minimize the effect.

2. Reductional Grouping

Carrot cell lines show various patterns of mitotic arrest simulating the colchicine effect. A variant of the previously described polyploidizing intraenvelope event (Section II,A,3) consists of two reduced prophase chromosome groups progressing to metaphase and then directly to an interphase stage; anaphase figures are missing completely but cytokinesis proceeds, giving two cells with reduced nuclei. T2 and A+ cell lines, which are highly embryogenic cell cultures, show these two types of division permanently during culture, and also in regenerated embryos up to the heart stage. In the T2 cell line the mechanism appears to work in the haploid–diploid range, separating mainly two groups of nine metaphase bipartite chromosomes. In the A+ line, instead, in that there is a high percentage of polyploids, the mechanism works by separating two (as in T2 line) or more prophase chromosome complements, according to their ploidy; the groups separated in prophase proceed directly into interphase after cell wall formation. Haploid cells constitute 40% of dividing cells, diploidy being restored by endoreduplication or by the precocious separation of the bipartite nine-sister chromatids, giving back a metaphase with 18 chromosomes, but at the 2C level. These cells may cycle again, after DNA synthesis, as a normal diploid cell, or directly undergo another reduction, with the same mechanism, to two haploid nuclei. A cytophotometric Feulgen/DNA absorption analysis in individual nuclei has confirmed the cytological data and the presence of the haploid population (Nuti Ronchi *et al.*, 1992a). Figure 2 shows the nuclear content of line T2

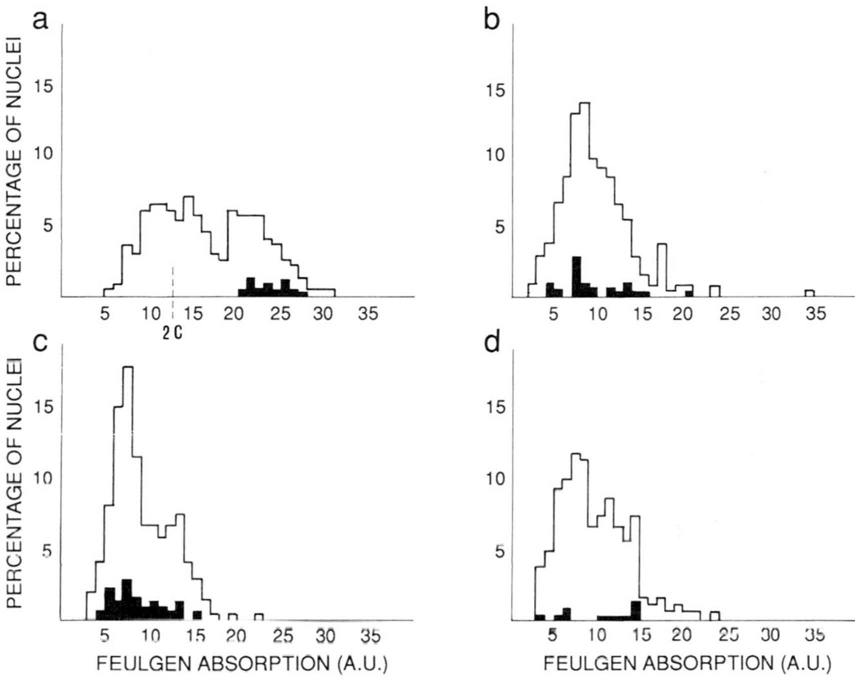

FIG. 2 Nuclear DNA content of line T2 cells, regenerated T2 embryos, and line A+ cells compared to carrot root tip meristem. Open bars: Feulgen/DNA absorption in arbitrary units (au) measured in individual nuclei of (a) root tip of cv. St. Valery meristems, (b) T2 cell line, (c) globular regenerated embryos of T2, and (d) cell line A+. Mitoses (solid bars) could be either haploid (n = 9, 2C, or 1C), diploid ($2n$ = 18, 4C), or segregated followed by centromere disjunction ($2n$ = 18, 2C). Besides the accumulation of cells at haploid values, an overall shift of DNA content to lower values is evident. [From Nuti Ronchi *et al.* (1992a). Reprinted by permission of Kluwer Academic Publisher.]

cells, T2 embryos, and line A+ cells compared to carrot root tip meristems. It is worth noting that, besides the accumulation of cells at haploid values, an overall shift of DNA content to lower values is evident.

The variant of mitosis found in the nonembryogenic carrot cell line E^2A^1C6 has been named *reductional grouping*. This line showed a high ploidy, reached by means of endopolyploidy, endomitosis, nuclear fusion and restitution nuclei, events that were all randomly present in the culture. As shown in Fig. 3, chromosome numbers even up to more than 150 were counted, but the distribution was in the range of multiples of the haploid number n=9, with a mode number $5n$ or $6n$, the haploid number being present in 5% of the metaphases. This variability was acquired by means of splitting, during prophase, of the polyploid nuclei into more groups of chromosomes that proceed to an arrested metaphase; in the same cell,

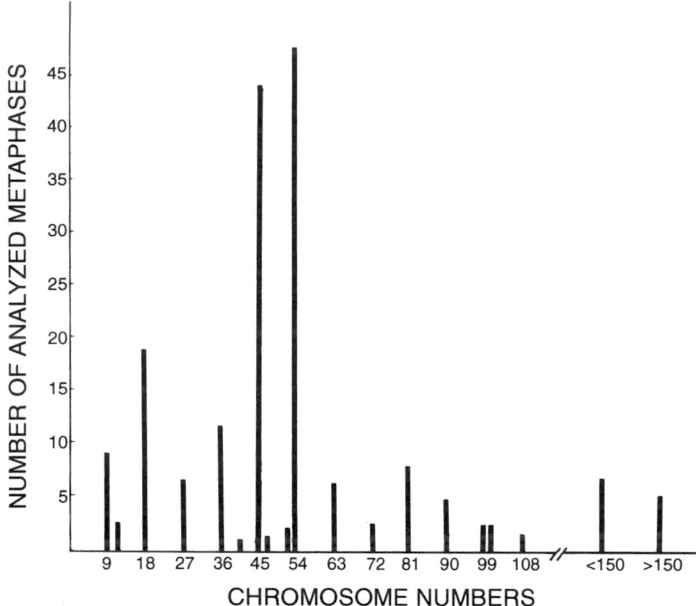

FIG. 3 Distribution of chromosome numbers of the carrot cell line E^2A^1C6. Different chromosome numbers in separate groups could be found in the same cell. [From Nuti Ronchi *et al.* (1992a). Reprinted by permission of Kluwer Academic Publisher.]

metaphases formed with groups of different chromosome numbers (multiples of 9) were commonly found. No anaphase followed but newly formed walls separated the nuclei in more cells with variable ploidy. The late formation of cell walls, that is, cytokinesis not strictly related to karyokinesis, is quite frequent in plants, and the norm in syncytial endosperm. The present evidence suggests that no spindle is formed, a fact also confirmed by the appearance of the metaphase chromosomes, which were similar to colchicine-induced arrested metaphase chromosomes.

It is important to stress that the reductional grouping mechanism has been described in the past to occur mainly during meiosis in plants, and in some cases the presence of multiple spindles was acknowledged. Owing to the extreme plasticity of the mitotic apparatus in plant cells, there is "the possibility that chromosomes separate by a mechanism that does not depend on an ordered bipolar mitotic apparatus, organizing separate spindles and moving chromosomes independently of each other" (Palevitz, 1993). However, the C-metaphase appearance of the E^2A^1C6 carrot line chromosomes and the lack of anaphase stages are not in accordance with the presence of a functional spindle.

It is interesting to note that events similar to the ones occurring in the carrot line have been described in protoplast cultures of *Brachycome dichromosomatica* and *Crepis capillaris,* with the formation of polynucleate cells with metaphases with different chromosome numbers, with ploidy ranging from $4n$ to $8n$, and often also endoreduplicated (Hahne and Hoffmann, 1986). The authors interpreted their results as defects during mitosis and cytokinesis, recognizing in endoreduplication the origin of the highly polyploid cells, and in the lack of cytokinesis the origin of the multinucleate cells. Also in this case, the presence of multiple metaphases with different chromosome numbers (but always in the range of multiples of the basic haploid value) in the same large cell and the aspect of the metaphase chromosomes similar to those induced by colchicine could suggest a mechanism of reductional grouping as in the carrot line.

On the other hand, the results of Hahne and Hoffmann (1986) have been obtained in protoplast cultures, a condition that may present a great difference in comparison with cell cultures. It is worth noting that the incapacity of protoplast to regenerate (Hahne and Hoffmann, 1984) has been suggested to be related to the lack of the network of cortical microtubules.

3. Split Spindle

The formation of split spindles has been reported to occur in plants *in vivo,* particularly during meiosis. The occurrence during mitosis has been studied in the interesting progeny of a colchicine-induced tetraploid ($4x=32$) of *Ribes nigrum* (Vaarama, 1949). The variant, split spindle, of the mitosis was established to be the cause of the gradual diminution of the high ploidy of the seed progeny of the colchicine-treated plants, with the induction of a high frequency of different somatic chromosome numbers. Besides the diploid number, which was the most frequent, all numbers divisible by four were more frequent than could be expected on a random distribution, including a high percentage of the same number 4. This fact suggested that the original basic number of the genus *Ribes* is 4 instead of 8, the recent species being a derived tetraploid. Two considerations may be inferred by this study, carried out long ago. The first concerns the similarity of this case to the one reported for carrot lines (Nuti Ronchi *et al.*, 1992a) and the *Brachycome* and *Crepis* protoplast culture (Hahne and Hoffmann, 1986). Because in *Ribes* the presence of the spindle was inferred only by the contemporary presence of more metaphases, with chromosomes defined supercontracted and resembling fairly closely the C-pairs of colchicine-treated cells, this case may also be included in the group of variants concerning the failure of spindle assembly. The second pertinent observation concerns the specific variation of the cell cycle, aimed to reduce chromosome number. The mechanisms operating with this objective may induce a "superreduction" of the chromosomes, extracting archaic haploid sets or single chromosomes into a separate group (like the original basic number 4 of *Ribes*). This is an important point to which we return again when discussing the mechanisms of somatic chromosome segregation and the occurrence in culture of twin metaphases or anaphases (split spindles) as processes of haploidization.

4. Failure of Cytokinesis

In plants, karyokinesis not followed by cytokinesis occurs particularly in organs and tissues that seem to have specific functions during development, mainly nutritional or for production of metabolites needed in neighbor cells or organs. For instance, syncytial endosperms are well known, as is the tapetum of the anthers in most angiosperm species.

It is controversial whether the above-described variants of mitosis, owing to arrested metaphases leading directly to restitution nuclei lacking anaphases, or the developmentally determined polynucleate endosperm, should be considered as variants of the mitotic cycle related to the failure of cytokinesis. This last stage, in fact, although not following directly after anaphase, is occurring in any case, even if delayed. On the other hand it may be discussed whether the delayed cellularization process, so common both *in vitro* and *in vivo,* may be considered comparable to a normal cytokinesis. In the process of wall formation described to occur in syncytial endosperm of *Stellaria media* the change from the free nuclear to cellular condition was accomplished by ingrowth of cell walls and freely forming walls, and often with modality similar to the reconstruction of protoplast walls (Newcombe and Fowke, 1973). As already discussed, the process of cellularization found in *N. glauca* after nuclear fragmentation (amitosis) followed the same patterns of wall formation occurring in syncytial endosperm (Nuti Ronchi *et al.,* 1973; Fasseas and Bowes, 1980; Nuti Ronchi and Terzi, 1988; Nuti Ronchi, 1990). In the amitotic division the ingrowth of existing cell walls and freely growing walls proceed through the amitotically derived mass of nuclei, held in the center of the cell by means of cytoplasmic bridges; this process allows the separation of new uninucleate cells within the mother cell (Nuti Ronchi and Terzi, 1988). As already discussed, ultrastructural studies have shown that, during the dedifferentiation *in vitro* of *P. vulgaris* cotyledons (Fasseas and Bowes, 1980) by means of a similar mechanism, no microtubules seemed to be involved in the process. It may be, therefore, that this cell wall formation could be highly deviating from the cytokinetic process normally occurring in the meristem.

Failure of cytokinesis and wall formation was also observed during the first days in culture of protoplasts of two *Nicotiana* species (Huang and Chen, 1988). Di- and multinucleates due to the failure of cytokinesis divided synchronously in the same cytoplasm, their spindles being oriented in such a way as to move to the same pole and fuse at anaphase, increasing the ploidy of the culture, which was started from haploid cells derived from pollen callus. No more haploid cells were present in the protoplast-derived culture after 30 days, and no haploid plant was regenerated. Even in this case, cell wall formation was delayed, but the spindles were functionally normal.

So far, few mutants in higher eukaryotes have been isolated showing karyokinesis not followed by cytokinesis, this last stage being inhibited experimentally only by means of treatments with specific drugs (Ghosh and Paweletz, 1993).

An intriguing question has emerged from advanced studies on the cell cycle with the mutant cdc16 of fission yeast related to the control of mitosis and cytokinesis (Chang and Nurse, 1993). The model suggested to explain the role of cdc16 is that it acts as a central switch controlling both cytokinesis and the end of mitosis, acting as a positive regulator of mitosis and a negative regulator of cytokinesis. Even if the relationship between cdc16, septation, and cdc2 kinase activity remains to be elucidated, it is proposed that a cell overexpressing cdc16 would arrest in mitosis with high cdc2 kinase activity, a spindle, and no septum. Mitotic spindle, in this model, appears necessary for the switch into cytokinesis, but it may be that only part of the spindle is required. It is fascinating to wonder, in comparing the fission yeast results to our plant deviant mitotic cycle, whether the lack of cytokinesis illustrated in *Nicotiana* and carrot culture may be ascribed to the supposed absence of the spindle. It is interesting to note that the thermosensitive line ts111 isolated from suspension cultures of Chinese hamster fibroblast showed mitotic aberrations similar to the ones described to occur in the carrot E^2A^1C6 line and in *Nicotiana* pith tissue: formation of giant cells, with more than 100 chromosomes, or cells with several metaphase figures, each with normal or reduced chromosomal number, whereas anaphases and cytokinesis did not occur (Hatzfeld and Buttin, 1975). Other mutants, with similar phenotype and blocked in cytokinesis when shifted to the nonpermissive temperature, were isolated in mammalian cells (Wang and Yin, 1976; Wang *et al.*, 1983). The similarity of the cell cycle variations, even if detected in different organisms and experimental conditions, once more confirm that the switch, coordinating subsequent events of the cell cycle, may be regulated by "checkpoint" controls highly conserved in many organisms (Lee and Nurse, 1988).

IV. Deregulation of the Cell Cycle: Meiosis

The use of the term *meiosis* for divisions of somatic cells that, in plant cell cultures, mimic the process of chromosome segregation and reduction normally occurring in the reproductive organs of sexually reproducing eukaryotes, may be open to criticism. Several other terms are probably more suitable, as, for instance, homologous chromosome segregation, perhaps more pertinent because it refers to somatic cells dividing in culture. The term meiosis has been chosen to define a mechanism of nuclear divisions occurring during the initial proliferation phase of carrot and other explant species because these events follow very closely the ones occurring during micro- and macrosporogenesis *in vivo*.

The choice was also compelled because the meiotic-like events occur in structures developing contemporaneously *in vitro,* similar to primitive reproductive organs; moreover, this mechanism may, like meiosis in plants, produce gametic-like cells competent to develop as proembryogenic masses.

Chromosome reduction has been considered the effect of the process of meiosis, occurring, in sexually propagating eukaryotes, in specific cells located in reproductive organs and originating haploid gametes. It is therefore conceivable that a similar process occurring, instead, in somatic tissue not related to sexual reproduction has never found general acceptance. This was also due to the paucity of its occurrence and the difficulty of its identification. Whereas the different mechanisms causing polyploidy have been generally recognized, the evidence for the occurrence of somatic reduction leading to haploid cells has been forgotten or ignored, even when it has been demonstrated beyond doubt. In fact, the mechanisms of chromosome reduction described to occur in plant cell cultures have also been induced *in vivo* long ago (Huskins, 1948), and since forgotten. On the other hand, the importance of somatic chromosome reduction in the biological process may be of great interest. In fact, new reported studies making use of the plant tissue culture have pointed to the need for continued investigation of these phenomena. Owing to the difficulty of accepting the possibility that somatic cells may perform, under particular conditions, the segregational process that is normally accomplished in reproductive cells, it is necessary to make a thorough search of the existing sources of information to define those that have demonstrated beyond doubt the occurrence of somatic reduction, ascertaining under which circumstances the event occurred. At present other data arising from plant tissue culture, somatic embryogenesis, transformed plant regeneration, and some puzzling questions put by the somaclonal variation are causing a renewed interest in the mechanism and possible role of somatic reduction. Unlike the results obtained in the past, more recent cytological data can be confirmed by genetic and molecular evidence, presently a feasible task with modern molecular techniques.

A. Homologous Chromosome Segregation

It may be pertinent to recall here all the investigations that, in the 1970s, have shown that the mean distance between homologous chromosomal sets in somatic cells is significantly shorter than that expected from random distribution, verifying previous cytological observations that homologous chromosomal sets are indeed associated in somatic resting nuclei (reviewed by Avivi and Feldman, 1980). Pretreatment with cold water (Feldman *et al.*, 1966) was the favorite pretreatment which, in preventing spindle formation, does not affect interphase chromosomal arrangement (Fig. 4); colchicine, however, which has a stronger effect in disrupting chromosomal distribution, may, under some conditions, preserve a natural association between homologous pairs. Such a close association induced by chemical, high or low temperatures, or other stress may therefore be the main cause of the spontaneous or induced somatic reduction events recorded in several plants.

1. Somatic Reduction in Plants

Evidence of the occurrence of somatic reduction has been produced occasionally over the years, and is either cytological or genetic. Already in 1948, Huskins observed in onion root tips the separation of chromosomes in two groups after treatment with sodium nucleate. Huskins could distinguish between meiotic-like phases, with apparent pairing and chiasmata, and a second event consisting of the separation of prophase chromosomes into two haploid groups. The data are illustrated by clear pictures of the phenomena and an interesting discussion on their meaning, in relation to evolutionary changes; the suggestion that the tomato could be the plant suitable to give other evidence of the phenomenon, especially if it were possible to grow its cells *in vitro,* was prophetic. In fact, in the following years, *Lycopersicon esculentum* would have given evidence of these events both *in vivo* and *in vitro*. Huskins found it difficult to explain which

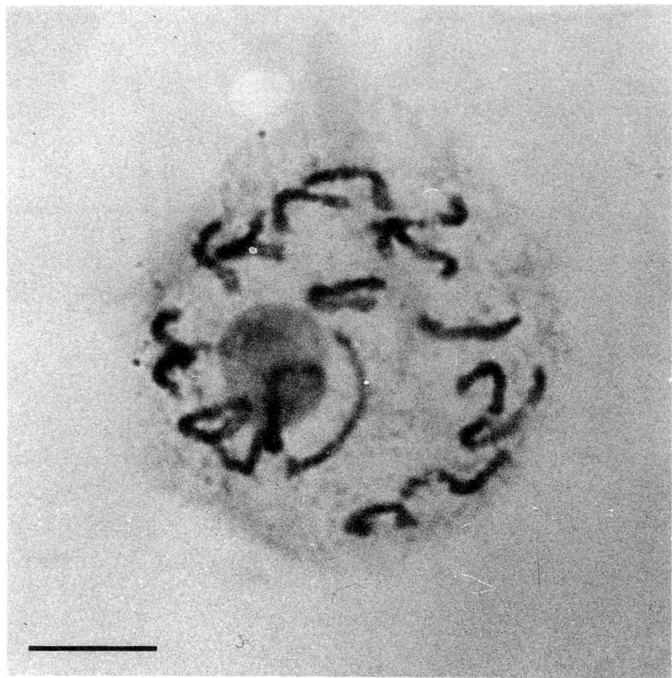

FIG. 4 Homologous chromosome associations in a carrot root tip cell after a 12-hr treatment at 0°C. Bar: 5 μm.

forces were operating to pull apart chromosomes; but he rightly identified in any sort of stress the primary event able to start the somatic reduction process. The Huskins results were confirmed in *Pterotheca falconeri* (Mehra, 1986), which has fewer chromosomes compared to *Allium:* several chemical substances, analogous to the ones used by Huskins, were shown to induce chromosome segregation of the clearly indentifiable homologous chromosome pairs of this species.

Wilson and Cheng (1949) and Patau (1950) also confirmed that homologous chromosomes in *Trillium* were separated into groups by a nonspecified "homologous repulsion."

Interesting cytological evidence of the possibility that cells of somatic tissues inside the plant could carry out meiotic process was also given by Battaglia (1947) and Battaglia and Dolcher (1947) in *Sambucus ebulus* and *Sambucus nigra,* respectively. In this case the tissues in which the reducing events occurred were strictly connected with sexual reproduction, because they were located in the glands situated in the basal portion of the stylar tissue. It is interesting to speculate on this closeness of the two tissues presenting the same concomitant phenomena: one, the true meiotic process ending with the embryo sac and egg formation; the other, a deregulated version of the meiotic cycle, ending with large cells whose thick hyaline walls recall an unsuccessful attempt to develop as a pollen cell.

A particular type of somatic reduction has been found in relation with colchicine treatments in *Sorghum*. Plants of *Sorghum vulgare* grown from seedlings treated with colchicine were found to mutate simultaneously in numerous traits (Franzke and Ross, 1952; Ross *et al.*, 1954). It was proposed that somatic chromosome reduction followed by doubling of the chromosomes occurred in the treated stem apex. To test the hypothesis of somatic reduction, marked chromosomes were obtained and used in proper crosses. Treatment of heterozygous *Sorghum* seedlings for a reciprocal translocation gave homozygous plants for the marker genes, confirming the occurrence of somatic reduction since the separation of members of the chromosome pairs was detected. It was therefore established that somatic reduction with subsequent doubling for at least the two pairs of chromosomes carrying the translocation was involved (Ross, 1962). It is interesting to note that, in *Sorghum,* the phenomenon of somatic reduction induced by colchicine was not always reproducible in different genotypes. A study was therefore carried out with the aim of explaining this fact and a differential reaction to colchicine treatment between genotypes was found, one particular variety being more reactive (24 reduced out of 43 treated plants) to the colchicine treatment (Atkinson *et al.*, 1957). Using the reactive genotype it was possible to define a laboratory method with the most suitable conditions (light, humidity, temperature, etc.) that assured the production of diplohaploid mutants, the growth conditions of the plants being therefore essential for the fulfilment of the process.

Colchicine as inducer of somatic reduction divisions was also found efficacious in soybean, causing the appearance of areas of leaf tissue apparently changed as a result of separation of the allelic genes for chlorophyll color in heterozygous material (Mark, 1961). Analogous results of leaf spotting were obtained after colchicine treatment of *Glycine max* seeds (Vig, 1971), and of axillary buds of flax (Dirks *et al.*, 1956).

Somatic reduction, although well demonstrated in the last cited works, seemed to play no significant role in cell differentiation and function, occurring occasionally only in particular genotypes, and therefore since then has ceased to evoke interest among plant scientists. In Section VII below, in discussing somaclonal variation and the meaning of polyploidy, this point will be resumed because it recalls similar findings related to somaclonal variation.

Somatic segregation has been demonstrated to occur not only in plant cells, but also in animal cells. The most famous paper related to this phenomenon (Martin and Sprague, 1969) described the parasexual cycle in cultivated human somatic cells, following the suggestion of Pontecorvo (1958) that a formal genetic analysis of the human species, based on recombination within cultivated somatic cells, could be possible. The paper shows somatic segregation for three different autosomes in two strains of human diploid fibroblasts derived from subjects known to be heterozygous for chromosomal variants. Ploidy-reducing processes and segregational events achieved by means of multipolar mitosis have also been demonstrated in mammalian cells by Pera and Reiner (1973).

2. Somatic Reduction in Plant Cultured Cells

The first report of the occurrence of somatic reduction in cultured plant cell concerns the carrot (Mitra *et al.*, 1960). Prophases with paired chromosomes and chiasmata and divisions suggesting abnormal reductional divisions forming haploid nuclei were observed together with proembryo-like structures; the authors reported that similar observations (somatic pairing and haploidy) have also been made in the cultured stem tissue of *H. gracilis*.

More evidence was given occasionally, over the years, of the presence in culture of cells with the haploid chromosome number: in *Gossypium* (cotton; Bajaj and Gill Manjeet, 1985), in *Nicotiana* species (Kovács, 1985), and in *Lathyrus sativus* L. (Lavania, 1982).

Except for Huskins and a few others, the evidence on meiosis-like events comes mostly from indirect proof, which is often quite reliable when based on genetic data. Cytological evidence is often lacking, or incomplete, and molecular data are nonexistent, because practically all references are old. In spite of this, pieces of information on the possibility of a somatic meiosis process are numerous, even if scattered over the years, and the general impression is that the phenomenon exists and may assume different features according to differences in plant species, genotypes, changing environments, and condition of stress. This last point appears to affect greatly the onset of the events and it is possibly the main reason that the stress induced by culturing tissues and cells, both of animal or plant origin, is turning out to be the most suitable inducer of the phenomenon. The major experiences are, at the moment, emerging from plant species: the excision of the tissue from the organism; its further chopping, normally also necessary (with its destruction of most of the cell-to-cell contacts and polarity); its transfer to a medium, either solid or liquid, whose composition is certainly far more unbalanced than the juice inside the plant; and again, the collision with the high concentration of growth factors, often also not natural hormones, not to mention the changes in light intensity and temperature; all are unnatural conditions cooperating to create the highest degree of stress. It is well known that plants have various defenses against stress, and that their stationary habit has provided them with peculiar qualities to tolerate even great changes in environment or extreme wounding or other damage. Monitoring stressful situations requires sensitive signaling, which is just now starting to be unraveled; but the most singular and extraordinary devices adopted to afford stress are certainly the rapid or programmed genomic changes, which are of common occurrence in higher plants (for further accounts of these strategies, see Walbot and Cullis, 1985; Walbot, 1985; Chasan and Walbot, 1993; Goldberg, 1988; Nuti Ronchi and Terzi, 1988; Nuti Ronchi, 1990).

The genomic changes occurring by means of variants of the division process, either by ploidy increase or diminution down to the haploid complement, is certainly part of these strategies. But the most astonishing aspect of the processes alternative to normal mitosis is their amazing capacity to increase the reservoir of genetic variation.

3. The Commitment to Somatic Embryogenesis in *Daucus carota*

Daucus carota was the first plant species that, when cultured *in vitro* (Steward *et al.*, 1958; Halperin and Wetherell, 1965), showed cellular structures reproducing exactly the very early phases of zygotic divisions, that is, closely paralleling the growth of carrot embryos, which are known to pass through a filamentous stage (Borthwick, 1931). Moreover, as already reported, the same authors (Mitra *et al.*, 1960) were also able to identify the presence, in carrot cultures, of different ploidy levels and also haploid cells. After these first, but precise, observations of the behavior in culture of carrot cells, the carrot system has been used by various laboratories all over the world, taking advantage of the fact that carrot cells could be cultured by suspension in plain medium and induced to perform embryogenesis simply by subtracting the growth factor (2,4-D, 2,4-dichlorophenoxyacetic acid) and decreasing cell density. A commitment event is necessary to induce embryogenic competence, which is highest in immature or mature embryos, seedlings, and floral structures (Bajaj and Gill Manjeet, 1985; Carman, 1990). The more suitable tissues, able to acquire the embryogenic competence, are not therefore the ones less differentiated, but the tissues more involved and connected with the important phase of the plant represented by reproduction; it seems that these tissues and these cells have more possibilities to pursue other reproductive pathways.

The hypothesis that cell commitment to somatic embryogenesis is, in carrot, preceded by a process similar to the meiotic process has been stimulated by the likeness of early stages of somatic embryo properties to those known to belong to the egg cell, such as isolation from other cells and the formation of a callose wall (Mackenzie *et al.*, 1967; Williams *et al.*, 1973). Another interesting likeness could be traced to the androgenesis process, in which an immature pollen grain, stimulated by suitable conditions and a proper stress input, may revert to a sporophytic phase and develop as an embryo. These considerations were logically connected to a third manner of forming embryos, that is, from the somatic cells in the living plant, a phenomenon defined as *homeotic transformation* (Leavitt, 1909; Sattler, 1988). Somatic embryos developing on the leaf margins in *Malaxis paludosa* (Taylor, 1967) are also to be considered homeotic, but other queer transformations are also possible and have been described to occur in nature: pollen grains that can be transformed in the embryo sac, and ovules that may contain pollen grains (for a review, see Meyer, 1966). The enormous capacity of plants to assume manifold developmental aspects according to necessity and conditions of stress make plausible the hypothesis that, to be competent for embryogenesis, a somatic cell should acquire a gametic condition by means of the same processes occurring in the reproductive organs. This idea was strongly supported by the previous studies of carrot lines expressing permanently in culture various types of variant of the mitotic cycle, some of them producing a high frequency of haploid cells that normally *in vivo* result from a meiotic event occurring in the reproductive organs.

With the intent of checking this hypothesis, experiments were planned aiming to perform a cytological analysis of the first divisions going on inside the 2- to 4-mm-long carrot hypocotyl explants (cultured for 20 days in liquid B5 medium supplemented with 2.2 μM 2,4-D) and in the derived cell suspension cultures.

Meiotic-like events were found to be present in different zones of the vascular strands, during the 20 days of culture, forming a population of segregated cells. Moreover, contemporaneously, structures similar to primitive reproductive organs developed along the vascular cylinder, as rough sporangium- and ovary-like forms or rudimentary inflorescence. The meiotic-like events occurring in the cultured hypocotyls or in the newly derived cultures showed some phases similar to the normal meiotic event; variants that could be compared to similar ones described in plants *in vivo* were also studied. Some of these variants were expressed permanently in suspension cultures of carrot, either very old or very young: in both cases the loss of embryogenic capacity was noticed.

B. Variant Meiotic-like Events

The switch of cells from a mitotic to a meiotic cycle has been extensively studied in budding yeast, in which the switching is dependent on genetic systems controlling the response to mating type and nutritional conditions. The use of mutants of yeast and other organisms has given evidence that M-phase PF (maturation promotion factor) complexes of p34 and cyclin regulate both the G_1-to-S and G_2-to-M transition in both the mitotic and meiotic cells in all eukaryotes (Murray, 1991). Homologs to the *cdc* yeast gene encoding the protein kinase p34 are found also in plants (Colasanti *et al.*, 1991).

Owing to the suitability of *Z. mays* for both cytological and genetical analysis, most of the work on genes affecting the course of the meiotic process have been carried out in this organism (Golubovskaya, 1979, 1989; Golubovskaya *et al.*, 1992, 1993). The comparison of the meiotic-like processes occurring in the cultured carrot cells with the ones extensively studied in other organisms presents many difficulties, because the process *in vitro* is often highly irregular and frequently, in the cultures, several phases appear simultaneously affected, making difficult the definition of the impaired mechanism.

1. Impairment of Entry into Meiosis

The study of this step may be helpful to understand the process that, in carrot hypocotyls explants, allows the passage of the resting cells directly into meiosis (Nuti Ronchi *et al.*, 1990; Giorgetti *et al.*, 1994).

In the ameiotic mutant identified in maize by Rhoades (1956) and designated *am,* cells do not enter meiosis and after two or three mitotic divisions, microsporocytes degenerate (Palmer, 1971). A study of the *ameiotic 1* gene and other mutant alleles (Golubovskaya *et al.*, 1993) indicate that the *am1* gene is involved in the commitment of the cells to meiosis and to crucial events related to prophase I and chromosome synapses. Male homozygous *am1* cells do not enter meiosis but undergo mitotic divisions; also, the female meiosis is not even started, the progression of the process is arrested at early prophase I, followed by chromatin and cell degradation and/or the formation of multinucleate cells. The arrest impedes any progression beyond the leptotene/early zygotene stages, when only short pieces of synaptonemal complex (SC) structures and axial elements were observed.

These results indicate that the *am1* gene controls the initiation of meiotic prophase in maize, probably in G_2, but it si difficult to say whether it also acts in G_1 or S. If we can transpose maize results to carrot events the normal expression of a gene comparable to *ameiotic 1* appears necessary for the transition from mitosis to meiosis, the critical step that a certain percentage of carrot cells is bound to take when dividing into the cultured hypocotyls. At the moment it does not appear easy to ascertain this point. Because the leptotene stage has been defined as the first critical stage for irreversible commitment of meiocites to meiosis (Stern and Hotta, 1967), either the hypocotyl cells enter the meiotic pathway directly from a mitotic G_2, or they are committed to the meiotic-like event during a previous mitosis. Unfortunately, the prophase stages of the involved cells are not clearly definable, as normally in meiosis with the resolution of a light microscope, but particularly so because of the small size of the carrot nuclei and chromosomes.

The histological analysis of the carrot hypocotyls with the aim of localizing the sites of the first meiotic-like event has revealed that the cells undergoing this type of division could enter the meiotic pathway soon after excision (even less than 7 hr) of the tissue from the seedling. These particular cells could soon be distinguished among the others because of a more conspicuous cell wall and a more dense cytoplasm; their location is also peculiar, situated close to the vascular tissue, especially the xylem elements. These cells are the most responsive to the cultural triggers; moreover, as shown by the microdensitometric measurements of DNA content per nucleus carried out during the first 20 days of culture on the hypocotyl cells, they appear to progress directly into meiosis without DNA synthesis, probably from a G_2 stage. This observation seems to exclude the second hypothesis, that hypocotyl cells are triggered to somatic meiotic-like divisions during a previous mitosis.

The peculiar aspect and behavior of this group of cells and their possible unique properties recall the concept of Newman (1965) of "continuing meristematic residue," also discussed by Steeves and Sussex (1989) in relation to the similarity with mammalian "stem cells." In the case of the carrot, and in all other plants in which similar phenomena occur, this interpretation seems pertinent to their function, which seems to be related to a rapid response to stress and to a developmentally oriented genome reprogramming. Moreover, these supposed stem cells appear to occur anywhere vascular and procambial cells are situated, and also in such tissues as leaves, and have been noticed in other plant species in culture (Lu and Vasil, 1981; Lipucci di Paola *et al.*, 1987; Barcelo *et al.*, 1991, 1992).

2. Pairing and Synaptonemal Complex

Whereas the early prophase stage is difficult to define in carrot hypocotyls, later phases are easily detected, showing pairing of the homologous chromosomes comparable to the pairing figures during carrot microsporogenesis, with eight bivalents and two univalents due to the precocious separation of one bivalent. Compared to the carrot meiosis, the bivalents appear slightly more despiralized, with chromosome arms pulled apart; the pairs are kept together at what appears to be chiasmata and therefore the phase is interpreted as early diakinesis. It is worth noting that, although some features are unquestionably meiotic, some more precocious events in prophase appear of dubious aspect. The entire problem has a resemblance to a cytological problem widely discussed in the past, and that could have some relevance in the context here discussed. The matter concerns a phenomenon, studied in *Lilium* genotypes (Walters, 1970), and also of frequent occurrence in several other plant species, where it is called "preleptotene chromosome contraction," and present also in mammals and other organisms, where it is known as "preleptotene prochromosomes," (Stahl and Luciani, 1971). In these cases a period of chromosome spiralization and contraction was observed between premeiotic interphase and leptotene, a feature resembling late mitotic prophase, even if the chromosomes also have some resemblance to chromosomes in late meiotic prophase (diakinesis); nevertheless, a normal meiotic cycle with a proper leptotene stage with normal pairing will closely follow. It was suggested that an irregularity in sequence of meiotic gene action, possibly an external condition of stress, allowed the transient expression of genes still related to the mitotic cycle. This point appears interesting, because it seems that, in species presenting this anomaly of the meiosis, the initiation of the meiosis itself is effective only after the microsporocytes have entered mitotic prophase. Going back to the carrot meiotic-like system, the transition from mitosis to meiosis may be precise, proceeding directly from mitotic interphase to leptotene-diakinesis; and this would explain the lack of clear early prophase figures. Or the transition may be less sharply defined, allowing a shift toward mitosis and back to meiosis, as it occurs in the case of *Lilium* (Walters, 1972; 1976). Also in this case it is opportune to recall the already cited paper of Moreno and Nurse (1994) about the fission yeast gene *rum1+*, which regulates the different potentialities of the early stages of the cell cycle at "checkpoints" at which it is decided to progress either to mitosis or to meiosis. On the other hand it is worth noting that the study of meiotic mutation in *Z. mays* already cited (Golubovskaya, 1979; Golubovskaya *et al.*, 1992) has allowed the detection of two genes, *am1* and *afd*, which affect the first division of meiosis, inducing the reversion of the highly differentiated cells, meiocytes, already undergoing meiosis, to mitosis. Only two monogenic mutations are thus required to induce such a remarkable upset; their influence not only on tissue culture and somatic embryogenesis, but also during development, is at the moment difficult to estimate. It is evident that, in eukaryotes, a genetic mechanism can switch cell division from mitosis to meiosis and vice versa; it is plausible that, under specific situations, this switch may occur also in somatic tissues.

In contrast to normal meiosis, in carrot hypocotyls a phase with a clear synapsis of the chromosomes is seldom visible; it is not evident whether this was due to the small sizes of the chromosomes or to the real lack of this phase. On the other hand it is well established that the SC is necessary for proper pairing and proper disjunction of the chromosomes. This is shown to be true even in carrot meiotic-like events, because proper pairing and proper chromosome disjunction are demonstrated by the data (Section V), proving that segregation of molecular markers has occurred owing to the meiotic event.

3. Chromosome Disjunction

The carrot meiotic-like process proceeds, after diakinesis, to what appears to be a normal metaphase I, where it is possible to recognize the nine pairs of homologous chromosomes. Whereas there is no evidence of the presence of the synaptonemal complex, the presence of the spindle is instead quite obvious, also emphasized by the unique appearance of the cytoplasm, which has a different density when compared to other cells and similar to the appearance of meiotic cells *in vivo*. Figure 9 shows, in cultured tomato hypocotyls, a meiotic-like metaphase. The pointed spindles characterize tomato meiosis (Hogan, 1987), whereas mitosis has a barrel-shaped spindle. It is interesting that spindles focused at the poles were also shown to occur in embryogenic protoplast cultures of white spruce *(Picea glauca)* and have also been detected in pollen generative cells of *N. tabacum* (Palevitz, 1993) and in the orchid *Phalaenopsis* (Brown and Lemmon, 1992).

Carrot cells correctly perform the chromosome disjunction of anaphase I (reductional), but this phase is rarely followed consequentially by the second equational division. Cytokinesis occurs with the formation of two reduced cells. The presence of two anaphases close to each other was not observed; instead, two nuclei in close proximity (comparable to a dyad) were seen to have undergone a simultaneous division by means of the mechanism previously illustrated, that is, prophase reduction, unthreading one from the other to give a tetrad. More frequently, these two cells appear to prosecute separately their further destiny, because the division by this last mechanism involved only one of the dyad cells, so that only two nuclei (and cells after cytokinesis) completely reduced to the haploid (1C) value were produced. At the moment it is unknown whether the other nucleus of the dyad undergoes the same, perhaps delayed, division, or enters a different pathway as a diploid 2C cell. It is difficult to compare this phase of carrot meiotic-like division with other organisms. A distinctive feature consists of the precocious centromere division, in carrot cells, at the end of anaphase I, a fact that appears to facilitate the further division of the nuclei by means of the mechanism of prophase reduction. But it is important to note that this event also has another notable consequence, because the disjunction of the sister chromatid centromeres returns the cell to the diploid number 18, even if the DNA content is still at the 2C value. As stated above, these dyad nuclei may therefore have multiple choices: (1) prophase reduction division (without DNA synthesis) to give 1C cells, (2) mitotic division directly through metaphase and anaphase with 18 chromosomes but at the 2C level to give 1C cells, or (3) DNA synthesis and mitotic division, entering the normal diploid cycle again. All these possibilities are probably exploited by the carrot cells, as confirmed by the cytological and microdensitometric data discussed below. It is also important to note, because this point will be recalled later, that in these different possibilities reside also the anomalies concerning the absence of cytokinesis, occurring at any time along the full course of the process. Restitution nuclei are the presumed results of these events, mainly first division or second meiotic division restitution nuclei, that moreover may also be obtained by means of other mechanisms (Werner and Poloquin, 1991); the anomalies involving the unsuccessful wall formation normally give unreduced nuclei.

It may be, therefore, tentatively asserted that the meiotic-like events occurring in the carrot hypocotyls are not of casual incidence, or just an odd curiosity referred to cultured cells; furthermore, they seem to occur not just in carrot, but any time a plant tissue (including vascular tissue) is excised and cultured *in vitro*.

Up to now, the phenomenon has been ascertained, always using the same cultural conditions as in the carrot, in several species such as tomato, sunflower, *Prunus* species, and apple tree (Nuti Ronchi *et al.*, 1992c; Blando *et al.*, 1991). Particularly interesting, in cultured tomato hypocotyls, is the high peak of divisions occurring in waves from the seventh hour to hr 18–20 of culture, all concerning the vascular strand parenchyma cells. Very frequent, in tomato, are divisions of the type prophase reduction, as shown in Fig. 5A, where it is possible to follow the sequence of the different phases, from the prophase nucleus, with a large nucleolus (both nucleolus and the whole prophase show the beginning of a narrowing in the equatorial region, foretelling the future division). Figures 5B–D shows the subsequent phases, up to the final formation of two nuclei, clearly composed each of 12 pairs of chromosomes. It is important to note that the nucleolus, in these divisions, is always present, submitted to the same internal division in two halves in concomitance with the

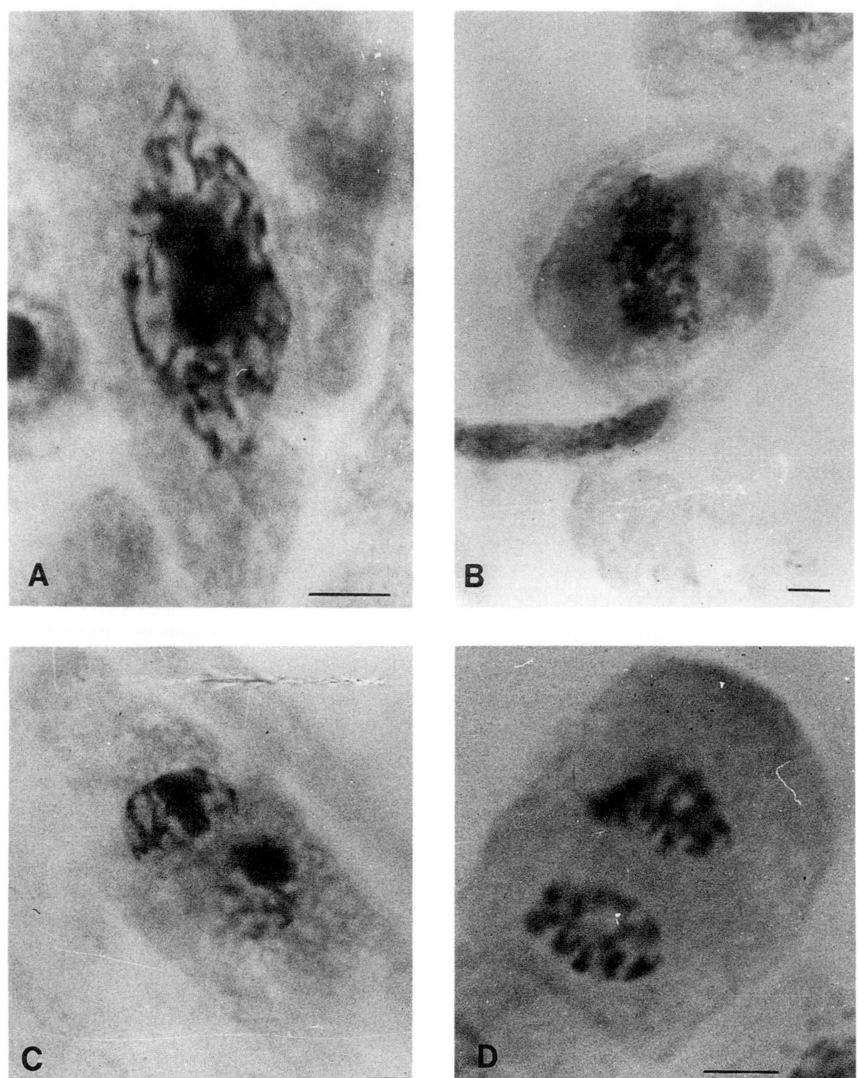

FIG. 5 Subsequent phases of a prophase reduction process in cells of a tomato hypocotyl explant cultured in the presence of 2,4-D for 24 hr. (A) Nucleolus and whole prophase showing an initial narrowing in the equatorial region, presaging the subsequent division; (B–D) subsequent phases, with the nucleolus still present, up to the final formation of two nuclei with 12 pairs of chromosomes. Bar: 5μm.

nucleus, during the successive phases. Figure 6 shows a tomato hypocotyl cell in diakinesis with 10 bivalent and 2 coupled chromosomes already separated.

Another interesting observation, concerning the course of the meiotic-like events in plants other than carrot, has been made in secondary polyploids, as for suspension cultures of *Malus domestica* and *Helianthus annuus* where multivalent chromosomal configuration in meiotic-like metaphases of cultured cells may predict the possible recovery of ancestral genomes. In fact, cells with reduced chromosomes numbers (below the haploid) are commonly found in the cultures of these plants; this may be interpreted, according to Jackson and Murray (1983), who obtained similar chromosome configurations with colchicine treatments, as the disruption of the normal attachment of genome of the nuclear membrane, allowing synapses of the ancestral genomes. Besides confirming the polyploid origin of these "diploid" species, the production, during meiosis-like events, of cells with the chromosome number reduced down to less than the diploid number, suggests the possibility of extracting from these cells, by means of a regeneration process, ancestral-type plants (Blando *et al.*, 1991; Nuti Ronchi *et al.*, 1992d; Nuti Ronchi and Giorgetti, 1994).

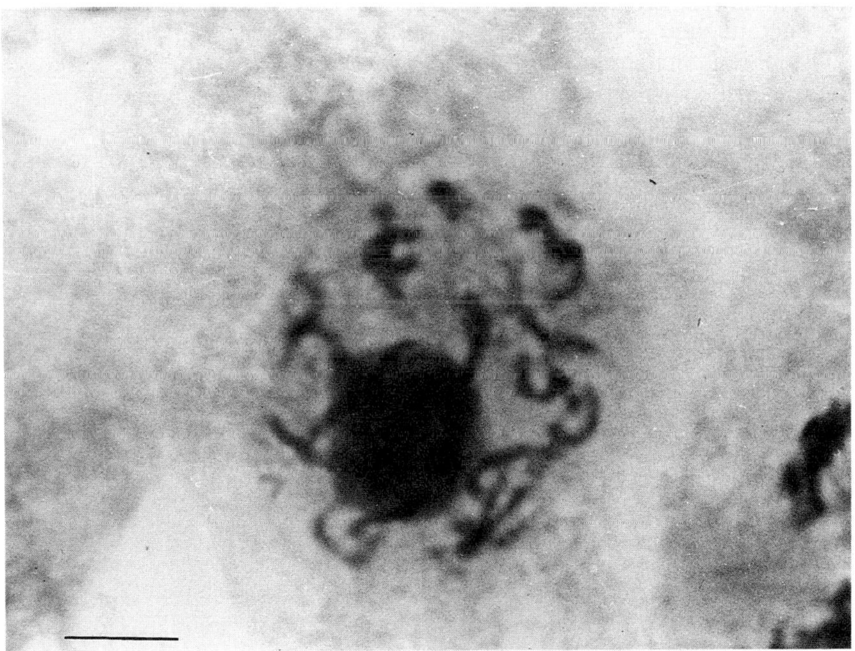

FIG. 6 Diakinesis-like phase in a somatic cell of a tomato hypocotyl explant cultured for 20 hr in the presence of 2,4-D. Bar: 5 μm.

4. Carrot Cell Lines Expressing Permanently Variant Meiotic Phases

Carrot suspension cultures have been described that have showed a number of deviant phenotypes already described and others besides (Nuti Ronchi *et al.,* 1992 a,b). Somatic meiotic-like aberrant divisions occur permanently in nonembryogenic W1 and W2 cell lines; they show, in general, the same phases occurring in the cultured hypocotyls, but with several variants due particularly to the presence or absence of the cell wall, and from the occurrence or nonoccurrence of DNA synthesis in one, few, or all nuclei of the dyads or tetrads (Fig. 7). High ploidy is generated by the frequent lack of cytokinesis, leading often to nuclear fusion, which may also be achieved through telophase fusion when spindles are too close. The final product of this cycle consists of round, large cells with a single nucleus situated in the peripheral position, near the cell wall, similar to immature uninucleate pollen cells; these large cells can go back in division, starting the cycle again, often complicated by a high ploidy. Interestingly enough, almost exactly the same course of events has been described in the already cited papers on the occurrence of meiotic events in somatic cells of *Sambucus* (Battaglia, 1947; Battaglia and Dolcher, 1947). The reference is important not only for the likeness of the phenomenon to the carrot events, but also because it is clear evidence that these events are not an "oddity" of the *in vitro* cultures, but may be expressed also in particular plant tissues, at different developmental stages or after specific stress conditions.

It is evident that the carrot cell lines W1 and W2 expressed permanently in culture a meiotic-like phenotype behaving like a sporogenous tissue. The deviation from normal processes seem to be frequent in occurrence, because they represent the most recurrent anomalies occurring in nonembryogenic cultures.

Similar cytogenetic anomalies were found to occur in aneuploid maize tissue culture regenerant meiosis: coenocytic microsporocytes that lacked cell walls between nuclei, or lacked cell wall formation after the first and/or the second meiotic division, were recorded. As in the carrot W1 and W2 cell lines the failure of cytokinesis formation was often followed by nuclear fusion (Rhodes *et al.,* 1986).

5. Multipolar Spindle

A meiotic irregularity that may be present, particularly in polyploid plants, concerns the distribution of chromosomes in more spindles, so that an unequal separation of chromosomes occurs at anaphase. This type of division has been demonstrated, *in vivo,* during meiosis, but may also be

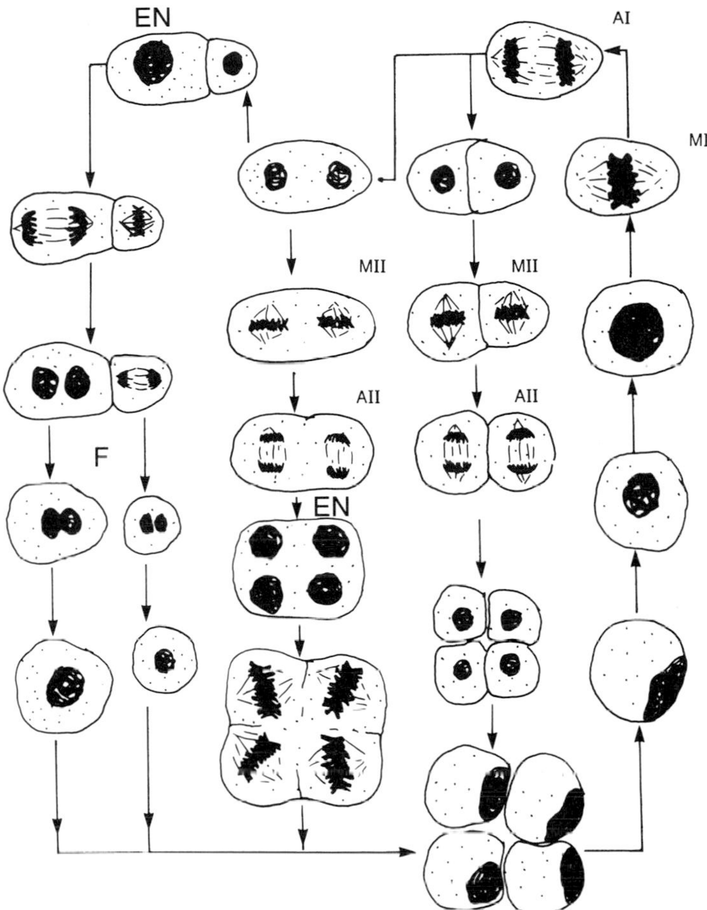

FIG. 7 Model of the parasexual cycle of carrot lines W1 and W2 as inferred from the cytological features of the nuclear divisions. From right to left: first column, nuclei go through a metaphase I stage (MI) followed by anaphase I (AI), which may (rarely) go through a metaphase II (MII), anaphase II (AII), tetrad, and back (after DNA replication) in the cycle. Alternatively, if the cell wall does not form after AI (third column) the following MII and AII phases produce a binucleate whose nuclei may fuse (fourth and fifth columns) and reenter the cycle. [From Nuti Ronchi *et al.* (1992b). Reprinted by permission of Kluwer Academic Publisher.]

present during the mitotic cycle. Section III,C has already discussed a variant of the mitosis which reduces polyploid nuclei to the diploid or even haploid level. The cytological mechanism of multipolar spindle allows a polyploid organism to produce gametes with extremely variable chromosome numbers. Because the tendency is for the formation of balanced genomes, some of these gametes will be functional. Examples in plants are found in *Rubus*. Thompson (1962) has called the process "complement fractionation," just for the possibility to separate independently operating groups of chromosomes within a cell, normally groups with a multiple of the basic *n* haploid number. Other examples are described in *Clarkia exilis* (Vasek, 1962), in *Agropyron cristatum* (Tai, 1970), and *Bromus* (Walters, 1958).

It should be noted that in these types of aberration it is also possible to obtain restitution nuclei with unreduced chromosomes. As explained above, cultured cells frequently show these anomalous behaviors, which appear necessary to reduce a high level of ploidy, but may also competitively restore a high level of ploidy. Particularly interesting is the case found with high frequency in *Nicotiana* species regenerated plants (Nuti Ronchi *et al.,* 1981). The case reported was the first to be studied, but other similar cases were found later, suggesting that this event occurs frequently in plants regenerated via organogenesis from suspension cultures of *Nicotiana*. The *N. glauca* plant at issue presented a constant proportion of cells with different ploidy ranging from 5*n* complements (which was present in 90% of the cells) to 4*n* (5%) 3*n* (3%), and 2*n* (2%). These different ploidy levels were constantly found in the same relative proportion in all tissues of the plant subject to analysis and during meiosis as well. No segregation of a particular chromosome number occurred within or among the developmental layers of the plant.

The more plausible explanation is that a self-perpetuating mechanism creating variability, originating and present in the cell culture phase, was transmitted to the regenerated plant, possibly by means of a regulated partition of the chromosome sets, so that a pentaploid cell could originate a triploid plus a diploid, and so on, fusion of groups, endoreduplication or C-metaphases restoring the pantaploid level (Nuti Ronchi *et al.,* 1981). The mechanism more suitable to give such a result should be a multipolar spindle, operating both in mitosis and meiosis. The high sterility of the regenerated *Nicotiana* plants (even if a small proportion of the gametes, being normal diploid, should have been fertile), did not allow investigators to ascertain if the defect was heritable.

Split spindles giving twin metaphases and anaphases also occur in cultured cell lines, even if at low frequency, and may therefore be considered as the meiotic anaphase II, reducing chromosome number to haploidy (Wilson and Cheng, 1949) when occurring in a diploid cell. Twin spindles are also frequent in polyploid mitoses, perhaps because of the difficulty in organizing a spindle large enough for all the numerous chromosomes.

V. Evidence for Consequences of Somatic Meiosis Events

A. Products of Somatic Meiosis Events

The products of a meiotic process are gametes, that is, from a single diploid cell four haploid progenies result. In plants, in the male organ (the anther), gametic cells undergo a maturation process that encompasses various progressive stages before reaching full maturation and the capacity to fertilize the female egg. Likewise, in the female organ (deep in the pistil, into the ovary), only one of the four haploid products of the meiosis initiates a more complex maturation path to secure what is necessary to welcome properly the male gametes and to comfortably accommodate the growth of the fruit of the fertilization process. The embryo sac has some features suitable for the growth of the embryo, and the three postmeiotic mitoses of the initial haploid cell enable the egg cell and others to accomplish further essential maternal functions. Pollen grain and embryo sac being the fulfillment of a meiotic process, it was presumed that similar cells could be found, as products of the somatic meiosis, in the cultured hypocotyl and in the derived cell lines. Pollen-like and embryo saclike cells were in fact found not only in carrot cultures, but also in tomato cultures, up to now unable to regenerate embryos. In fact, whereas in carrot both these two reproductive-like cells are able to continue development, having acquired embryogenesis competence [expressed in proembryogenic masses (PEMs)], tomato cells instead undergo a degradative process, unable to divide further.

In carrot, pollen-like cells are similar to the uninucleate immature pollen found *in vivo;* after a first asymmetrical division giving two unequal cells, proliferation proceeds to form PEMs, one single cell of PEMs being competent to develop, under the proper cultural conditions, as one single embryo. The embryo sac may also be identified in the large vacuolate cell whose single nucleus may divide, forming eight nuclei, or seven for the fusion of two nuclei, as in the plant *in vivo*. The already cited early paper of the Steward group (1958) showed, by means of observations of living carrot cells with the phase-contrast microscope, the capacity of very large (up to 300 μm long and 100–150 μm in diameter) vacuolated cells to divide, giving primitive filamentous forms similar to the earliest stages of zygotic embryos. These data have since been forgotten, and large vacuolated cells in suspension cultures of carrot have up to now been considered unable to divide and therefore nonembryogenic (De Vries *et al.*, 1988), instead of the very small cells demonstrated by several authors (Komamine *et al.*,

1992; Dudits *et al.*, 1991) to be the embryogenic ones. *In vivo* observations (Nuti Ronchi and Giorgetti, 1995, in press) have showed that the disagreement between the old and new data was due to the fact that the small embryogenic cells are derived from the primitive filamentous embryonal structure, growing from or out of the large cells that behave as an embryo sac. This filamentous embryonal structure may itself become a PEM, originating directly from the large cell. This behavior is favored when culture growth and embryogenesis are accomplished at high temperature (32°C). Small PEM cells may, by rounding up their cell walls, break loose into the medium. Proembryogenic masses, therefore, besides giving rise to the somatic embryos, may set free more small cells to perpetuate the process. Two important consequences come from these observations: each regenerated embryo may be considered of secondary origin, because it develops from cells belonging to an unsuccessful attempt to form an embryo; the second consideration is that every single embryo may be considered derived from a single cell (with itself possibly derived from a segregating process), an important condition from a genetic point of view and a favorable circumstance under which to obtain genetic proofs of the events involved in the commitment to somatic embryogenesis. This last point has been one of the most controversial questions concerning plant tissue cultures and their capacity to regenerate faithful copies of the donor plant. Is plant cell totipotency really to ascribed to all healthy living cells of plants, as several scientists still believe, or do only some cells respond to the attempt to regenerate plants, even if almost all may divide in cultures to produce callus, that is, a proliferating amount of heterogeneous cells of various shapes and degrees of cytodifferentiation. The data just reported give a clear suggestion that, to acquire the competence for somatic embryogenesis, a specific commitment of particular single cell is needed.

It is interesting, the fact that an indirect confirmation of the real identity of the embryo saclike cell is found in *Prunus* (Nuti Ronchi *et al.*, 1994) where two characteristic features, specific for these species, have been used as markers to recognize these cells in culture: that is, the presence of a haustorium and the longer persistence of the callose cell wall, identifiable with the aniline blue-induced fluorescence under ultraviolet light (Pimienta and Polito, 1983).

As already stated above, the surprising event that occurs in concomitance with the meiotic ones is the development in culture, from the hypocotyls or other organs of all plants tested, of structures similar to reproductive organs and inflorescence. The histological study of these forms has allowed investigators to identify primitive ancestral features that, however, even if unable to develop to normal size and forms, are nevertheless competent enough to express (as detected by means of Northern blots and *in situ* hybridization with floral organ-specific probes) homologous or nonhomologous floral genes (Gasser *et al.*, 1989; Doman *et al.*, 1990; Pnueli *et al.*, 1991) that are detected, *in vivo* in the plants, in specific cell layers of mature reproductive organs (Pitto *et al.*, 1992, 1993). This evidence stresses the likeness of the structures developing in the hypocotyls to homeotic mutant forms that display different types of organ transformation. The structures may therefore be defined as homeotic, according to the definition that homeosis is "the assumption by one part of an organism of likeness to another part" (Sattler, 1988). This hypothesis is confirmed in tomato, because one tomato gene, TM 8 (flower specific with distinguishable temporal expression), is expressed in cultured tomato hypocotyls from day 4 to day 15. TM 8 has been isolated from meristems of *anantha*, a recessive mutation of tomato with arrested floral development (Lifschitz, 1988; Pnueli *et al.*, 1991). The TM 8 gene, together with others, is part of the MADS family, members of which share a homologous putative DNA-binding domain with three organisms (plant, yeast, and human), and plays a regulatory role in determining the cell developmental fate. TM 8 is expressed, in tomato plants, mainly in flower meristems both of the normal plant and of the *anantha* mutant. Other genes, shown by *in situ* hybridization to be expressed in tomato hypocotyls, are MON 9617 and MON 9612, which are expressed in tomato flowers only in the transmitting tissue of the style (Gasser *et al.*, 1989; Budelier *et al.*, 1990). Besides confirming the floral nature exploited by means of *in vitro* cultures in the tomato hypocotyls, these data reveal that, although extremely primitive and underdeveloped, the structures may express genes very tightly regulated both spatially and temporally within the pistil. More detailed molecular research is required to explain the meaning and the extension of the "morphic translocation," as Leavitt (1909) would have called this homeotic phenomenon, which is the replacement of a vegetative structure by a reproductive one.

B. Analysis of Nuclear DNA Content during Somatic Embryogenesis Commitment

The haploid condition of the meiotic cells was detected by chromosome counting and by cytophotometric DNA absorption analysis of nuclei from the hypocotyls at different days of culture, from the hypocotyl-derived suspension cultures and from the different stages of regenerated embryos. In the cultured hypocotyls a gradual increase of the haploid population appears, up to the beginning of the suspension cultures, the latter and the regenerated embryos cycling between the 1C- and 2C-type constitution (Nuti Ronchi *et al.*, 1990; Giorgetti *et al.*, 1995, in press). This investigation has allowed another interesting discovery: a progressive, nongeometric reduction of DNA content per nucleus (besides the one due to the haploid constitution) occurred all during the period of culture, persisting in the regenerated embryos until the heart stage. This trend overlaps exactly with the one shown in Fig. 2, concerning the variant cell lines in which prophase reduction was demonstrated. DNA content per nucleus was somehow progressively recovered at the torpedo and plantlet stages. Modulation of DNA content in development and during the dedifferentiation phases of plant cells has been frequently described (Parenti *et al.,* 1973; Bassi, 1990). But most surprising is the loss of DNA sequences during the commitment to embryogenesis and meiotic phases and their subsequent reattainment when meristems start the activity at the seedling stage. These data have been confirmed by means of molecular analysis (Geri *et al.,* 1992). The authors have identified, by means of C_0t analysis, the highly repetitive, medium and unique carrot genome sequences; these have been used as probes, in parallel with ribosomal cistrons (belonging to the medium repetitive sequences and often amplified in culture and in plants after stressful events), and genes of small multigenic families [actin, ubiquitin, and chalcone synthase (CHS)], in slot-blot hybridization experiments. The result of the subsequent densitometric scanning and statistical elaboration of the data show a trend that matches exactly with that of cytophotometric analysis. Whereas the highly repeated sequences did not appear too affected, the ribosomal cistrons decreased in parallel with the medium repetitive sequences, but in contrast to what usually happens in culture. The high reduction of CHS sequences is surprising; it has been observed to occur progressively during all the stages, not yet recovering normal values at the torpedo and plantlet phases. For further demonstration the authors have analyzed the exact copy number in the various stages so far studied, in comparison with *in vivo* plant tissue. The copy number per haploid genome for the CHS gene varied from 30 copies in leaves to 4 in the hypocotyls on day 20 of culture; the copies were up to 6 at the plantule stage.

Similar selective chromatin diminution in response to environmental conditions has been found during cell culture and plant generation of *Scilla sciberica* (Deumling and Clermont, 1989). Even in this species, CHS was drastically reduced in copy number; the authors interpreted such an excessive and specific diminution as a necessary prerequisite to render the cells of *S. siberica* "omnipotent," the large amount of DNA present in the nuclei of the original plant constituting a hindrance to plant regeneration.

Because the reprogramming appears to be performed mainly in concomitance with the meiotic-like process, and the sequences are progressively lost at the same time that the segregating events occur, it is more likely that the sequence modulation precedes, rather than follows, the meiotic-like process. It is worth saying that meiotic-like figures and haploid metaphases have also been detected, even if only rarely, in the regenerated somatic embryos up to the early torpedo stages; this concomitant occurrence of the two processes (the meiotic one and the specific DNA sequence diminution) and their extension to the regenerated somatic embryos is puzzling. It is the opinion of the author that all processes that can be detected in cultured cells, even if often irregular, are also part of similar processes normally occurring, during development or reproduction, in plants *in vivo*. It is therefore tempting to propose a hypothesis concerning the possibility that cells reaching meiosis in plants may also, before or during the process, undergo a loss of some unnecessary DNA sequences, or of hindrance, for the proper completion of the meiotic division; the recovery of the full or partial DNA complement could occur, also progressively, any time from meiosis to the early stages during embryo development in the seed, or even later during some adult phases of the plant. The occurrence of reducing division, also during the early somatic embryo phases of growth, suggests an unstable genome, poised between mitosis and meiosis, a situation that may persist up to the moment both shoot apical and root tip meristems start to function. In plants *in vivo*, few data are available on DNA content of single nuclei in the transition of micro or macrosporocytes from the premeiotic cells stage to the pollen or egg cells, probably due to the difficulty of measurement in these phases of the meiosis. Sparrow has shown the exclusion from nuclei, during the early stages of meiotic prophase, of some material that was interpreted as chromatin because it was Feulgen positive and absorbed at the same wavelength as nucleic acid (Sparrow and Hammond, 1947). The phenomenon was present in eight genera of plants, characterized by larger than average chromosomes. Similar observations have been made in many species (V. Nuti Ronchi, unpublished observations), but no conclusive proofs are available. Some data seem to indicate that the nucleus in meiotic prophase has a functional dynamic that could, in some way, be interpreted as a phenomenon of DNA sequence extrusion. For instance, nuclear vacuoles appearing in the prophase nucleus of pollen mother cells, reported by some authors (Rodriguez-García *et al.*, 1988; Sheffield *et al.*, 1979), which up to now have not yet been satisfactorily explained, could be related to an event of extrusion from the nucleus of surplus copy number sequences, maybe of impediment for the pairing and crossing-over process.

Very few additional data have been published on DNA content in the nuclei of zygotic embryos in early developmental stages, because the references concern mostly germinating seeds or seedlings. A significant difference in the Feulgen-staining DNA content was found in *He. annuus* seeds, depending on the head portion in which seeds have developed. The data show a rapid drop in DNA content from early mitotic prophases of anthers and pistils to those in embryos at the heart stage, early or late cotyledonary stages remaining unchanged in each embryo during further development and seed germination. The decrease in DNA content occurs during meiosis and/or embryogenesis, but the content differs significantly, the embryos developing at the center of the head having less DNA than those developing at its periphery (Cavallini *et al.*, 1989). The authors rightly interpreted the difference as due to diversification in the microenvironment in which embryogenesis takes place, because reproduction events occur on the head in centripetal succession. On the other hand, it is the author's opinion that these findings represent confirmation that, similar to what occurs in carrot cultures, in some plants the genomic reprogramming going on through meiosis may include a phase of DNA sequence elimination, possibly to limit the consequences of inequal crossing over, followed by their resynthesis, when needed, during development. This point is, at the moment, the object of experimental evaluation. It may also be pertinent to recall that the already cited major aptitude of the immature embryos of various plant species to respond to the *in vitro* culture and to regenerate somatic embryos may be due just to this DNA instability, possibly poised between mitosis and meiosis and at any rate open to new possibilities of development. The frequent incidence of haploid mitoses in cultured embryos favors this suggestion (Singh, 1986; Bajaj and Gill Manjeet, 1985; Linacero and Vazquez, 1992).

Nuclear changes during development have been well documented, and certainly whichever kind of stressful conditions may provide rapid genomic changes, inducing variations in the nuclear DNA content (Walbot and Cullis, 1985).

As rightly discussed by Deumling and Clermont (1989), special "recognition sequences" must be postulated to effect the correct joining of the resynthesized DNA sequences in the chromosomes. On the other hand, a decrease in chromatin is known as a regular step during differentiation processes (Tobler, 1986) or for adaptation to environmental stress as in flax (Durrant, 1962; Cullis, 1983).

C. Genetic and Molecular Analysis of Somatic Segregation Process

The cytological data demonstrating the occurrence of meiotic-like and prophase reduction processes and interpreted as a prerequisite for the acquisition of somatic embryogenesis competence needed to be confirmed, so that the genetic consequences of the segregation event could be demonstrated at DNA level. The author's group has analyzed at the molecular level the possibility of segregation events related to the acquisition of embryogenic competence.

The experiment was carried out using two pure carrot lines, their sexual hybrid, and the cell lines obtained by differentiating 10 embryos regenerated from the sexual hybrid. Two molecular approaches were carried out, in order to reveal molecular variants between parental lines, both present in the sexual hybrid, and segregating when somatic embryogenesis was carried out in the sexual hybrid.

The patterns of the restriction length polymorphism (RFLP), relative to codominant expression, and of the randomly amplified polymorphic DNA (RAPD), relative to a dominant/null type of expression were used. The results showed segregation in 17 of 40 analyzed patterns for RAPD analysis (real frequency twice as much, for its dominant/null type of expression), and in 24 of 30 analyzed patterns in the RFLP analysis. Two of the 10 tested embryo cell lines showed no segregation for any marker, including those of the RAPD analysis; these two embryos may probably be derived from processes like those discussed above and frequently occurring in plants as nuclear restitution due to failure of karyokinesis or of cytokinesis and nuclear fusion; a multicellular origin of the two embryos may not be excluded (Giorgetti *et al.*, 1995, in press). Because cosegregation was not observed for any pair of markers, we have seven nonsyntenic markers that cover the majority of the nine haploid chromosome carrot complement.

Segregational events, either meiotic-like or by means of the prophase reduction mechanism, have been observed in cell cultures of many other plants with almost the same modality already described to occur in the carrot; among others, tomato, sunflower, *Malus*, and some other *Prunus* species were thoroughly analyzed. Using the same simple protocol of culture (B5 medium and 2.2 μM 2,4-D) meiotic-like division and development, along the vascular strands, of reproductive structures could be observed. As already explained, up to now only carrot could be analyzed for the genetic proof of the segregation events, because carrot remains the only plant species for which the single-cell origin of the somatic embryos has been so well characterized. Other plants, although producing gametic-like cells, are unable to regenerate somatic embryos because these cells show a precocious degeneration. The plants regenerating by means of organogenesis normally arise from more cells (as shown by the frequent incidence of sectors or mosaics for different markers), of which a variable number (only few or all) among these participating to form the shoots may have been submitted previously to the segregation process; these plants, therefore, will randomly present the effect of the phenomenon, and only when a proper marker is available. Evidence for this point may be drawn from the analysis of regenerated plants and of their offspring. It is plausible that a large portion of the genetic variation known as somaclonal may be attributed to these segregating mechanisms occurring early in culture as a prerequisite of plant regeneration in culture. On this point it is worth noting that these data appear to confirm the suggestions of Barbier and Dulieu (1980), that most variability of the culture is determined by the donor plant, because it is in the explant early cell divisions where the first cell cycle variants take place.

In conclusion, the meiotic process, in light of these data, acquires a new meaning, as an essential part of an entire process leading to the expression of reproductive potentialites, and a direct answer to conditions of mishap and stress threatening the survival of the organism. If confirmed to be of general occurrence, the evolutive significance of these phenomena, where, in the span of a few days, the efficacious sources of variability of the segregational processes are coupled with the attempt to develop a phenotype fit for the new environment, will be of great interest.

VI. Chromosome Aberration

An important point that may be connected with the mitotic or meiotic cycle concerns the occurrence and evolution, with time, of events that affect changes in number, behavior, and structure of the chromosomes. The cause of changes in chromosome numbers, related to variants of the mitotic or meiotic cycle, has already been discussed (see Sections III and IV). Similarly, the behavior of chromosomes is deeply affected by the way the spindle operates, as discussed above, and by molecular motors such as cyclins. The changes in structure of the chromosomes are of course also connected with the S phase of the cell cycle, but very important and interesting is the relationship between chromosome breakage and heterochromatin. Some of these points concerning the possible involvement of events related directly or indirectly to the cell cycle are discussed here.

A. Chromosome Distribution

Chromosome aberrations occurring in cultured plant cells and variously affecting the competence to regenerate new plants either via somatic embryogenesis or organogenesis have been one of the most discussed subjects; their origin is differently interpreted by the authors, who, however, agree to attribute mutagenic potentialities to the medium components and to the presence of the growth factors. The synthetic hormone 2,4-D was considered active, but only under specific conditions (Nuti Ronchi *et al.*, 1976), in *Nicotiana* species pith tissue cells in an extremely unbalanced auxin condition, or in combination treatments with kinetin. Moreover, regression analysis showed a linear relationship between the number of transfers (up to day 720 of culture) and both frequencies of aberrant anaphases and chromatid breaks in freshly excised pith tissue cultured on a medium supplemented with 2,4-D. When, instead, chromosome numbers were considered, the variability was higher during the first months of culture, compared to 2-year-old cell lines, when a stable pentaploid, no longer organogenic population was established (Nuti Ronchi *et al.*, 1981).

Some published data (Bennici and D'Amato, 1978) have reinforced the idea that an aberrant division process, such as amitosis, could be responsible for most of the aneuploidy frequently found both in cultures and in regenerated plants. In the experience of the author it is unlikely that amitosis, which is a precise division process in which single or coupled full *n* chromosome complements are separated without the apparent presence of the spindle, may provoke the loss of single or a few chromosomes. Amitosis induces a high variability of the ploidy level (Martini and Nuti Ronchi, 1974), whereas single chromosome loss has more chances to occur in the course of an anaphase with a perturbed spindle (a frequent possibility) or when crowded with chromosomes in a polyploid nucleus (Ashmore and Shapcott, 1989).

The frequent occurrence of various chromosome numbers, often ranging from haploids to hexaploids, or more, in the same culture or also in the same regenerated plant, as in the already cited *Nicotiana* case (Nuti Ronchi *et al.*, 1981), may be more plausibly attributed to a variant mechanism of the mitotic cycle, such as the ones already described. Only a competent and thorough cytological analysis may help to ascertain the true variant mechanism operating in each case. Changes in chromosome number may also be found in aneusomatic plants *in vivo*, this instability being transmitted to the cultures derived from these plants. But chromosomal instability appears to be more frequent in seedlings (and often related to seed age) and early developmental stages, suggesting a selective mechanism favoring normal cells in the course of development (Nuti Ronchi and Martini, 1962). Again these findings suggest that *in vivo*, the early stages of growth of the embryo are more unstable and subject to environmental conditions than the mature plant, being therefore similar to the unbalanced somatic carrot embryos. The presence of haploid mitoses in calluses derived from cultured embryos (Singh, 1986; Bajaj and Gill Manjeet, 1985; Linacero and Vazquez, 1992) also suggests events similar to the ones occurring in carrot, and seems to confirm the unstable condition of the early developmental stages. Moreover, a large number of different chromosomal aberrations is often present together with the change in ploidy level, this last fact again confirming that the causal events are possibly due to a chromosome set partition similar to the mechanisms illustrated to occur in *Nicotiana* and *Daucus carota*.

Correct distribution of the chromosomes is an important part of the cell cycle for the production of normal gametes during meiosis and normal division in mitosis. Some of the aberrations shown by plant cell lines concerns the distribution of chromosomes and particularly the production of aneuploids mainly by nondisjunction or chromosome loss. In plants little is known about the genes involved in the production of these anomalies, whereas in the last few years much progress has been made, using lower eukaryotes, to reveal the dynamic of the spindle and of the complex sets of motile events and the proteins required for the proper distribution of the chromosomes. This type of work would be extremely useful to help understand the still mysterious type of movement that allows separation of the chromosomes during the prophase reduction mechanism. The apparent reciprocal repulsion forces could be mediated by microtubules or other motor proteins; it would not be surprising that these forces could be related to archaic primitive structures and functions. Some proteins, identified in *Saccharomyces cerevisiae* and localized to the spindle microtubules that lie between the poles, may have a role in producing a pole separating force from within the spindle (Hoyt *et al.*, 1992; Roof *et al.*, 1992). These components of the mitotic spindle are required for a motility event, but are apparently also essential structural elements (Saunders and Hoyt, 1992); these authors conclude by proposing that spindle structure is generated and maintained by opposing mechanical forces.

These studies and the manifold diversities of the mutants that show the impairment of the functions regulated by these genes related to the cell cycle prompt several suggestions linked with variants found in plant tissue cultures. For instance, some of these motor proteins may be connected with other mechanisms of nuclear division, such as amitosis, possibly related to long forgotten organisms and to genes no longer in function, recalled again to operate when needed by cultured cells in selecting the fittest phenotype. These are only a few hints on the possibility that the discovery of novel kinesin proteins may provide, allowing a better understanding of the basis of chromosome movement in mitosis and meiosis. Identification of homologous genes in plants would help us to understand the origin of the large number of aberrations that often accompany the cycling cells in culture.

B. Heterochromatin and Chromosome Breakage

One of the causes often attributed to chromosomal breakage events is the heterochromatin. Johnson *et al.* (1987), discussing the role of heterochromatin in breakage events, suggested that a perturbation involving heterochromatic regions was induced by tissue culture, and similar effects were invoked by other authors (Phillips *et al.*, 1990; Karp and Bright, 1985; Eizenga, 1989). It is well known that induced or spontaneous chromosome breakage is preferentially occurring in specific regions involving the centromere or the centromeric regions, and often also in telomeres where heterochromatin is present. The effect may be compared with the ones induced by chemical or physical mutagens, which were shown to induce nonrandomly distributed aberrations in plants mainly located in heterochromatic regions composed of repetitive DNA (Nuti Ronchi *et al.*, 1986a.b; Natarajan and Ahnström, 1969; Natarajan and Raposa, 1975). The high specificity of the breakage sites was suggested to be related to a high affinity of some zones for the inducing agents, either because of the presence of fragile sites, or because of the composition of specific DNA sequences possibly connected with the nuclear matrix. Repetitive DNA sequences situated at the anchorage sites of the scaffold of the nuclear matrix could be more exposed during the DNA replication process (Sparvoli *et al.*, 1976; Goldberg *et al.*, 1983). Transposing these results to cultured plant cells, it may be hypothesized that several events may work in cultured cells, separately or in concomitance, to give alterations in the chromosomal structure and/or function: DNA turnover, gain and loss of DNA sequences, altered mitotic cycles due to failure of nuclear envelope breakdown, or other mishaps due to several possible variants of the cell cycle. But even variants of the meiotic-like events, which are supposed to occur mainly at the initial culture times, and the possible effects of transposable elements, may be included among the causal events.

The involvement of heterochromatin in the release of somaclonal variation has also been discussed by several authors (Lapitan *et al.*, 1988; Lee *et al.*, 1988; Eizenga, 1989; Benzion and Phillips, 1988).

In several cases the influence of heterochromatin was attributed to the well-known delayed DNA replication characteristic of the sequences belonging to this type of nuclear component (Lima-de-Faría, 1969). The chromosomal aberrations supposed to derive from this late DNA replication during interphase are mainly chromosome bridges leading to translocations, deficiencies, and duplications (Johnson *et al.*, 1987; Lee and Phillips, 1988).

Eleven agronomic traits related to yield, earliness, tillering, and plant height were examined to evaluate the somaclonal variation in plants regenerated from immature embryo cultures of triticale lines differing in telomeric heterochromatin on chromosome arms 7RL (Drira cultivar) and 6RS (Rosner cultivar) (Bebeli *et al.*, 1988; 1993a). Significant somaclonal variation was observed in the R2 arm for several of the traits studied; the lines possessing telomeric heterochromatin released more variation. It is interesting to note that, instead, in a similar study carried out on rye

(Secale cereale L.) lines differing in the presence or absence of telomeric heterochromatin 7RL, 7RS, and 5R, the somaclonal variation observed was more pronounced in the lines lacking telomeric heterochromatin (Bebeli *et al.,* 1993b). The variance could be due to several factors, related to the fact that triticale, being a man-made cereal resulting from crossing wheat to rye (Kaltsikes, 1974), may react to the *in vitro* culture differently from the genotypes constituted by the single rye genome. Therefore the possible effect on cell cycle variation in relation to the late replication attribute of heterochromatin may have different consequences in rye as compared to triticale.

That heterochromatin may be of great influence in mutation induction and chromosome breakage and structure may be inferred from the demonstration, by means of *in situ* nick translation of meiotic chromosomes, that homologous heterochromatic regions of the grasshopper *Atractomorpha similis* show a different staining, which could be explained by a differential endonuclease cleavage (de la Torre *et al.,* 1993). The authors tentatively suggest that homologous heterochromatic regions are able to incorporate, during their evolution, different mutations that, chiasmata not being commonly formed within heterochromatic segments in this species (John and King, 1985), would probably be preserved separately. If similar situations occur in plants, in which such a large number of changes in repetitive sequences and heterochromatin blocks has been demonstrated, it should influence variability even more.

Heterochromatin changes were detected in anther-derived doubled haploids of tobacco *(N. tabacum)* (Dhillon *et al.,* 1983). Further studies, although confirming a 41% increase of nuclear DNA in tobacco double haploids, were unable to detect substantial enlargement of preexisting heterochromatic segments (Reed *et al.,* 1989).

Another type of *in vitro* technique, namely protoplast culture, has been reported to yield extensive karyotypic changes in the majority of the cells, chromosomal aberrations ranging from numeral variations to structural anomalies (Hahne and Hoffman, 1984, 1986; Karp *et al.,* 1987). The reported presence of megachromosomes among several other aberrations, detected in 4% of the analyzed cells, appears interesting (Karp *et al.,* 1987). In the detailed study of some *Nicotiana* hybrids, Burns and Gerstel (1969) found that when a heterochromatic block from *N. otophora* is introduced into *N. tabacum* genomes, spontaneous chromosome breakage and formation of megachromosomes occurs. Spontaneous deletion of the distal part of the heterochromatin block, obtained by means of specific crosses, stopped both chromosome breakage and formation of megachromosomes. Atypical conditions arrived at by hybridization, exposure to radiation or disease, or in tumors have allowed the detection of similar phenomena in such different organisms as plants, cestodes, mice, and insects (reviewed in Gerstel and Burns, 1967). Cultured plant cells often show megachromosomes (Evans and Gamborg, 1982; Kovács, 1985).

The discovery of these abnormally long chromosomes and of unusually extensive chromosome aberrations in protoplast cultures do not appear to be a casual phenomenon. This and the other cases just reported are consequences of conditions of stress, and they may also be related to the heterochromatin connection with the nuclear matrix. It is the opinion of the author that the process of cell wall digestion, which allows protoplast recovery and culture, may perturb these specific chromatin–nuclear envelope associations, promoting chromosome breakage, possibly in specific sites (data on this point is lacking); when heterochromatin is broken, or shifted from its normal position on the nuclear envelope, a process of nuclear amplification may be stimulated that results in the formation of megachromosomes. Specific chromatin–nuclear association may also be involved in *Nicotiana* hybrid megachromosomes: if genomes of different species do not properly fit into their normal postion in the nucleus, they may possibly be more prone to an amplification process. This hypothesis also foresees that the rebuilding of the protoplast cell wall is achieved following a completely new program, possibly connected with the reprogramming of the nucleus. The perfection of the chromosome breaks and the synthesis of any possible DNA amplification, or the occurrence of other processes connected with the rebuilding of the cell wall, may possibly also be related to the expression of the regenerative capacities. It is in fact well known that in the carrot, for instance, the embryogenic capacity is dominant in protoplast fusion experiments (Sung *et al.*, 1984).

Burns and Gerstel (1969), discussing the presence, in cytologically unstable *Nicotiana* hybrids, of dicentric chromosomes undergoing the breakage–fusion–bridge cycles, compared their results with the ones obtained in *Z. mays*. In this species similar phenomena, involving regions undergoing breaks, rouse transposable elements to move in the chromosomes (McClintock, 1944). Dicentric chromosomes undergoing a permanent breakage–fusion–bridge cycle have been studied in carrot cell suspension (Nuti Ronchi *et al.*, 1976; Toncelli *et al.*, 1985). Somatic embryos regenerated from these cultures showed the same aberrant divisions, which therefore did not preclude the differentiation phase; but further growth after the torpedo stage could continue only if healing of the broken dicentric ends or dicentric loss occurred. These data also recall the McClintock (1956) results, because they share some important elements with the *Z. mays* system of transposable elements: the passage through a meiotic-like process, the traumatic element of a permanent breakage–fusion–bridge cycle, the development of the early embryo stages, notwithstanding the chromosomal aberration, and the possibility of further growth of the regenerates if only healing of the chromosome broken ends or dicentric loss occurred. There are not, of course, any markers to verify the likeness of the two processes, but the comparison marks once again how cells in culture behave with a surprising analogy to the plant system, any apparent oddity reflecting exactly what plants have devised, in the course of time, to endure any kind of environmental situation.

It is pertinent to discuss now the process of DNA modulation, either diminuition or amplification, this last process having been first detected just in plants, during the first dedifferentiation hours of *N. glauca* pith tissue in culture (Nuti Ronchi *et al.*, 1973; Parenti *et al.*, 1973; Durante *et al.*, 1982). This phenomenon, which has since been found in other plant systems (Zheng *et al.*, 1987; Deumling and Clermont, 1989; Arnholdt-Schmitt, 1993), is not discussed in depth here, but it is interesting to point out some suggestions, related to the cell cycle, that are emerging both from plant tissue culture and also from the analysis of regenerated plants. One initial observation comes from data showing that, in cultured plant cells, a diminution of DNA specific sequence content per nucleus often occurs with a regeneration process (Geri *et al.*, 1992; Deumling and Clermont, 1989; Arnholdt-Schmitt, 1993; Cecchini *et al.*, 1992; Altamura *et al.*, 1991; Natali *et al.*, 1993).

A second point concerns the fact that the DNA variations are often accompanied by a changed pattern of DNA methylation that may even be related to the cell cycle and phase changes as well. The study of *N. glauca* during the early dedifferentiation stages of pith tissue culture has shown a selective DNA synthesis of specific sequences (satellite and/or repetitive). Experiments with hypomethylation-inducing drug, such as azacytidine, suggested the presence, in cultured *Nicotiana* species, of a period of DNA synthesis during the cell cycle, in which some particular sequences may undergo variations in the endogenous methylation pattern; subsequent changes in gene expression and in the differentiative pathways to regeneration, habituation (hormonal autonomy), or other cytodifferentiation events, if submitted to the proper cultural conditons, are induced (Durante *et al.*, 1989). Any perturbation during this time may affect the methylation pattern, inducing modification of the plant genome and leading to a process of demethylation and therefore allowing the expression of sequences related to a differentiation process phase. Rapid genomic variations due to *de novo* methylation and to an increase in the number of copies of repeated DNA sequences were observed during the first hours of culture of carrot root explants (Arnholdt-Schmitt, 1993).

The correlation between hypomethylation and gene expression has been made in several systems. In plant cells (Durante *et al.*, 1989), demethylation may induce tumorous transformation.

It is interesting to note the suggestions coming from the results obtained by the study of the variations in the pattern of DNA methylation on carrot cell lines and during somatic embryogenesis (Lo Schiavo *et al.*, 1989).

In discussing experiments with demethylating drugs showing that the level of auxin, both natural or synthetic, may change the 5-methylcytosine level in the cultured carrot cell DNA from 15 to 70% of total cytosine, Lo Schiavo *et al.* (1989) suggest the presence of two levels (or ''patterns'') of methylation. One pattern is necessary for the maintenence of the differentiative state: hypomethylation by means of drugs disorganizes this pattern, that is, arrests embryogenesis. The other type of methylation occurring as a response to auxin, is lost in the absence of its causative hormone and may disorganize, in concomitance with a proper signal, the differentiative patterns, for example, those involving a change of program (e.g., versus a tumorous phenotype). Whatever the real meaning and effect of the variations in the DNA methylation patterns, in the plant system the presence or absence of the hormone seems always to be involved; and again, these variations may also contribute to the overall variability of the cultured cells.

VII. Polyploidy

The mechanisms operating, separately or in conjunction, both *in vivo* and *in vitro* to change the nuclear ploidy toward lower or higher levels have been schematised (Fig. 8) (Nuti Ronchi and Terzi, 1988). The scheme shows how, covering different pathways, cells may choose to reach the fittest chromosomal complements needed for developmental purposes or for resistance to stress. Variously crossing each other, these different pathways allow sudden changes and overturning of tendency not only from the highest ploidy state to haploidy, but also to deviant odd ploidy levels, aneuploid complements or aberrant chromosome constitutions.

As we have already discussed, polyploidy is always reached by means of two DNA doublings between which mitosis was completely omitted, or aborted, or took place but chromosome separation was impeded by the failure of the nuclear envelope to dissolve, or by a failure of spindle formation or of cytokinesis to occur; although displaying great variety in detail, this event appears, in plants, to be the most frequent answer to different types of stimuli, both endogenous and exogenous. The evolutionary importance of the polyploidy, and therefore of the mechanisms devised to attain this aim, has been recognized and widely elucidated by some authors (Dobzansky, 1964; Stebbins, 1959). Similar devices can be found following the evolution of a cell population in culture (Patau and Das, 1961; Nuti Ronchi and Terzi, 1988; Pijnacker *et al.,* 1989). In a fast-growing cell line, where very frequent medium changes keep a high rate of cell division, diploidy level is maintained under control, cells with higher chromosome number being present but rarely dividing. The function of

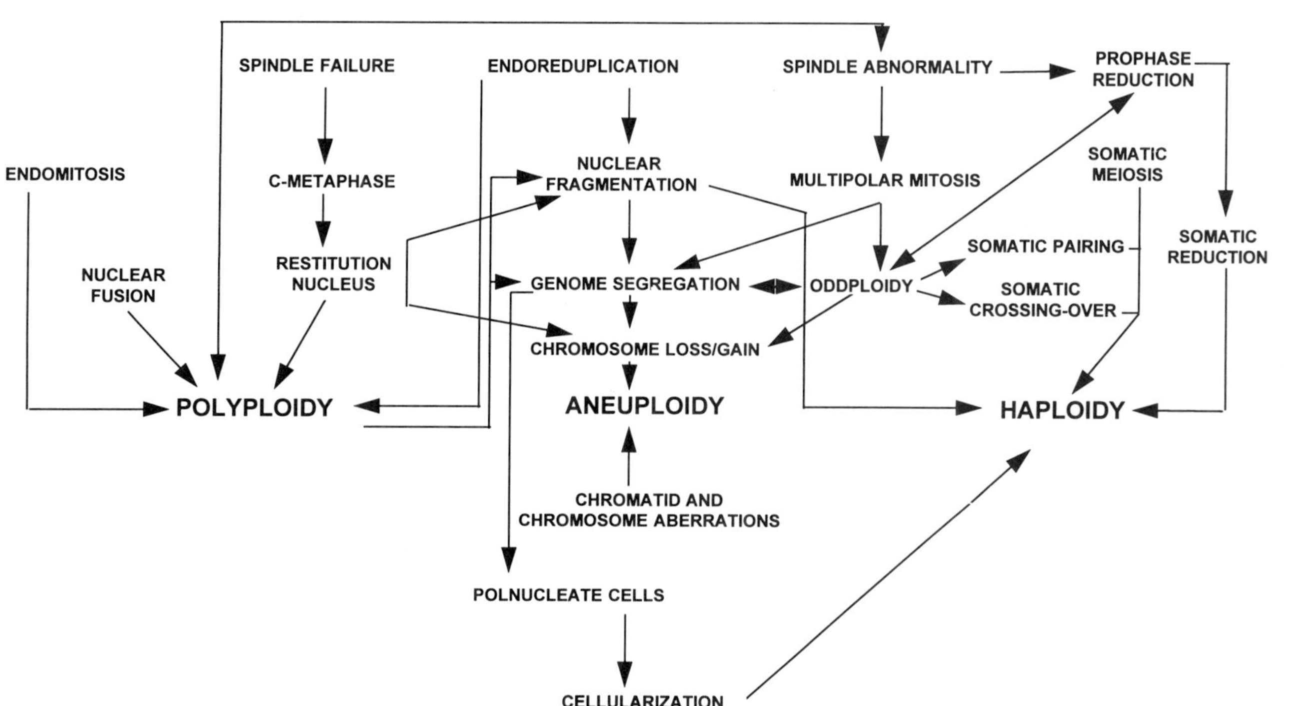

FIG. 8 Different pathways leading to higher degrees of polyploidy and, through various mechanisms, to aneuploidy and haploidy. [From Nuti Ronchi and Terzi (1988). Copyright © 1988 John Wiley & Sons. Reprinted by permission of Wiley-Liss, a division of John Wiley & Sons, Inc.]

these cells is possibly different, according to the circumstances and to the micro- or macroenvironments surrounding them; as a rule, they may perform the duties of nurse cells, providing some metabolites required by other sister cells for growth. But these larger cells are also the fittest to survive mishaps, are able to enter division, and also adopt some of the reducing devices previously described, when most of the other diploid cells have ceased to divide because of lack of nourishment or due to other stress. Also in plants, *in vivo* polyploidy may carry out similar nutritional functions, and in some tissues this concept is extremely clear, as, for instance, in the tapetal tissue of the anthers or in the suspensor or glandular cells; but where polyploidy is certainly extremely useful, with the full possibility offered by its manifold ways to obtain the fittest chromosome complement orders, is when events such as wounding or physical–chemical attacks perturb the state of the tissue or cell lines. The strategies displayed to survive the situation may be found in Fig. 8 and may be different according to the environment and to the final objective, whether it be survival, defence, stimulus to a reprogrammed development, or wound healing. As already discussed, the response of plant tissues consisting only of parenchyma cells, such as the pith tissue, to excision will be different from the response of tissue including some vascular strands. In the first case the large parenchyma cells undergo a series of endoreduplication cycles before going back to the diploid values by means of an amitotic process; this mechanism of cellularization provides a fast way to overcome the drawback given by the largesse of the cells, filling them in a single event with many reduced nuclei, even of different ploidy, and later on with many small cells (Nuti Ronchi *et al.*, 1973; Nuti Ronchi and Terzi, 1988). In *N. glauca* cell suspension cultures, the innermost cells of this cellularized large cell immediately undergo a differentiation to a tracheid cell, starting thus the formation of the first vascular structure that, by canalizing the solutes, will allow bud regeneration (Nuti Ronchi, 1981).

When the explants include vascular strands, as in carrot hypocotyls, the strategies are different. Cells undergo an extensive DNA reduction, down to the haploid level, by means of mechanisms as somatic meiosis-like and prophase reduction processes. These reduction mechanisms also concern the endoreduplicated cells, which are numerous in tomato hypocotyls. The aim, in this case, is the production of reproductive structures that, mimicking the gametes, could contribute to the survival of the mother plant genes by means of somatic embryogenesis. These mechanisms exploit the totipotency of plant cells, releasing the amount of variability employed, also *in vivo,* to overcome the disadvantage of immobility.

Another interesting potentiality offered by the mechanisms of reduction of the ploidy level is the genome partition event. As in all the cases of chromosomal reduction, a nonrandom arrangement of chromosomes in the nucleus is required, so as to allow the separation of the chromosomes belonging to the same genome basic number *n*, behaving as a unit rather than an individual during cell division. This is particularly evident in allopolyploids undergoing genome partition, as shown in polyploid oats *(Avena sativa)*, in which the spontaneous production of a diploid sector $2n=14$, representing an entire genome of one of the diploid progenitors, was found (Ladizinksy and Fainstein, 1978). Extraction of specific genomes from allopolyploids has been made in wheat, in which the tetraploid genome was obtained from the hexaploid wheat (Kerber, 1964). The possibility of extracting specific genomes from allopolyploids or secondary polyploids exists, for evolutionary study but also for breeding purposes, with the help of *in vitro* culture reducing mechanisms and subsequent plant regeneration.

Several chemicals have been shown to increase the frequency of these events of segregation of entire genomes from hybrids and allopolyploids or homozygous tissue from heterozygous plant. The substances used range from colchicine to other chemicals well known for their mutagenic effect, such as alkylating agents. Colchicine was the most used, and some species appeared more suitable than others to induce segregation of entire genomes or somatic crossing over in *Sorghum*, tomato, and *G. max*. The colchicine effect is probably related to the homologous chromosome association supposed to be nonrandom and brought about, and maintained in the interphase nuclei, by a given location of the heterochromatic regions on the nuclear envelope. Mitotic spindle (and therefore microtubules) and nuclear membrane appear to be involved in this arrangement of the chromosomes in the interphase nucleus, an arrangement that, once disrupted by the drug, may lead to cytological consequences that justify the hypothesis of an association of homologous chromosomes. Avivi and Feldman (1980) have exhaustively reviewed this topic; their proposed hypothesis as to the mechanisms involved in determining the specific pattern of arrangement within a given set being, in the opinion of the author of this article, widely supported by the evidence on the variants of mitosis and meiosis described before. The Avivi and Feldman hypothesis suggests a specific attachment of centromeres and telomeres of every chromosome to definite sites on the nuclear envelope; supposing that the centromere regions are responsible for positioning the homologous chromosomes close to one another, the mitotic apparatus being therefore concerned in the matter, any physical–chemical agents or stress affecting the microtubular system may act on the nuclear envelope to modify the structure; in any case, the association and pairing of the chromosomes may be greatly affected. Because the passage from mitosis to meiosis involves only few genes (only two for Golubovskaya, 1979), and genes affecting the structure of the spindle are known (Avivi *et al.*, 1970a,b; Avivi and Feldman, 1973), a similar hypothesis may be the most probable and open to experimental confirmation.

VIII. Concluding Remarks

The leitmotiv of the previous discussion may be surely identified as the concept of plant genome plasticity surviving whatever stress arises. When considered in detail and in comparison to the manner of division of other organisms, the behavior of plant cells in culture may appear fanciful, and some deducted conclusions hazardous. But it has been shown that all these different ways to act are also performed in the plant *in vivo*, both in specific tissues or under conditions of stress. When plant cells are concerned, manifold possibilities are the rules, and this article has attempted to give an overview of all these different opportunities cultured cells may choose among, and on which selection operates.

The main points arising from the previous discussion may be summarized as follows.

A. Plant Cell Totipotency

Totipotency should no longer be defined as a vague attribute of any healthy plant cell, but as a specific competence, acquired by somatic cells by means of processes that may be compared to the ones occurring in the reproductive organs, in reprogramming nucleus and cytoplasm to generate gametic-like cells. It is suggested that the cells undergoing this reprogramming process belong to a particular cell population, preexisting in strict connection with the xylem in plant vascular strands, and having unique properties, transmissible to their derivatives, which can perpetuate them all along the development of the plant. These cells respond to the concept of apical initials put forth by Newman (1965), who also called them "continuing meristematic residue"; Steeves and Sussex (1989), discussing shoot apical initials, recalled this concept, recognizing "their striking similarity to certain cells that play an important role in animal development" and are designating them as stem cells. It is not possible at the moment to be more precise as to the origin of these cells and their possible relationship with apical initials, but surely their functions appear to be involved in the response to any sort environmental stimuli, either hormonal or physical–chemical.

The concept of stem cells has been recalled in a discussion on the relationship between the relative quantities of DNA in individual nuclei of stem and leaf epidermal cells of *Arabidopsis*. Endopolyploid nuclei up to the 16C level occur in various vegetative tissues, but not in floral structures of *Arabidopsis* (Galbraith *et al.*, 1991; Dickson *et al.*, 1992). A definite relationship between cell size and ploidy level in this model plant has been demonstrated, but some 2C cells are present and continually contributing to form new 2C cells, the rough proportion between different ploidy levels remaining the same. The authors conclude that the apparently undifferentiated 2C cells constitute a line of "stem" cells in leaf epidermis (Melaragno *et al.*, 1993). It is also opportune to cite some data that in the opinion of the author, may be interpreted as another example of stem cells comparable to the ones already described in cultured hypocotyls. Organogenesis and somatic embryogenesis may occur, in monocotyledon plants, with more success when sectors of the youngest leaves were cultured. Nuclear DNA measurements have always demonstrated, during the first days of culture, when cell division was more active, the presence of a cell population peaking mostly in 2C, the cells in 4C being almost absent (Karlsson and Vasil, 1986; Taylor and Vasil, 1987; Wernicke and Milkovits, 1984; Hesemann and Schroeder, 1982). This paradox may be solved only if the cell population, which divides precociously in culture, is composed by cells having a 2C nuclear content, and is cycling with mechanisms of chromosome reduction like those illustrated to occur in carrot "stem" cells. The data on this subject appear particularly interesting, showing that the spatial and temporal expression of the *Arabidopsis cdc2a* gene is not coupled with cell proliferation but always precedes it (Hemerly *et al.*, 1993). This fact suggests that the *cdc2a* gene expression may reflect a state of competence to divide. Also in *Arabidopsis,* the activity of this gene is present in the same cells that in carrot seem to behave as stem cells, namely in the parenchyma cells of the protoxylem and phloem, during both development or under different environmental conditons of stress. The authors point out the fact that fully differentiated cells do not switch off *cdc2a* expression; these cells therefore potentially retain "a pool of the essential kinase that can be used if necessary." This, in the author's opinion, seems exactly the right definition of a stem cell.

However, the concept of totipotency should not be considered as related only to the production of gametic-like cells. Totipotent cells are part of a more complex process that mainly concerns the transformation of a vegetative structure to a reproductive meristem, leading to the expression of floral and reproductive genes, including the meiotic process. Only gametic-like cells may, possibly, if given the proper conditions, prosecute growth to give a bipolar embryo; a return to vegetative status may lead, instead, to shoot development. The important questions is, what triggers these cells, even assuming that they are the equivalent of stem cells, and express *cdc2* gene and have a 2C DNA content, to start a developmental pathway when submitted to specific culture conditions? It is clear that the hormone action is the trigger, both endogenous or exogenous hormones being involved, because induction of somatic embryogenesis competence may occur in carrot in the absence of exogenous hormones (Smith and Krikorian, 1990); explant tissue consisting of zygotic embryos instead of hypocotyls, and hormone action replaced by low to medium pH, have been used in these experiments.

A model suggests that it is a question of hormone receptors; if the cells have, or could be induced to produce, the proper receptor, complete embryogenesis will follow; if the receptors are not the proper ones, organogenesis or callus will result (Filippini *et al.,* 1992; Guzzo *et al.,* 1994).

B. Cytological Features

The meiotic-like process is initiated in the specific 2C cells as soon as the stressful situation begins, segregating directly by making use of one of the reducing mechanisms previously described, without any DNA synthesis. Evidence of the occurrence of this event is given at the cytological, histological, and molecular level. Moreover, even if the ultrastructural analysis is lacking, some specific cytological features support the interpretation of meiotic cells: the presence of a pointed spindle, as shown in Fig. 9, which represents a metaphase I-like spindle in cultured tomato hypocotyls. Pointed spindles during meiosis have been demonstrated to be formed in *L. esculentum* (tomato), unlike during mitosis when a barrel-shaped spindle is present (Hogan, 1987). Other specific cytological features concern the presence of a callose cell wall on the *Prunus* embryo saclike cells and the presence of haustorium. Finally, RFLP and RAPD patterns have confirmed, at the molecular level, the segregation events.

C. Heat Shock Proteins and Gametogenesis

Indirect support for the present interpretation that stresses conditions are the main inducers of events related to reproductive pathways may also be found in some heat shock protein (hsps), whose expression is stage specific. It has been shown, from studies in a number of evolutionarily divergent eukaryotes, that the expression of specific members of the low molecular weight hsp families often accompanied gametogenesis (Kurtz

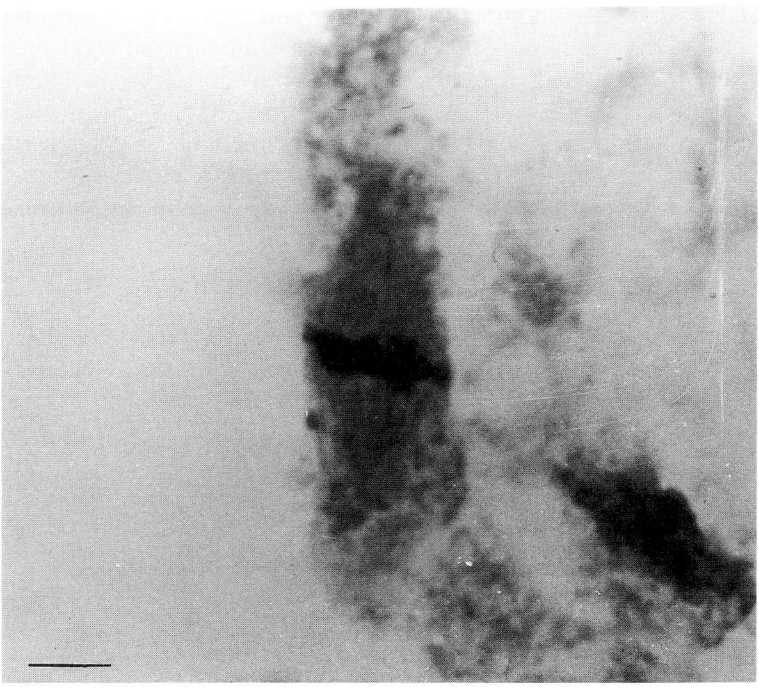

FIG. 9 Meiotic-like metaphase I in a somatic cell of tomato hypocotyl explant cultured in the presence of 2,4-D for 20 hr. A pointed spindle characteristic of meiotic tomato metaphases is evident. Bar: 5μm.

et al., 1986). Studies with *Z. mays* and *Lilium* (Bouchard, 1990) have reported enhanced expression and/or accumulation of 18-kDa hsp mRNAs in the male gametophyte in the absence of heat stress.

The work in *Z. mays* has confirmed the previous results showing that the stage-specific expression of particular 18-kDa hsp genes results from gene-specific regulation during microsporogenesis and gametophyte development rather than from an overall activation of the heat shock or stress response (Atkinson *et al.*, 1993). The finding that specific, small hsp mRNAs can be correlated with meiosis and gametogenesis in organisms from three eukaryotic kingdoms (yeast, *Drosophila*, and plants), implies that some small hsps may serve a specific and, perhaps, common function in the normal development of events related to the reproduction process. It may therefore be justified and not endanger the hypothesis proposed in the present article. The response of plant tissue to the *in vitro* culture stress may involve the induced expression of similar genes, connected with the floral reproduction, giving rise to the series of events described in detail in the present article.

This last statement needs, of course, a proper scientific demonstration, but some results may be interpreted in this direction. In soybean (Gee *et al.,* 1991), the expression of the auxin-responsive transcript GH3 is restricted largely to specific tissues within organs of the developing flowers in the absence of added auxin, whereas the exposure of plant parts of exogenous auxin results in activation of these genes within the vascular tissues of most plant organs. These data, comparing GH3 and small auxin up RNA (SAUR) transcript expression in normal and 2,4-D-treated soybean seedlings and flowers, suggest the existence of different classes auxin-induced genes (and of auxin receptors) as postulated in carrot (Filippini *et al.,* 1992). The results could also be extrapolated to confirm that auxin-responsive genes, such as GH3, which are tightly regulated in the whole plant, may respond to exogenous 2,4-D activating a floral response, as shown in cultured hypocotyls.

If this hypothesis is shown to be well grounded, it would be possible to conclude that ancient device specific for surviving the danger of extinction, and possibly for support of evolutionary needs, is still persistent in plants, perhaps also in other organisms.

D. Evolutive Aspect

The variants of the mitotic cycle and the primitive homeotic reproductive structures developing in the cultured tissues of different plant species may offer some hints of extreme interest for studies on the evolution of mitosis and of the mitotic spindle, and of the homeotic ancestral reproductive structures as well. It is of course not an easy task to survey this subject, because most important ultrastructural and molecular details of the phenomena discussed here are still lacking. Moreover, most of the mechanisms and functions of microtubules and of the spindle are still unknown, and therefore the distinction between their static and dynamic properties is unclear.

It is suggested here that some unusual patterns of nuclear division, as discussed above, may represent primitive mitotic systems that selective pressure has discharged but that may be called on again to function when needed in cell culture conditions. As rightly observed by Heath (1980), discussing variant mitoses in lower eukaryotes as possible indicators of the evolution of mitosis, "it is highly likely that the selection pressure operating in the mitotic system will differ substantially from those operating on the entire organism.'' It is therefore possible to find a ''primitive'' mitotic system in a cell that is otherwise ''advanced'' and vice versa.

Kubai (1975), exhaustively reviewing the evolution of the mitotic spindle, has posed the question as to whether there are any examples in which microtubules concerned in nuclear reproduction seem to play solely (or predominantly) a cytoskeleton role. The idea was that ''a recounting of the evolutionary past of the mitotic processes of chromosome movement would entail tracing a gradual shift from membranes to microtubules as the primary forces for chromosome movement'' (Kubai, 1975). For instance, certain fungal meiotic divisions in which more than one spindle per nucleus is observed favors a system in which entire sets of chromosomes are physically constrained to an appropriate site. The designation of *spindle pole body* (SPB) concerns the form found to be associated with the nuclear envelope to the polar regions of dividing fungal nuclei. In the case of fungi, Kubai (1975) pointed out findings consistent with the idea that an intimate connection of kinetochore and SPBs may coordinate movement of entire sets of chromosomes and the lack of association between sister kinetochores at ''metaphase.''

Some of the variants of the cell cycle described above should be of help to trace back their possible role as intermediates in the evolution of the mitotic apparatus. But it should also be considered that any condition of stress (including lack of hormonal balance) may act to deviate the mitotic process in the plant. These considerations pose the question of the molecular triggers of these mechanisms working also *in vivo* (e.g., amitosis in *Tradescantia* stem, chromosome segregation in *Sorghum,* etc.) and how selection works to master them.

The same question is important concerning the homeotic structures developing in the cultured plant tissues; the detection, on these primitive ancestral reproductive organs, of genes that have normally an exclusive expression on specialized floral tissues and structures of the plant, may allow a further inquiry, in the comparison with the normal inflorescence, on possible evolutionary differences of genes and/or functions.

E. Evidence from Regenerated Plants

A final remark must be made on some characteristic features presented by regenerated plants that could be interpreted as due to the effect of the phenomena previously described to occur either during *in vitro* culture or during the commitment to regenerate.

Besides the already-discussed changes in ploidy level and in chromosome structure that could be transmitted to regenerated plants, other differences at the genic or molecular level were also noticed in normal-appearing regenerants.

1. Evidence of Genome Segregation

One of the first reports, referring to cereal grains, is important because it suggests that in these species a segregation mechanism may also operate in cultured cells and tissue. Orton (1980) has shown, with an analysis of specific isozyme activities in plants regenerated from tissue culture of hybrids of *Hordeum vulgare* × *H. jabatum,* that a quantitative segregation of the two genomes had occurred; this author suggests the occurrence of a meiosis-like reduction division, such as the one observed in somatic tissue by Wilson and Cheng (1949). Alternatively, the segregation of genomes in interspecific hybrids as observed in oats (Ladizinsky and Fainstein, 1978) or multipolar anaphases segregating whole parental genomes (Tai, 1970) was postulated. It is, however, important to note that genome separation occurs, throughout the cell cycle, in hybrids between widely diverged species (Heslop-Harrison and Bennet, 1990), as shown by *in situ* hybridization of interphase nuclei with two probes specific for the two genomes (Schwarzacher *et al.,* 1989). The position of different parental genomes in the interphase nucleus depends, possibly, at least in part, on the nuclear envelope attachments or the nuclear skeleton functions. It is worth noting that spindle poisons can disrupt genome separation into the nucleus. This last finding seems to confirm the hypothesis already posed before when discussing the high incidence of chromosome breakage in heterochromatic sites, that any perturbation may change specific chromatin–nuclear envelope association, promoting chromosome breakage, but also differential positioning of whole genomic sets or associations of homologous pairs. The high incidence of chromosomal aberrations reported by Orton (1980) that always occurs in these interspecific *Hordeum* hybrid cultures may have been originated by culture condition-induced changes of specific chromatin–nuclear envelope association.

Evidence of a segregation process was also given in rye regenerates, in which 1 haploid plant and 3 other mosaic diploid–haploid plants, in a total of 97 regenerated plants, were found (Linacero and Vazquez, 1992). Somatic segregation was also suggested to explain the high rate of recessive mutations (Evans, 1989); 18 of 19 regenerated mutant plants breeding true for at least 3 generations were reported in tomato (Gavazzi *et al.,* 1987).

2. Single Recessive Mutations

Most of the analyses of first and second self-pollination generations of plants regenerated from tissue cultures reveal variant phenotypes segregating as if controlled by single, recessive genes. The reported mutation frequencies are high, ranging from 0.3 to 1.3 per regenerated plant. Most data come from maize (Edallo *et al.*, 1981; McCoy and Phillips, 1982; Armstrong and Phillips, 1988 Lee *et al.*, 1988). But other species also showed similar frequencies, as in tomato (Evans and Sharp, 1983; Gavazzi *et al.*, 1987) and oats (Dahleen, reported by Phillips *et al.*, 1990). In this last case the significant variation from control concerned, in each regenerated plant, at least one of the eight quantitative traits measured and found in more than 80% of the regenerated plants. Because each quantitative trait is controlled by many genes it appears difficult to envisage a mechanism capable of mutating so many quantitative traits in a single experiment.

Another interesting case reported by Phillips *et al.*, (1990) concerns the performance in maize, of inbred A188 regenerant-derived families in testcrosses with the inbred that was the explant source. In most of these crosses the yield (which was about 50% lower in the regenerant-derived families than the uncultured sources) is completely restored to normal level. The question these authors posed about this heterosis effect is how completely recessive gene mutations may be induced by tissue culture in the lower yielding regenerant-derived line.

These puzzling results have been attributed to different causes, such as DNA methylation, transposable elements activation, and structurally altered chromosome; even if these last genetic changes may contribute to the above-mentioned results, the more plausible genetic explanation for somaclonal variation is offered by the meiotic-like or segregational events, shown to occur early in culture and supposed to be a prerequisite for the commitment to regeneration capacity.

3. Variation in Genetic Recombination Frequency

Another result of somaclonal variation not easily explained is the fact that regenerated plants have generally higher chiasma frequencies and more distally localized chiasmata than the donor plants (Puolimatka and Karp, 1993; Singsit *et al.*, 1990; Molnár-Láng *et al.*, 1991). The variation is not always in favor of an increase in the recombination rate; in tomato a decreased recombination rate and map distance were also found, the differences depending on the distance of the studied loci from the centromere (Compton and Veilleux, 1991).

DNA physical modification such as amplification, increases in repeated DNA sequences, and rearrangements may be involved in the variation of the genetic recombination frequencies. In fact, the data showing in carrot that a large phenomenon of DNA modulation occurs when cells undergo the reprogramming process, a supposed prerequisite for the acquisition of the regeneration capacity, widely justify these modification of DNA recombination rates. For instance, the dramatic variation in copy numbers of the CHS gene (Geri *et al.,* 1992), with the increase in adult tissue and decrease down to only two copies in cultured cells, probably generates wide perturbations both in the resynthesis process and insertion sites. The phenomenon recalls the DNA changes already shown to occur in flax genotrophs (Durrant, 1962), and studied at molecular level by Cullis (1983), changes that were shown to be heritable and induced by different specific environments (Walbot and Cullis, 1985).

The aim of this article has been to ascertain, considering the old and new information on the sources of variability operating in plant cell cultures, the role they play in creating new possibilities for human needs and how the variants arising *in vitro* may affect regenerated plants. The variability is certainly great, because the deviant events often occur in conjunction and/or in succession, quickly responding to even slight changes in the microenvironments and to selective pressure. A better understanding of the mechanisms involved will make it possible, in future, to master the processes for their possible utilization. But the major contribution will be the one given to solve scientific problems at genetic, molecular, and developmental levels.

Acknowledgments

The author thanks M. Terzi and R. Vergara for critical reading of the manuscript; and L. Pitto, C. Geri, and L. Giorgetti for stimulating discussions. The author's group work quoted herein was supported by the National Research Council of Italy (Special Project RAISA, subproject No. 2, Paper No. 1585) and by the Italian Ministry of Agriculture and Forestry "Sviluppo Tecnologie Innovative" Program.

References

Afon'kin, S. Yu. (1989). Induced and spontaneous polyploidisation in large amebas. *Int. Rev. Cytol.* **115,** 231–266.

Altamura, M. M., Cavallini, A., Cionini, G., Monacelli, B., Pasqua, G., and Cionini, P. G. (1991). Cytophotometry of nuclear DNA in buds and flowers of *Nicotiana tabacum. Cytobios.* **67,** 85–93.

Armstrong, C. L., and Phillips, R. L. (1988). Genetic and cytogenetic variation in plants regenerated from organogenic and friable, embryogenic cultures of maize. *Crop Sci.* **28,** 363–369.

Arnholdt-Schmitt, B. (1993). Rapid changes in amplification and methylation pattern of genomic DNA in cultured carrot explants *Daucus carota* L. *Theor. Appl. Genet.* **85,** 793–800.

Ashmore, S. E., and Shapcott, A. S. (1989). Cytogenetic studies of *Haplopappus gracilis* in both callus and suspension cultures. *Theor. Appl. Genet.* **78,** 249–259.
Atkinson, B. G., Raizada, M., Bouchard, R. A., Frappier, R. H., and Walden, D. B. (1993). The independent stage-specific expression of the 18-kDa heat shock protein genes during microsporogenesis in *Zea mays* L. *Dev. Genet.* **14,** 15–26.
Atkinson, G. H., Franzke, C. J., and Ross, J. G. (1957). Differential reaction of two varieties of sorghum to colchicine treatment. *J. Hered.* **48,** 259–264.
Avivi, L., and Feldman, M. (1973). The mechanism of somatic association in common wheat, *Triticum aestivum* L. IV. Further evidence for modification of spindle tubulin through the somatic association genes as measured by vinblastine binding. *Genetics* **73,** 379–385.
Avivi, L., and Feldman, M. (1980). Arrangement of chromosomes in the interphase nucleus of plants. *Hum. Genet.* **55,** 281–295.
Avivi, L., Feldman, M., and Bushuk, W. (1970a). The mechanism of somatic association in common wheat, *Triticum aestivum* L. II. Differential affinity for colchicine of spindle microtubules of plants having different doses of the somatic association suppressor. *Genetics* **65,** 585–592.
Avivi, L., Feldman, M., and Bushuk, W. (1970b). The mechanism of somatic association in common wheat, *Triticum aestivum* L. *Genetics* **66,** 449–461.
Bajaj, Y. P. S., and Gill Manjeet, S. (1985). In vitro induction of genetic variability in cotton (*Gossypium* spp.) *Theor. Appl. Genet.* **70,** 363–368.
Band, R. N., and Mohrlok, S. (1973). Observations on induced amitosis in *Acanthamoeba Exp. Cell Res.* **79,** 327–337.
Barbier, M., and Dulieu, H. L. (1980). Effects génétiques observés sur des plantes de tabac regenerées a partir de cotyledons par culture in vitro. *Ann. Amélior. Plant.* **30,** 321–326.
Barcelo, P., Lazzeri, P. A., Martin, A., and Lörz, H. (1991). Competence of cereal leaf cells. I. Patterns of proliferation and regeneration capability in vitro of the inflorescence sheath leaves of barley, wheat and tritordeum. *Plant Physiol.* **77,** 242–251.
Barcelo, P., Lazzeri, P. A., Martin, A., and Lörz, H. (1992). Competence of cereal leaf cells. II. Influence of auxin, ammonium and explant age on regeneration. *Plant Physiol.* **139,** 448–454.
Bassi, P. (1990). Quantitative variations of nuclear DNA during plant development: A critical analysis. *Biol. Rev.* **65,** 185–225.
Battaglia, E. (1947). Divisione eterotipica in cellule somatiche di *Sambucus ebulus*. *Nuovo G. Bot. Ital.* **54,** 744–753.
Battaglia, E., and Dolcher, T. (1947). "Eumeiosi" e "Meiosi apomeotipica" seguita da "Meiosi a diplounivalenti" nel tessuto somatico di Sambuco. *Nuovo G. Bot. Ital.* **54,** 632–641.
Bayliss, M. W. (1974). Origin of chromosome number variation in cultured plant cells. *Nature (London)* **246,** 529–530.
Bayliss, M. W. (1980). Chromosomal variation in plant tissue in culture. Perspectives in plant cell and tissue culture. *Int. Rev. Cytol. Suppl.* **11A,** 113.
Bebeli, P. J., Karp, A., and Kaltsikes, P. J. (1988). Plant regeneration and somaclonal variation from cultured immature embryos of sister lines of rye and triticale differing in their content of heterochromatin. *Theor. Appl. Genet.* **75,** 929–936.
Bebeli, P. J., Kaltsikes, P. J., and Karp, A. (1993a). Field evaluation of somaclonal variation in rye lines differing in telomeric heterochromatin. *J. Genet. Breed.* **47,** 15–22.
Bebeli, P. J., Kaltsikes, P. J., and Karp, A. (1993b). Field evaluation of somaclonal variation in triticale lines differing in telomeric heterochromatin. *J. Genet. Breed.* **47,** 249–258.
Bennici, A., and D'Amato, F. (1978). In vitro regeneration of Durum wheat plants. 1. Chromosome numbers of regenerated plantlets. *Z. Pflanzenzuecht.* **81,** 305–311.

Benzion, G., and Phillips, R. L. (1988). Cytogenetic stability of maize tissue cultures: A cell line pedigree analysis. *Genome* **30,** 318–325.

Blando, F., Giorgetti, L., Tonelli, M., and Nuti Ronchi, V. (1991). Cytological characterization of cell suspension cultures of fruit trees. *Acta Hortic.* **300,** 377–380.

Borthwick, H. A. (1931). Development of the macrogametophyte and embryo of *Daucus carota. Am. J. Bot.* **18,** 784–796.

Bouchard, R. A. (1990). Characterization of expressed meiotic prophase repeat transcript clones of *Lilium:* Meiosis specific expression, relatedness and affinities to small heat-shock proteins genes. *Genome* **33,** 68–79.

Breton, A. M., and Sung, Z. R. (1982). Temperature-sensitive carrot variants impaired in somatic embryogenesis. *Dev. Biol.* **90,** 58–86.

Broek, D., Bartlett, R., Crawford, K., and Nurse, P. (1991). Involvement of p34cdc2 in establishing the dependency of S phase on mitosis. *Nature (London)* **349,** 388–393.

Brown, R. C., and Lemmon, B. E. (1992). Pollen development in orchids. 4. Cytoskeleton and ultrastructure of the unequal pollen mitotis in *Phalaenopsis. Protoplasma* **167,** 183–192.

Budelier, K. A., Smith, A. G., and Gasser, C. S. (1990). Regulation of a stylar transmitting tissue. *Mol. Gen. Genet.* **224,** 183–192.

Burns, J. A., and Gerstel, D. U. (1969). Consequences of spontaneous breakage of heterochromatic chromosome segments in *Nicotiana* hybrids. *Genetics* **63,** 427–431.

Caligo, M. A., Nozzolini, M., and Nuti Ronchi, V. (1985). Proline and serine affect polarity and development of carrot somatic embryos. *Cell Differ.* **17,** 193–198.

Carman, J. G. (1990). Embryogenic cells in plant tissue cultures: Occurrence and behavior. *In Vitro Cell Dev. Biol.* **26,** 746–753.

Cavallini, A., Zolfino, C., Natali, L., Cionini, G., and Cionini, P. G. (1989). Nuclear DNA changes within *Helianthus annuus.* L.: Origin and control mechanisms. *Theor. Appl. Genet.* **77,** 12–16.

Cecchini, E., Natali, L., Cavallini, A., and Durante, M. (1992). DNA variations in regenerated plants of pea. (*Pisum sativum* L.). *Theor. Appl. Genet.* **84,** 874–879.

Chang, F., and Nurse, P. (1993). Finishing the cell cycle: Control of mitosis and cytokinesis in fission yeast. *Trends Genet.* **9,** 333–335.

Chasan, R., and Walbot, V. (1993). Mechanisms of plant reproduction: Questions and approaches. *Plant Cell* **5,** 1139–1146.

Colasanti, J., Tyers, M., and Sundaresan, V. (1991). Isolation and characterization of cDNA clones encoding a functional p34cdc2 homologue from *Zea mays. Proc. Natl. Acad. Sci. U.S.A.* **88,** 3377–3381.

Compton, M., and Veilleux, R. E. (1991). Variation for genetic recombination among tomato plants regenerated from three tissue culture systems. *Genome* **34,** 810–817.

Conard, A. (1928). Sur la structure et l'origine des noyaux polymorphes et fragmentes de la tige de *Tradescantia virginica* L. *Mem. Cl. Sci., Acad. R., Belg., Collect.* **9,** 8.

Cullis, C. A. (1983). Environmental induced DNA changes in plants. *CRC Crit. Rev. Plant Sci.* **1,** 117–131.

D'Amato, F., and Avanzi, M. G. (1948). Reazioni di natura auxinica ed effetti-rizogeni in *Allium cepa* L. Studio cito-istologico sperimentale. *Nuovo G. Bot. Ital.* **55,** 161–213.

de la Torre, C., and Gimenez-Martin, G. (1975). The nucleolus in the induced amitosis. *J. Exp. Bot.* **26,** 713–721.

de la Torre, J., López-Fernandéz, C., Herrero, P., and Gosálvez, J. (1993). In situ nick translation of meiotic chromosomes to demonstrate homologous heterochromatin heterogeneity. *Genome* **36,** 268–270.

Deumling, B., and Clermont, L. (1989). Changes in DNA content and chromosomal size during cell culture and plant regeneration of *Scilla siberica. Chromosoma* **97,** 439–448.

De Vries, S. C., Booij, H., Meyerink, P., Huisman, G., Wilde, H. D., Thomas, T. L., and

Van Kammen, A. B. (1988). Acquisition of embryogenic potential in carrot cell-suspension cultures. *Planta* **176,** 196–204.
Dhillon, S. S., Wernsman, E. A., and Miksche, J. P. (1983). Evaluation of nuclear DNA content and heterochromatin changes in anther-derived dihaploids of tobacco (*Nicotiana tabacum*) cv. Coker 139. *Can. J. Genet. Cytol.* **25,** 169–173.
Dickson, E. E., Arumuganathan, K., Kresovich, S., and Doyle, J. J. (1992). Nuclear DNA content variation within the Rosaceae. *Am. J. Bot.* **79,** 1081–1086.
Dirks, V. A., Ross, J. G., and Harpstead, D. D. (1956). Colchicine-induced true-breeding chimeric sectors in flax. *J. Hered.* **47,** 229–233.
Dobzansky, T. (1964). "Genetics and the Origin of the Species." Columbia Univ. Press, New York.
Domon, C., Evrard, J.-L., Herdenberger, F., Pillay, D. T. N., and Steinmetz, A. (1990). Nucleotide sequence of two anther-specific cDNAs from sunflower (*Helianthus annuus* L.). *Plant Mol. Biol.* **15,** 643–646.
Dudits, D., Bögre, L., and Gyorgyey, J. (1991). Molecular and cellular approaches to the analysis of plant embryo development from somatic cells in vitro. *J. Cell Sci.* **99,** 4475–4484.
Durante, M., Geri, C., Buiatti, M., Baroncelli, S., Parenti, R., Nuti Ronchi, V., Martini, G., Collina Grenci, Grisvard, J., and Guillé, E. (1982). DNA heterogeneity and genetic control of tumorogenesis in *Nicotiana* tumorous and non-tumorous genotypes. *Dev. Genet.* **3,** 25–32.
Durante, M., Cecchini, E., Natali, L., Citti, L., Geri, C., Parenti, R., and Nuti Ronchi, V. (1989). 5-Azacytidine-induced tumorous trasformation and DNA hypomethylation in *Nicotiana* tissue cultures. *Dev. Genet.* **10,** 298–303.
Durrant, A. (1962). The environmental induction of heritable change in flax. *Heredity* **17,** 27–61.
Dutta, A., and Stillman, B. (1992). cdc2 family kinases phosphorylate a human cell DNA replication factor, RPA, and activate DNA replication. *EMBO J.* **11,** 2189–2199.
Edallo, S., Zucchinali, C., Perenzini, M., and Salamini, F. (1981). Chromosomal variation and frequency of spontaneous mutation associated with in vitro culture and plant regeneration in maize. *Maydica* **26,** 39–56.
Eizenga, G. C. (1989). Meiotic analysis of tall fescue somaclones. *Genome* **32,** 373–379.
Evans, D. A. (1989). Somaclonal variation—genetic basis and breeding applications. *Trends Genet.* **5,** 42–45.
Evans, D. A., and Gamborg, O. L. (1982). Chromosome stability of cell suspension cultures of *Nicotiana* spp. *Plant Cell Rep.* **1,** 104–107.
Evans, D. A., and Sharp, W. R. (1983). Single gene mutations in tomato regenerated from tissue culture. *Science* **221,** 949–951.
Fasseas, C., and Bowes, B. G. (1980). Ultrastructural observations on proliferating storage cells of mature cotyledons of *Phaseolus vulgaris* L. cultured in vitro. *Ann. Bot.* (*London*) [N.S.] **46,** 143–152.
Feiler, H. S., and Jacobs, T. W. (1990). Cell division in higher plants: A cdc2 gene, its 34-kDa producing kinase activity in pea. *Proc. Natl. Acad. Sci. U.S.A.* **87,** 5397–5401.
Feldman, M., Mello-Sampayo, T., and Sears, E. R. (1966). Somatic association in *Triticum aestivum*. *Proc. Natl. Acad. Sci. U.S.A.* **56,** 1192–1199.
Filippini, F., Terzi, M., Cozzani, F., Vallone, D., and Lo Schiavo, F. (1992). Modulation of auxin-binding proteins in cell suspensions. II. Isolation and initial characterization of carrot cell variants impaired in somatic embryogenesis. *Theor. Appl. Genet.* **84,** 430–434.
Franzke, C. J., and Ross, J. G. (1952). Colchicine-induced variants in *Sorghum*. *J. Hered.* **43,** 107–115.
Galbraith, D. W., Harkins, K. R., and Knapp, S. (1991). Systemic endopolyploidy in *Arabidopsis thaliana*. *Plant Physiol.* **96,** 985–989.

Gaponenko, A. K., Petrova, T. F., Iskakov, A. R., and Sozinov, A. A. (1988). Cytogenetics of in vitro cultured somatic cells and regenerated plants of barley (*Hordeum vulgar* L.) *Theor. Appl. Genet.* **75,** 905–911.

Gasser, C. S., Budelier, K. A., Smith, A. G., Shah, D. M., and Fraley, R. T. (1989). Isolation of tissue-specific cDNAs from tomato pistils. *Plant Cell* **1,** 15–24.

Gavazzi, G., Tonelli, C., Todesco, G., Arreghini, E., Raffaldi, F., Vecchio, F., Barbuzzi, G., Biasini, M. G., and Sala, F. (1987). Somaclonal variation versus chemically induced mutagenesis in tomato (*Lycopersicon esculentum* L.) *Theor. Appl. Genet.* **74,** 735–738.

Gee, M. A., Hagen, G., and Guilfoyle, T. (1991). Tissue-specific and organ-specific expression of soybean auxin-responsive transcripts GH3 and SAURs. *Plant Cell* **3,** 419–130.

Geri, C., Giorgetti, L., Turrini, A., and Nuti Ronchi, V. (1992). Modulation of DNA sequences in relation to the commitment of carrot embryogenesis. *Atti AGI* **38,** 139–140.

Ghosh, S., and Paweletz, N. (1993). Mitosis: Dissociability of its events. *Int. Rev. Cytol.* **144,** 217–258.

Giorgetti, L., Vergara, M. R., Evangelista, M., Lo Schiavo, F., Terzi, M. (1994). On the occurrence of somatic meiosis in embryogenic carrot cell cultures. *Mol. Gen. Genet.* (in press).

Giuliano, G., Lo Schiavo, F., and Terzi, M. (1984). Isolation and developmental characterization of temperature-sensitive carrot cell variants. *Theor. Appl. Genet.* **67,** 179–183.

Goldberg, G. I., Collier, I., and Cassel, A. (1983). Specific DNA sequences associated with the nuclear matrix in synchronized mouse 3T3 cells. *Proc. Natl. Acad. Sci. U.S.A.* **80,** 6887–6891.

Goldberg, R. B. (1988). Plants: Novel developmental processes. *Science* **240,** 1460–1467.

Golubovskaya, I. N. (1979). Genetic control of meiosis. *Int. Rev. Cytol.* **58,** 247–290.

Golubovskaya, I. N. (1989). Meiosis in maize: mei-genes and conception of genetic control of meiosis. *Adv. Genet.* **26,** 149–192.

Golubovskaya, I. N., Avalkina, N. A., and Sheridan, W. F. (1992). Effects of several meiotic mutants on female meiosis in maize. *Dev. Genet.* **13,** 411–424.

Golubovskaya, I. N., Grebennikova, Z. K., Avalkina, N. A., and Sheridan, W. F. (1993). The role of the ameiotic 1 gene in the initiation of meiosis in subsequent meiotic events in maize. *Genetics* **135,** 1151–1166.

Guzzo, F., Baldan, B., Mariani, P., Lo Schiavo, F., and Zerzi, M. (1994). *J. Exp. Bot.* **279,** 1427–1432.

Hahne, B., and Hoffmann, F. (1984). Dimethyl sulfoxide can initiate cell divisions of arrested callus protoplasts by promoting cortical microtubule assembly. *Proc. Natl. Acad. Sci. U.S.A.* **81,** 5449–5453.

Hahne, B., and Hoffmann, F. (1986). Cytogenetics of protoplast cultures of *Brachycome dichromosomatica* and *Crepis capillaris* and regeneration of plants. *Theor. Appl. Genet.* **72,** 244–251.

Halperin, W., and Wetherell, D. F. (1965). Ontogeny of adventive embryos in wild carrot. *Science* **147,** 756–758.

Hamaguchi, J., Tobey, R. A., Pines, J., Crissman, H. A., Hunter, T., and Bradbury, E. M. (1992). Requirement for p34cdc2 kinase is restricted to mitosis in the mammalian cdc2 mutant FT210. *J. Cell Biol.* **117,** 1041–1053.

Handschumacher, R. E., and Welch, A. D. (1960). Agents which influence nucleic acid metabolism. In: "The Nucleic acids" (E. Chargaff and J. N. Davidson, eds.), Vol. 3, pp. 453–526. Academic Press, New York.

Hatzfeld, J., and Buttin, G. (1975). Temperature-sensitive cell cycle mutants: A Chinese hamster cell line with a reversible block in cytokinesis. *Cell (Cambridge, Mass.)* **5,** 123–129.

Heath, B. (1980). Variant mitoses in lower eukaryotes: Indicators of the evolution of mitosis? *Int. Rev. Cytol.* **64,** 1–80.

Hemerly, A. S., Ferreira, P., de Almelda Engler, J., Van Montagu, M., Engler, G., and Inzé, D. (1993). cdsc2a expression in *Arabidopsis* is linked with competence for cell division. *Plant Cell* **5,** 1711–1723.

Hesemannn, C. U., and Schroeder, G. (1982). Loss of nuclear DNA in leaves of rye. *Theor. Appl. Genet.* **62,** 325–328.

Heslop-Harrison, J. S., and Bennet, M. D. (1990). Nuclear architecture in plants. *Trends Genet.* **6,** 401–405.

Hogan, C. J. (1987). Microtubule patterns during meiosis in two higher plant species. *Protoplasma* **138,** 126–136.

Hoyt, M. A., He, L., Loo, K., and Saunders, W. S. (1992). *S. cerevisiae* genes required for cell cycle arrest in response to loss of microtubule function. *J. Cell Biol.* **118,** 109–120.

Huang, H. C., and Chen, C. C. (1988). Genome multiplication in cultured protoplasts of two *Nicotiana* species. *J. Hered.* **79,** 28–32.

Huskins, C. L. (1948). Segregation and reduction in somatic tissue. *J. Hered.* **39,** 311–318.

Jackson, R. C., and Murray, B. G. (1983). Colchicine induced quadrivalent formation in *Helianthus:* Evidence of ancient polyploidy. *Theor. Appl. Genet.* **64,** 219–222.

John, B., and King, M. (1985). Pseudoterminalization, terminalization, and non-chiasmate modes of terminal association. *Chromosoma* **92,** 89–99.

John, P. C. L., Sek, F. G., and Lee, M. G. (1989). A homolog of the cell cycle control protein p34cdc2 participates in the division cycles of *Chlamydomonas*, and a similar protein is detectable in higher plants and remote taxa. *Plant Cell* **1,** 1185–1193.

Johnson, S. S., Phillips, R. L., and Rines, H. W. (1987). Possible role of heterochromatin in chromosome breakage induced by tissue culture in oats (*Avena sativa*). *Genome* **29,** 439–446.

Jonak, C., Pay, A., Bögre, L., Hirt, H., and Heberle-Bors, E. (1993). The plant homologue of MAP kinase is expressed in a cell cycle-dependent and organ-specific manner. *Plant J.* **3,** 611–617.

Kallak, H., and Yarvekylg, L. (1977). Nuclear behavior in callus cells: Morphology and division. *Biol. Plant.* **19,** 48–52.

Kaltsikes, P. S. (1974). Methods for triticale production. *Z. Pflanzenzuecht.* **71,** 264–286.

Karlsson, S. B., and Vasil, I. K. (1986). Growth, cytology and flow cytometry of embryogenic cell suspension cultures of *Panicum maximum* Jacq and *Pennisetum purpureum* Schum. *J. Plant Physiol.* **123,** 211–227.

Karp, A., and Bright, S. W. J. (1985). On the causes and origins of somaclonal variation. *Oxford Surv. Plant Mol. Cell Biol.* **2,** 199–234.

Karp, A., Wu, Q. S., Steele, S. H., and Jones, M. G. K. (1987). Chromosome variation in dividing protoplasts and cell suspensions of wheat. *Theor. Appl. Genet.* **74,** 140–146.

Kerber, E. R. (1964). Wheat: Reconstitution of the tetraploid component (AABB) of hexaploids. *Science* **143,** 253–255.

Kirschner, M. (1992). The cell cycle then and now. *Trends Biochem. Sci.* **17,** 281–285.

Komamine, A., Kawahara, R., Matsumoto, M., Sunabori, S., Toya, T., Fujiwara, A., Tsukahara, M., Smith, J., Fukuda, H., Nomura, K., and Fujimura, T. (1992). Mechanisms of somatic embryogenesis in cell cultures: Physiology, biochemistry and molecular biology. *In Vitro Cell Dev. Biol.* **28P,** 11–14.

Kovács, E. I. (1985). Regulation of karyotype stability in tobacco tissue cultures or normal and tumorous genotypes. *Theor. Appl. Genet.* **70,** 548–554.

Kubai, D. F. (1975). The evolution of the mitotic spindle. *Int. Rev. Cytol.* **43,** 167–227.

Kurtz, S., Rossi, J., Petko, L., and Lindquist, S. (1986). An ancient developmental induction: Heat-shock proteins induced in sporulation and oogenesis. *Science* **231,** 1154–1157.

Ladizinsky, G., and Fainstein, R. (1978). A case of genome partition in polyploid oats. *Theor. Appl. Genet.* **51,** 159–160.

Lapitan, N. L. V., Sears, R. G., and Gill, B. S. (1988). Amplification of repeated DNA sequences in wheat X rye hybrids regenerated from tissue cultures. *Theor. Appl. Genet.* **75,** 381–388.

Larkin, P. J., and Scowcroft, W. R. (1981). Somaclonal variation—a novel source of variability from cell cultures for plant improvement. *Theor. Appl. Genet.* **60,** 197–214.

Lavania, U. C. (1982). Chromosomal instability in *Lathyrus sativus* L. *Theor. Appl. Genet.* **62,** 135–138.

Leavitt, R. G. (1909). A vegetative mutant, and the principle of homeosis in plants. *Bot. Gaz. (Chicago)* **47,** 30–68.

Lee, M., and Nurse, P. (1988). Cell cycle control genes in fission yeast and mammalian cells. *Trends Genet.* **4,** 287–290.

Lee, M., and Phillips, R. L. (1987). Genetic variants in the progeny of regenerated maize plants. *Genome* **29,** 122–128.

Lee, M., and Phillips, R. L. (1988). Genetic variation in progeny of regenerated maize plants. *Annu. Rev. Plant Physiol. Plant Mol. Biol.* **39,** 413–437.

Lee, M., Geadelmann, J. L., and Phillips, R. L. (1988). Agronomic evaluation of inbred lines derived from tissue cultures of maize. *Theor. Appl. Genet.* **75,** 841–849.

Libbenga, K. R., and Torrey, J. G. (1973). Hormone-induced encoreduplication prior to mitosis in cultured pea root cortex cells. *Am. J. Bot.* **60,** 293–299.

Lifschitz, E. (1988). Molecular markers for the floral program. *Flower. Newsl.* **6,** 16–20.

Lima de Faría, A. (1969). DNA replication and amplification in heterochromatin. *In* "Handbook of Molecular Cytology" (A. Lima de Faría, ed.), pp. 227–325. North-Holland Publ., Amsterdam and London.

Linacero, R., and Vazquez, A. M. (1992). Genetic analysis of chlorophyll-deficient somaclonal variants in rye. *Genome* **35,** 428–430.

Lipucci di Paola, M., Fossi, D., Tognoni, F., and Nuti Ronchi, V. (1987). Adventitious bud induction in vitro from juvenile leaves of *Cupressus arizonica* Green. (1987). *Plant Cell Tissue Organ Cult.* **10,** 3–10.

Lo Schiavo, F., Giuliano, G., and Sung, Z. R. (1988). Characterization of a temperature-sensitive carrot cell mutant impaired in somatic embryogenesis. *Plant Sci.* **54,** 157–164.

Lo Schiavo, F., Pitto, L., Giuliano, G., Torti, G., Nuti Ronchi, V., Marazziti, D., Vergara, R., Orselli, S., and Terzi, M. (1989). DNA methylation of embryogenic carrot cell cultures and its variation as caused by mutation, differentiation, hormones and hypomethylating drugs. *Theor. Appl. Genet.* **77,** 325–331.

Lo Schiavo, F., Giuliano, G., de Vries, S. C., Genga, A., Bollini, R., Pitto, L., Cozzani, F., Nuti Ronchi, V., and Terzi, M. (1990). A carrot cell variant temperature sensitive for somatic embryogenesis reveals a defect in the glycosylation of extracellular proteins. *Mol. Gen. Genet.* **223,** 385–393.

Lu, C., and Vasil, I. K. (1981). Somatic embryogenesis and plant regeneration from leaf tissue of *Panicum maximum* Jacq. *Theor. Appl. Genet.* **59,** 275–280.

Mackenzie, A., Heslop-Harrison, J., and Dickinson, H. G. (1967). Elimination of ribosomes during meiotic prophase. *Nature (London)* **215,** 997–999.

Mark, C. L. (1961). Colchicine-induced genetic changes in soybean leaf tissue. *Proc. South Dakota Acad. Sci.* **40,** 219–225.

Martin, G. M., and Sprague, C. A. (1969). Parasexual cycle in cultivated human somatic cells. *Science* **166,** 761–763.

Martini, G., and Nuti Ronchi, V. (1974). Microdensitometric and autoradiographic analysis of cell proliferation in primary cultures of *Nicotiana glauca* pith tissue. *Cell Differ.* **3,** 239–247.

Martini, G., and Nuti Ronchi, V. (1977). In vivo observation of nucleolar extrusion in cultured callus cells of a non-tumorous *Nicotiana glauca* × *Nicotiana langsdorffii hybrid. Protoplasma* **91,** 409–415.

McClintock, B. (1944). Maize genetics. *Year Book, Carnegie Inst. Washington* **43,** 127–135.
McClintock, B. (1956). Mutation in maize. *Year Book, Carnegie Inst. Washington* **55,** 323–332.
McCoy, T. J., and Phillips, R. L. (1982). Chromosome stability in maize (*Zea mays*) tissue cultures and sectoring in some regenerated plants. *Can. J. Genet. Cytol.* **24,** 559–565.
Mehra, P. N. (1986). Induced meiotic reductions in root-tips. IV. Concluding remarks. *Cytologia* **51,** 467–472.
Melaragno, J. E., Mehrotra, B., and Coleman, A. W. (1993). Relationship between endopolyploidy and cell size in epidermal tissue of *Arabidopsis*. *Plant Cell* **5,** 1661–1668.
Meyer, V. G. (1966). Flower abnormalities. *Bot. Rev.* **32,** 165–218.
Miller, R. H. (1980). Amitosis and endocytogenesis in the fruit of *Malus sylvestris*. *Ann. Bot.* (*London*) [N. S.] **46,** 567–571.
Mitra, J., Mapes, M. O., and Steward, F. C. (1960). Growth and organized development of cultured cells. IV. The behavior of the nucleus. *Am. J. Bot.* **47,** 357–368.
Molnár-Láng, M., Galiba, G., Kovács, G., and Sutka, J. (1991). Changes in the fertility and meiotic behaviour of barley (*Hordeum vulgare*) × wheat (*Triticum aestivum*) hybrids regenerated from tissue cultures. *Genome* **34,** 261–266.
Moreno, S., and Nurse, P. (1994). Regulation of progression through the G1 phase of the cell cycle by the *rum1* + gene. *Nature* (*London*) **367,** 236–242.
Murray, A. W. (1991). Remembrance of things past. *Nature* (*London*) **349,** 367–368.
Nagl, W. (1971). Teilweise Umstimmung von Wurzelspitzen-Mitosen zu Endomitosen durch Phytohormone. *Z. Naturforsch., B: Anorg. Chem., Org. Chem., Biochem., Biophys., Biol.* **26B,** 1390–1391.
Nagl, W. (1974). The *Phaseolus* suspensor and its polytene chromosomes *Z. Pflanzenphysiol.* **73,** 1–44.
Nagl, W. (1993). Induction of high polyploidy in *Phaseolus* cell cultures by the protein kinase inhibitor, K-252a. *Plant Cell Rep.* **12,** 170–174.
Natali, L., Cavallini, A., Cionini, G., Sassoli, O., Cionini, P. G., and Durante, M. (1993). Nuclear DNA changes within *Helianthus annuus* L.: Changes within single progenies and their relationship with plant development. *Theor. Appl. Genet.* **85,** 506–512.
Natarajan, A. T., and Ahnström, J. R. (1969). Heterochromatin and chromosome aberrations. *Chromosoma* **28,** 48–61.
Natarajan, A. T., and Raposa, T. (1975). Heterochromatin and chromosome aberrations. A comparative study of three mouse cell lines with different karyotype and heterochromatin distribution. *Hereditas* **80,** 83–90.
Newcombe, W., and Fowke, L. C. (1973). The fine structure of the change from free-nuclear to cellular condition in the endosperm of the chickweed *Stellaria media*. *Bot. Gaz.* (*Chicago*) **134,** 236–240.
Newman, I. V. (1965). Pattern in the meristems of vascular plants. III. Pursuing the patterns in the apical meristem where no cell is a permanent cell. *J. Linn. Soc. London, Bot.* **59,** 185–214.
Nuti Ronchi, V. (1981). Histological study of organogenesis in vitro from callus cultures of two *Nicotiana* species. *Can. J. Bot.* **59,** 1969–1977.
Nuti Ronchi, V. (1990). Cytogenetics of plant cell cultures. *In* "Plant Tissue Culture: Applications and Limitations" (S. S. Bhojwani, ed.), pp. 276–300. Elsevier, Amsterdam.
Nuti Ronchi, V., and Giorgetti, L. (1995). The cell's commitment to somatic embryogenesis. *Biotechnol. Agric. For.* **30,** 1–19. Springer Verlag, Berlin Heidelberg. (in press).
Nuti Ronchi, V., and Martini, G. (1962). Germinabilità, sviluppo delle plantule e frequenza delle aberrazioni cromosomiche in rapporto all età del seme del frumento. *Caryologia* **15,** 293–302.
Nuti Ronchi, V., and Terzi, M. (1988). Induced aneuploidy in plant tissue culture.

In "Aneuploidy. Part B: Induction and Test Systems" (A. Sandberg and B. K. Vig, eds.), pp. 19–38. Alan R. Liss, New York.

Nuti Ronchi, V., Avanzi, S., and D'Amato, F. (1965). Chromosome endoreduplication (endopolyploidy) in pea root meristems induced by 8-azaguanine. *Caryologia* **18,** 599–617.

Nuti Ronchi, V., Bennici, A., and Martini, G. (1973). Nuclear fragmentation in dedifferentiating cells of *Nicotiana glauca* pith tissue grown in vitro. *Cell Differ.* **2,** 77–85.

Nuti Ronchi, V., Martini, G., and Buiatti, M. (1976). Genotype-hormone interaction in the induction of chromosome aberrations: Effect of 2,4-dichlorophenoxyacetic acid (2″4-D) and kinetin on tissue cultures of *Nicotiana* spp. *Mutat. Res.* **36,** 67–72.

Nuti Ronchi, V., Nozzolini, M., and Avanzi, L. (1981). Chromosomal variation in plants regenerated from two *Nicotiana* spp. *Protoplasma* **109,** 433–444.

Nuti Ronchi, V., Bonatti, S., and Turchi, G. (1986a). Preferential localization of chemically induced breaks in heterochromatic regions of *Vicia faba* and *Allium cepa* chromosomes. I. Exogenous thymidine enhances the cytologic effects of 4-epoxyethyl-1,2-epoxy-cyclohexane. *Environ. Exp. Bot.* **26,** 115–126.

Nuti, Ronchi, V., Bonatti, S., Durante, M., and Turchi, G. (1986b). Preferential localization of chemically induced breaks. II. 4-Epozyethyl-1,2-epoxy-cyclohexane interacts specifically with highly repetitive sequences of DNA in *Allium cepa*. *Environ. Exp. Bot.* **26,** 127–135.

Nuti Ronchi, V., Caligo, M. A., Martini, G., and Luccarini, G. (1986c). Polarity acquisition and progression during carrot and alpha-alpha somatic embryo development. *In* "Progress in Developmental Biology," Part A, pp. 189–193. Alan R. Liss, New York.

Nuti Ronchi, V., Giorgetti, L., and Tonelli, M. G. (1990). The commitment to embryogenesis: A cytological approach. *In* "Progress in Plant Cellular and Molecular Biology" (H. J. J. Nijkamp, L. H. W. Van Der Plas, and J. Van Aartrijk, eds.), pp. 437–442. Kluwer Academic Publishers, New York.

Nuti Ronchi, V., Giorgetti, L., Tonelli, M., and Martini, G. (1992a). Ploidy reduction and genome segregation in cultured carrot cell lines. I. Prophase chromosome segregation. *Plant Cell Tissue Organ Cult.* **30,** 107–114.

Nuti Ronchi, V., Giorgetti, L., Tonelli, M., and Martini, G. (1992b). Ploidy reduction and genome segregation in cultured carrot cell lines. II. Somatic meiosis. *Plant Cell Tissue Organ Cult.* **30,** 115–120.

Nuti, Ronchi, V., Giorgetti, L., Tonelli, M., Belloni, P., and Martini, G. (1992c). Analogies of chromosome reducing events in somatic cultured cells and in microsporogenesis. *In* "Angiosperm Pollen and Ovules" (D. L. Mulcahy, G. Bergamini-Mulcahy, and E. Ottaviano, eds.), pp. 245–247. Springer-Verlag, New York.

Nuti Ronchi, V., Giorgetti, L., Geri, C. C., Pitto, L., Vergara, M. R., and Martini, G. (1992d). Cytological, anatomical and morphological aspects of somatic embryogenesis. *In* "Regulation of Plant Somatic Embryogenesis" (M. Griga and E. Teyklova, eds.), pp. 67–72. Research Institute of Technical Crops and Legumes, Sumperk, Czechoslovakia.

Nuti Ronchi, V., *et al.* (1994). Submitted for publication.

Orton, T. G. (1980). Chromosomal variability in tissue cultures and regenerated plants of *Hordeum*. *Theor. Appl. Genet.* **51,** 161–174.

Palevitz, B. A. (1993). Morphological plasticity of the mitotic apparatus in plants and its developmental consequences. *Plant Cell* **5,** 1001–1009.

Palmer, R. G. (1971). Cytological studies of ameiotic and normal maize with reference to premeiotic pairing. *Chromosoma* **35,** 233–246.

Parenti, R., Guillé, E., Grisvard, J., Durante, M., Giorgi, L., and Buiatti, M. (1973). Transient DNA satellite in dedifferentiating pith tissue. *Nature* (*London*) **246,** 227–229.

Patau, K. (1950). A correlation between separation of the two chromosome groups in somatic reduction and their degree of homologous segregation. *Genetics* **35,** 128–132.

Patau, K., and Das, N. K. (1961). The relation of DNA synthesis and mitosis in tobacco pith tissue cultured in vitro. *Chromosoma* **11,** 553–572.

Pera, F., and Reiner, B. (1973). Studies of multipolar mitoses in euploid tissue cultures. I. Somatic reduction to exactly haploid and triploid chromosomal sets. *Chromosoma* **42,** 71–86.

Phillips, R. L., Kaeppler, S. M., and Peschke, V. M. (1990). Do we understand somaclonal variation? *In* "Progress in Plant Cellular and Molecular Biology" (H. J. J. Nijcamp, L. H. V. Van Der Plas, and J. Van Artijk, eds.), pp. 131–141. Kluwer Academic Publishers, New York.

Pijnacker, S., Ramulu, S., Dijkhuis, P., and Ferwerda, M. A. (1989). Flow cytometric and karyological analysis of polysomaty and polyploidization during callus formation from leaf segments of various potato genotypes. *Theor. Appl. Genet.* **77,** 102–110.

Pimienta, E., and Polito, V. S. (1983). Embryo sac development in almond (*Prunus dulcis* (mill.) D. A. Webb) as affected by cross-self and non-pollination. *Ann. Bot.* (*London*) [N. S.] **51,** 469–479.

Pitto, L., Miarelli, C., Giorgetti, L., Colella, C., Luccarini, G., Nuti Ronchi, V. (1992). Occurrence of floral homeotic structures in carrot, tomato, and helianthus hypocotyls in liquid cultures. *Atti AGI* **38,** 107–108.

Pitto, L., Giorgetti, L., Miarelli, C., Luccarini, G., Evangelista, M., Colella, C., Nuti Ronchi, V. (1993). Hormone induced floral reprogramming in cultured hypocotyls: Induction of gametic-like cells and expression of floral homeotic mad genes. *Atti AGI* **39,** 111–112.

Pnueli, L., Abu-Abeid, M., Zamir, D., Nacken, W., Schwarz-Sommer, Zs., and Lifschitz, E. (1991). The MADS-box gene family in tomato: Temporal expression during floral development, conserved secondary structures and homology with homeotic genes from *Anthirrinum* and *Arabidopsis*. *Plant J.* **1,** 255–266.

Pontecorvo, G. (1958). "Trends in Genetic Analysis." Columbia University Press, New York.

Puolimatka, M., and Karp, A. (1993). Effect of genotype on chromosome variation in tissue culture of inbred and outbred rye. *Heredity* **71,** 138–144.

Reed, S. M., Burns, J. A., and Wernsman, E. A. (1989). Cytological observations of the heterochromatin of *Nicotiana tabacum* L. androgenic doubled haploids. *J. Hered.* **80,** 74–76.

Rhoades, M. M. (1956). Genic control of chromosomal behavior. *Maize Genet. Coop. News Lett.* **30,** 38–42.

Rhodes, C. A., Phillips, R. L., and Green, C. E. (1986). Cytogenetic stability of aneuploid maize tissue cultures. *Can. J. Genet. Cytol.* **28,** 374–384.

Rizzoni, M., and Palitti, F. (1973). Regulatory mechanisms of cell division. I. Colchicine-induced endoreduplication. *Exp. Cell Res.* **77,** 453–458.

Rodríguez-García, M. I., Majewska-Sawka, A., and Fernandez, M. C. (1988). Why do nuclear vacuoles appear in the prophasic nucleus of pollen mother cells? Facts and hypotheses. *In* "Sexual Reproduction in Higher Plants" (M. Cresti, P. Gori, and E. Pacini, eds.), pp. 163–168. Springer-Verlag, New York.

Roof, D. M., Meluh, P. B., and Rose, M. D. (1992). Kinesin related proteins required for assembly of the mitotic spindle. *J. Cell Biol.* **118,** 95–108.

Ross, J. G. (1962). Proof of somatic reduction after colchicine treatment using marked chromosomes. *Manit. Med. Rev.* **42,** 536–539.

Ross, J. G., Franzke, C. J., and Schuh, L. A. (1954). Studies in colchicine-induced variants in *Sorghum*. *J. Bot.* **39,** 625–633.

Sattler, R. (1988). Homeosis in plants. *Am. J. Bot.* **75,** 1606–1617.

Saunders, W. S., and Hoyt, M. A. (1992). Kinesin-related proteins required for structural integrity of the mitotic spindle. *Cell* (*Cambridge, Mass.*) **70,** 451–458.

Scheffield, E., Cawood, A. H., Bell, P. R., and Dickinson, H. G. (1979). The development of nuclear vacuoles during meiosis in plants. *Planta* **146,** 597–601.

Schröder, R., and Kurth, W. (1972). Untersuchungen zur Bestimmung morphometrischer Korrektur-faktoren und Messungen von Zelldichte, Mitose-und Amitoserate am gesunden, menschlichen Grosshirn-Hemisphärenmark. *Microsc. Acta* **73,** 205–216.

Schwarzacher, T., Leitch, A. R., Bennet, M. D., and Heslop-Harrison, J. S. (1989). *In situ* localization of parental genomes in a wide hybrid. *Ann. Bot.* (*London*) [N. S.] **64,** 315–324.

Singh, B. D., Harvey, B. L., Kao, K. N., and Miller, R. A. (1972). Karyotypic changes and selection pressure on *Haplopappus gracilis* suspension cultures. *Can. J. Genet. Cytol.* **14,** 65–70.

Singh, R. J. (1986). Chromosomal variation in immature embryo derived callus of barley (*Hordeum vulgare* L.). *Theor. Appl. Genet.* **72,** 710–716.

Singsit, C., Veilleux, R. E., and Sterret, S. B. (1990). Enhanced seed set and crossover frequency in regenerated potato plants following anther and callus culture. *Genome* **33,** 50–56.

Smith, D. L., and Krikorian, A. D. (1990). Somatic embryogenesis of carrot in hormone-free medium: External pH control over morphogenesis. *Am. J. Bot.* **77,** 1634–1647.

Sparrow, A. H., and Hammond, M. R. (1947). Cytological evidence for the transfer of desoxyribose nucleic acid from nucleus to cytoplasm in certain plant cells. *Am. J. Bot.* **34,** 439–445.

Sparvoli, E., Galli, M. G., Mosca, A., and Paris, G. (1976). *Exp. Cell. Res.* **97,** 74–82.

Stahl, A., and Luciani, J. M. (1971). Individualisation d'un stade préleptotène de condensation chromosomique au début de la méiose chez l'ovocyte foetal humain. *C. R. Hebd. Seances Acad. Sci.* **272,** 2041–2044.

Stebbins, G. L. (1959). "Variation and Evolution in Plants." Columbia Univ. Press, New York.

Steeves, T. A., and Sussex, I. M. (1989). "Patterns in Plant Development," 2nd Ed. Cambridge University Press, Cambridge.

Stern, H., and Hotta, Y. (1967). Chromosome behavior during development of meiotic tissue. *In* "The Control of Nuclear Activity" (L. Goldstein, ed.), pp. 47–76. Prentice-Hall, Englewood Cliffs, NJ.

Steward, F. C., Mapes, M. O., and Smith, J. (1958). Growth and organized development of cultured cells. I. Growth and division of freely suspended cells. *Am. J. Bot.* **45,** 693–703.

Sung, Z. R., Fienberg, A., Chorneau, R., Borkird, C., Furner, I., Smith, J., Terzi, M., LoSchiavo, F., Giuliano, G., Pitto, L., and Nuti Ronchi, V. (1984). Developmental biology of embryogenesis from carrot. *Plant Mol. Rep.* **2,** 3–14.

Tai, W. (1970). Multipolar meiosis in diploid crested wheatgrass *Agropyron cristatum*. *Am. J. Bot.* **57,** 1160–1169.

Taylor, M. G., and Vasil, I. K. (1987). Analysis of DNA size, content, and cell cycle in leaves of Napier grass (*Pennisetum purpureum* Schum.). *Theor. Appl. Genet.* **74,** 681–686.

Taylor, R. L. (1967). The foliar embryos of *Malaxis paludosa*. *Can. J. Bot.* **45,** 1553–1556.

Terzi, M., Giuliano, G., Lo Schiavo, F., and Nuti Ronchi, V. (1982). Studies of plant cell lines showing temperature-sensitive embryogenesis. *In* "Embryonic Development. Part B: Cellular Aspects" (M. M. Burger, ed.), pp. 521–534. Alan R. Liss, New York.

Thompson, M. M. (1962). Cytogenetics in *Rubus*. III. Meiotic instability in some higher polyploids. *Am. J. Bot.* **49,** 575–582.

Tobler, H. (1986). The differentiation of germ and somatic cell lines in nematodes. *In* "Results and Problems in Cell Differentiation" (W. Hennig, ed.), Vol. 13, pp. 1–69. Springer-Verlag, New York.

Toncelli, F., Martini, G., Giovinazzo, G., and Nuti Ronchi, V. (1985). Role of permanent dicentric systems in carrot somatic embryogenesis. *Theor. Appl. Genet.* **70,** 345–348.

Vaarama, A. (1949). Spindle abnormalities and variation in chromosome number in *Ribes nigrum*. *Hereditas* **35,** 136–162.

Vasek, F. C. (1962). "Multiple spindle"—a meiotic irregularity in *Clarkia exilis*. *Am. J. Bot.* **49,** 536–539.

Vig, B. K. (1971). Increase induced by colchicine in the incidence of somatic crossing over in *Glycine max*. *Theor. Appl. Genet.* **41,** 145–149.

Walbot, V. (1985). On the life strategies of plants and animals. *Trends Genet.* **1,** 165–169.

Walbot, V., and Cullis, C. A. (1985). Rapid genomic change in higher plants. *Annu. Rev. Plant Physiol.* **36,** 367–396.

Walters, M. S. (1958). Aberrant chromosome movement and spindle formation in meiosis of *Bromus* hybrids: An interpretation of spindle organization. *Am. J. Bot.* **45,** 271–289.

Walters, M. S. (1970). Evidence on the time of chromosome pairing from the preleptotene spiral stage in *Lilium longiflorum* "Croft." *Chromosoma* **29,** 375–418.

Walters, M. S. (1972). Preleptotene chromosome contraction in *Lilium longiflorum* "Croft." *Chromosoma* **39,** 311–332.

Walters, M. S. (1976). Variation in preleptotene chromosome contraction among three cultivars of *Lilium longiflorum*. *Chromosoma* **57,** 51–80.

Wang, R. J., and Yin, L. (1976). Further studies in a mutant mammalian cell line defective in mitosis. *Exp. Cell Res.* **101,** 331–336.

Wang, R. J., Wissinger, W., King, W. J., and Wang, G. (1983). Studies on cell division in mammalian cells. VII. A temperature-sensitive cell line abnormal in centriole separation and chromosome movement. *J. Cell Biol.* **96,** 301–306.

Werner, J. E., and Peloquin, S. J. (1991). Occurrence and mechanisms of $2n$ egg formation in $2x$ potato. *Genome* **34,** 975–882.

Wernicke, W., and Milkovits, L. (1984). Developmental gradients in wheat leaves—response of leaf segments in different genotypes cultured in vitro. *J. Plant Physiol.* **115,** 49–58.

Williams, E. G., Heslop-Harrison, J., and Dickinson, H. G. (1973). The activity of the nucleolus organising region and the origin of cytoplasmic nucleotides in meiocytes of *Lilium*. *Protoplasma* **77,** 79–93.

Wilson, G. B. V., and Cheng, K. (1949). Segregation and reduction in somatic tissue. *J. Hered.* **40,** 3–6.

Zheng, K. L., Castiglione, S., Biasini, M. G., Biroli, A., Morandi, C., and Sala, F. (1987). Nuclear DNA amplification in cultured cells of *Oryza sativa* L. *Theor. Appl. Genet.* **74,** 65–70.

Ecto-ATPases: Identities and Functions

Liselotte Plesner
Department of Biophysics, University of Aarhus, DK-8000 Aarhus C, Denmark

Ecto-ATPases are ubiquitous in eukaryotic cells. They hydrolyze extracellular nucleoside tri- and/or diphosphates, and, when isolated, they exhibit E-type ATPase activity, (that is, the activity is dependent on Ca^{2+} or Mg^{2+}, and it is insensitive to specific inhibitors of P-type, F-type, and V-type ATPases; in addition, several nucleotide tri- and /or diphosphates are hydrolysed, but nucleoside monophosphates and nonnucleoside phosphates are not substrates). Ecto-ATPases are glycoproteins; they do not form a phosphorylated intermediate during the catalytic cycle; they seem to have an extremely high turnover number; and they present specific experimental problems during solubilization and purification. The T-tubule Mg^{2+}-ATPase belongs to this group of enzymes, which may serve at least two major roles: they terminate ATP/ADP-induced signal transduction and participate in adenosine recycling. Several other functions have been discussed and identity to certain cell adhesion molecules and the bile acid transport protein was suggested on the basis of cDNA clone isolation and immunological work.

KEY WORDS: Ecto-ATPase, E-type-ATPase activity, (Ca^{2+} or Mg^{2+}) ATPase, T-tubule Mg^{2+}-ATPase, Apyrase, Cell adhesion molecules, Bile acid transport, Adenosine recycling, P-glycoprotein.

I. Introduction

A. Delimitation of Subject

Information about the enzymes that hydrolyze extracellular nucleoside tri- and/or diphosphates on the cell surface may be found in the literature under many different names. These include Ecto (Ca^{2+}- or Mg^{2+}-dependent)-nucleoside triphosphatase, nucleotide phosphohydrolase, adenylpyrophosphatase, ATP pyrophosphohydrolase, ATP diphosphohydrolase, ATPDase, ATP-DPH, apyrase, ATP-diphosphatase, adenosine diphosphatase, nucleoside-diphosphatase, nucleoside-diphosphate phosphohydrolase, nucleotidase, ATPase, ADPase, NTPase, or NDPase. The ecto-ATPase of the skeletal muscle transverse tubules was called T-tubule Mg^{2+}-ATPase, and frequently in work on a variety of tissues the ecto-ATPases were referred to as basal, basic, or nonspecific Mg^{2+}-ATPases. Also, different EC numbers have been used: EC3.6.1.3, 3.6.1.5, 3.6.1.6, or 3.6.1.15 (see Section VII).

In the following, the term *Ecto-ATPases* is used for enzymes that hydrolyze extracellular nucleoside tri- and/or diphosphates. Furthermore, an ecto-ATPase is assumed to be responsible if an isolated enzyme activity is (1) dependent on Ca^{2+} or Mg^{2+}, (2) insensitive to specific inhibitors of P-type, F-type, and V-type ATPases, and (3) hydrolyzes several nucleoside tri- and/or diphosphates, but nucleoside monophosphates and nonnucleoside phosphates are not substrates. An activity that exhibits these characteristics is called *E-type ATPase activity*.

This article does not deal with the physiological effects of extracellular nucleotides that act as transmitters, cotransmitters, neurotransmitters, and growth factors. This fast-growing field is covered in numerous reviews and symposium proceedings (e.g., Su, 1983; Burnstock and Kennedy, 1986; Gordon, 1986; Burnstock, 1988; White, 1988; Seifert and Schulz, 1989; Kennedy, 1990; Olsson and Pearson, 1990; Dubyak, 1991; Dubyak and Fedan, 1991; von Kugelgen and Starke, 1991; Bean, 1992; El-Moatassim *et al.*, 1992; Vassort *et al.*, 1992; Dubyak and El-Moatassim, 1993).

In contrast, only two reviews have been published on the enzymes that catalyze the hydrolysis of these important agonists (Banerjee, 1981 and Pearson, 1985). This contrast may reflect the fact that the ecto-ATPases represent a field of research that has so far been only partly successful: while elegant studies were done on the ecto-ATPase activity of intact cells in cultures and in fixed tissue sections, biochemical and structural work was hampered by specific experimental problems presented by these enzymes during isolation. These particular experimental problems are discussed in Section I,C.

B. Survey of Suggested Identities and Functions

Following the identification of ATP in 1929, a number of ATPases were shown to have either mechanochemical functions or to serve as pumps. They hydrolyzed intracellular ATP, and the dogma developed that ATP was strictly intracellular.

Yet the presence on the cell surface of enzymes capable of hydrolyzing ATP and ADP were described in cell cultures and tissues, for example, yeast (Rothstein and Meier, 1948), intestinal mucosa cells (Rothstein *et al.,* 1953), mammalian nerve cells (Abood and Gerard, 1954), Ehrlich ascites cells (Acs *et al.,* 1954), erythrocytes (Herbert, 1956), and intact muscle (Marsh and Haugaard, 1957). The name "ecto-ATPase" was used for the first time in 1957 by Engelhardt, who studied nucleated erythrocytes (Engelhardt, 1957). The same year Wachstein and Meisel (1957) introduced the histochemical method for demonstration of nucleoside phosphatase activity in tissue samples for microscopy. The fixed samples were incubated with substrate and lead; the released P_i formed lead phosphate, and on immersion in dilute ammonium sulfide solution the sites of enzymatic activity were revealed by deposits of black lead sulfide. In their original work, Wachstein and Meisel (1957) observed intense ATPase activity at the canalicular membrane of liver cells. Using this method, several authors soon reported the presence of ATPase activity at cell membranes of a variety of cells. Whereas special treatments to permeabilize the cells were often needed to show ATP hydrolysis by intracellular ATPases P_i release at the cell surface, catalyzed by the ecto-ATPase, was readily demonstrated. In addition, ecto-ATPase activity sustained the fixation procedures better than did the intacellular ATPases, and ecto-ATPase activity was not significantly inhibited by lead. Further studies, including electron microscopy, were initiated to determine whether the reaction product, P_i, was formed on the extra or intracellular side of the cell membrane. From these studies it was established that lead phosphate was precipitated on the outside of the cell membrane, and it was the contention of this group of researchers that they had visualized the activity of an ecto-ATPase (Essner *et al.,* 1958; Novikoff *et al.,* 1962a). Moreover, it was shown by Novikoff *et al.* (1961), who tested several rat and mouse tissues, that Ca^{2+} could replace Mg^{2+} without reduction of ATPase staining intensity, which was not influenced either by the presence or absence of Na^+ and ouabain; in addition, most cell membranes stained to the same degree with several nucleoside triphosphates and ADP; that is, in this study the E-type ATPase characteristics were demonstrated by the histochemical method. Soon these characteristics were confirmed by Wallach and Ullrey (1962), who investigated in detail substrate specificity and divalent cation dependency of the ecto-ATPase activity of Ehrlich ascites carcinoma cells in suspension.

Thus, the ecto-ATPases and the Na^+,K^+-ATPase (Skou, 1957) were in fact identified at about the same time. Yet there is a striking difference in the body of knowledge available on the two systems, and papers claiming the discovery of a "new" ATPase with typical E-type activity are still being published.

In the past 30–40 years, there have been several suggestions regarding the function of the ecto-ATPase.

1. By electron microscopic histochemistry it was shown that the ecto-ATPase activity of HeLa cells was intense at microvilli, but minimal where cell surfaces were smooth. This localization was considered to be suggestive of a **mechanochemical** function of the enzyme (Epstein and Holt, 1963a), and the possibility that the ecto-ATPase was an actomyosin-like protein engaged in cell motility and/or adhesion was discussed by many authors (e.g., Gropp *et al.,* 1958; Wallach and Ullrey, 1962; White and Krivit, 1965; B. M. Jones, 1966; Chambers *et al.,* 1967; Jones and Kemp, 1970a,b; Ronquist and Ågren, 1975; Ohnishi and Kimura, 1976; Lambert and Christophe, 1978). Results that showed that extracellular ATP distinctly influenced the volume of Ehrlich cells (Stewart *et al.,* 1969; Hempling *et al.,* 1969) and isolated kidney tubule cells (Rorive and Kleinzeller, 1972) were interpreted to support this idea (see Section II,A,3, for further examples).

2. Trams and Lauter (1974), who may have been the first to suggest that the Ca^{2+}- or Mg^{2+}-dependent ATPase of the plasma membrane was an ectoenzyme and ubiquitous in eukaryotic cells, suggested its role in **nonsynaptic information transfer.** Trams (1974) had observed a transient increase in membrane permeability with extrustion of K^+ and possibly cytoplasmic ATP on addition of micromolar concentrations of ATP to the medium of certain cultured cell lines. The ecto-ATPase would be needed to control or limit this biphasic change or signal (Trams and Lauter, 1978a).

3. Ecto-ATPase activity has been **associated with virus and cancer cell function.** Novikoff *et al.* (1962b) studied by histochemistry and electron microscopy the myeloblastosis avian tumor virus in thymus gland of leukemic chicks and found an intense ecto-ATPase activity on the outermost membrane of the virus and on the membrane of the blastlike cells from which the virus emerged by budding. In similar experiments Epstein and Holt (1963b) demonstrated that the outer viral limiting membrane of herpes simplex virus exhibited a high ecto-ATPase activity that was carried off from the HeLa host cells. Banerjee studied the E-type ATPase activity of several viruses, and found that in some it was surprisingly high, for example in avian myeloblastosis virus, where 33 U/mg of virus protein was measured (Banerjee and Racker, 1977; Banerjee, 1978) (One unit (U) is the amount of enzyme that will catalyze the release of one micromole of inorganic phosphorous per min at 37°C.) Banerjee (1981) speculated that the ecto-ATPase activity of some viruses may modify the permeability properties of cells that respond to an external source of ATP.

Karasaki and Okigaki (1976) found that the ecto-ATPase activity of cultured hepatoma cells could be correlated to their tumorigenicity in neonatal Wistar rats, and concluded that the ecto-ATPase activity may provide a reliable index of oncogenic potential of different liver epithelial cells. Ohnishi and Kimura (1976) noticed that the ecto-ATPase activity of cultured hepatoma cells was located intensively on the surface of cell contact, and evidence was presented that the ecto-ATP activity was correlated to the ability of the cell to overcome contact inhibition (Yamaguchi and Ohnishi, 1977). Other researchers reported results that did not agree with the proposals of Karasaki, Ohnishi, and co-workers, but studies by A. F. Knowles and co-workers may have solved his controversy and shed new light on the possible role of ecto-ATPases in malignant transformation (see Section II,G).

4. Secretion of E-type ATPase-containing microvesicles (diameter, 150 nm) into the oviduct was described by Rosenberg *et al.*(1977) who coined the term **exo-ATPase** for the enzyme in the microvesicles. Similar E-type ATPase-containing microvesicles were present in the lumen of pancreas acini (Beaudoin *et al.*, 1986) and in prostatic fluid (Ronquist and Brody, 1985). Although it is not known to what extent this type of microvesicular secretion is found in other animal cells, it is an interesting possibility that the ecto-ATPase associated with the microvesicles may play a role in secretory processes (Beaudoin *et al.*, 1986).

5. Several researchers investigated the possibility that ecto-ATPase or E-type ATPase would have **pumping functions,** for example, mediating Ca^{2+} transport (Lambert and Christophe, 1978; LeBel *et al.*, 1980) or bicarbonate transport (Martin and Senior, 1980) in pancreas, proton transport in liver (Luu-The *et al.*, 1987), and Ca^{2+} entry and Mg^{2+}transport across the cell membrane in cardiac cells (Dhalla and Zhao, 1989). But the fate of the energy derived from the hydrolysis of extracellular nucleoside phosphates remained an open question, as no transport was found to be associated with the ecto-ATPase activity. In the vast literature on pump isolations, the role assigned to this enzyme was the one of "unspecific basal Mg^{2+}-ATPase," that is, the abundant activity that must be eliminated when purifying the established pumps (: P-type, V-type, and F-type ATPases).

For a while considerable amounts of work were focused on the fact that only a minor part of the Ca^{2+}-stimulated ATPase activity of plasma membranes was catalyzed by the Ca^{2+} pump. A series of papers appeared with almost identical titles, stating that the Ca^{2+}-transporting ATPase and the Ca^{2+} -or Mg^{2+}-dependent nucleoside phosphatase were separate enzymes (Pershadsingh and McDonald, 1980; Ghijsen *et al.*, 1984; Lin, 1985a,b; Minami and Penniston, 1987; Birch-Mashin and Dawson, 1988; Enyedi *et al.*, 1988; Gandhi and Ross, 1988; Kelley *et al.*, 1990; Yoshida *et al.*, 1990; Sun *et al.*, 1990). While the latter had E-type ATPase activity, the plasma membrane Ca^{2+}-transporting ATPase needed both Ca^{2+} and Mg^{2+} for activity, was specific for ATP, and was inhibited by the P-type inhibitor, vanadate. No phosphorylated intermediate could be identified corresponding to the E-type ATPase activity, which in most cases was inactivated by detergents, resistant to proteolysis, and often activated by concanavalin A. (Further references are given in the respective sections below.) The identity between the enzyme that catalyzed this E-type activity and the ecto-ATPase was gradually accepted.

6. At this time biological responses to extracellular ATP had been documented in virtually every major organ and/or tissue system that had been studied, and the question was asked whether the ecto-ATPases might be **purinoceptors.** Several studies have addressed this question and found evidence against the suggested identity. The $P_{2\times}$ receptor from rat vas deferens was solubilized and separation of receptor and ATPase activity was obtained by sucrose density gradient centrifugation (Bo *et al.*, 1992). In porcine vascular smooth muscle a spectrum of nucleotide analogs was used to show that the structure–activity relationship of the ecto-ATPase did not resemble the structure–potency relationship of the purinoceptor (Cusack *et al.*, 1983; Martin *et al.*, 1985; Pearson *et al.*, 1985). Similar results were obtained in guinea pig smooth muscle (Welford *et al.*, 1986, 1987). Although no evidence could be found for identity between ecto-ATPases and purinoceptors, the ecto-ATPases necessarily play important roles in purinoceptor function, regulating, or certainly affecting, the concentrations of agonists and antagonists around the purinoceptors. Thus Welford *et al.* (1987) conclude that susceptibility to degradation does limit the potency of ATP and its degradable analogs.

7. Extracellular ATP can be utilized by kinase activities that appear to be localized on the surface membrane of some cells, possibly acting as effectors of ATP-mediated signal transduction (for references, see Dubyak and El-Moatassim, 1993). Another role for the ecto-ATPases could therefore be to **regulate ectokinase substrate concentration.** The inherent experimental difficulties in attempts to identify ectokinase activity in cells that also contain ecto-ATPase activity were reported by De Souza and Reed (1991), who observed phosphorylation of intracellular proteins within minutes after the addition of 20 μM [^{32}P]ATP to PC12 cells, a cell line derived from rat pheochromocytoma. Incubating the cells with $^{32}P_i$ of comparable specific activity gave the same overall protein labeling pattern, and the simultaneous addition of [^{32}P]ATP and nonradioactive P_i reduced the incorporation of radioactivity into the proteins. The results thus suggest that the phosphorylation of intracellular proteins by extracellular [^{32}P]ATP required the initial extracellular hydrolysis of [^{32}P]ATP and subsequent incorporation of the $^{32}P_i$ in the intracellular ATP pool.

In 1991 Saborido *et al.* (1991a) reported the ectoorientation of T-tubule Mg^{2+}-ATPase. As T-tubule membranes represent more than 80% of the skeletal muscle cell surface and as the E-type ATPase is the most prominent enzyme activity of the membranes, this new member represents a substantial addition to the ecto-ATPase family. Unfortunately, however, it remains another branch with unknown function.

8. A possible role for T-tubule and other ecto-ATPases in **adenosine**

recycling has been suggested: as ATP, ADP, and AMP cannot reenter the cell (Henderson and Paterson, 1973; Trams and Lauter, 1974), several authors pointed to the important role of the combined action of ecto-ATPases and 5′-nucleotidase to produce the hydrolysis end product, adenosine, in case ATP had leaked from a cell. Adenosine is not degraded outside the cell, and it is readily recaptured by specific transporters into the cell, where it is valuable for resynthesis of nucleotides. In the canalicular membrane of rat liver the specific Na^+-dependent adenosine transporter was shown to be closely associated with the ecto-ATPase (Moseley *et al.*, 1991; Che *et al.*, 1992). As substantial amounts of ATP have been found to leak from working muscle cells, T-tubule ecto-ATPase may be needed for adenosine recycling in this tissue. Cytosolic ATP (as opposed to ATP sequestered within intracellular granules or vesicles) can be released in several tissues from intact cells activated by physiological or pathological stimuli (Trams, 1974; Pearson and Gordon, 1979). In certain blood vessels secondary release of cytosolic ATP from intact endothelial and smooth muscle cells may constitute about 90% of the total ATP released on neural stimulation (Westfall *et al.*, 1990), and this noncytolytic release of cytoplasmic ATP may be mediated by specific transporter proteins. Abraham *et al.* (1993) have shown that cells with increased levels of the multidrug resistance (*mdr*1) gene product, P-glycoprotein, release ATP to the medium in proportion to the concentration of the protein in their plasma membrane. As indicated by measurements of whole-cell and single-channel currents with patch-clamp electrodes, ATP may directly permeate the P-glycoprotein channel as a charge carrier. Overexpression of another ABC-transporter (traffic ATPase), the cystic fibrosis transmembrane regulator protein, was also correlated with enhanced flux of ATP from transfected cells (Prat *et al.*, 1992). If in this way the electrochemical gradient for ATP is used to transport xenobiotics and/or some endogenous intracellular products of metabolism to the cell surface, then the ecto-ATPase would play a role in the transport mechanism, as part of the necessary adenosine recycling system.

9. Attention has been focused on the possible **cell adhesion function** of the ecto-ATPases: in 1989 the amino acid sequence of the ecto-ATPase from rat liver was deduced from analysis of cDNA clones and genomic clone by Lin and Guidotti. The deduced amino acid sequence consists of 519 amino acids, encompassing an extracellular catalytic domain, a transmembrane region that anchors the glycoprotein in the lipid bilayer, and a cytoplasmic domain that contains two potential tyrosine phosphorylation sites (Tyr^{488} and Tyr^{513}) and a cyclic AMP-dependent serine phosphorylation site (KRPTS, amino acids 499–503). The amino acid sequence was found to share substantial homology with human biliary glycoprotein 1 (BGP 1) (Hinoda *et al.*, 1988). This protein is a member of the CEA

subgroup of the carcinoembryonic antigen (CEA) gene family, which is a subfamily in the immunoglobulin superfamily and proposed to be involved in Ca^{2+}-independent cell adhesion.

Increasing evidence indicates structural and functional association between various cell adhesion molecules (CAMs) and membrane ATPases. A prominent example is the β_2 subunit of the Na^+,K^+-ATPase, which is homologous to the cell adhesion molecule of glia (AMOG) (Gloor *et al.*, 1990). The ectoenzyme 5′-nucleotidase that hydrolyzes the product of the ecto-ATPase reaction, AMP, was modulated by extracellular matrix proteins laminin and fibronectin, and a possible bifunctional role as ectoenzyme and cell adhesion molecule was discussed (Olmo *et al.*, 1992, and references therein).

It is a problem to researchers, whose research grows to include the CAMs, that most of these proteins have several different names, as apparently they have each been "discovered" several times. In addition, virtually every CAM has multiple isoforms that are expressed in different tissues and/or at different times in development, and it adds to the confusion that well-defined functions have not been established for many of the molecules. Therefore in the following some information is given on each of the CAMs shown to have amino acid sequences identical to the one reported for rat liver ecto-ATPase (HA4, pp120, and C-CAM) and on those that were associated with the ecto-ATPases based on different evidence (T-cadherin and NCAM).

HA4 is a protein localized predominantly in the apical (bile canalicular) domain of the hepatocyte, and the pathway of newly synthesized HA4 to the apical membrane was studied along with other integral membrane proteins during hepatocyte plasma membrane biogenesis (Bartles *et al.*, 1987). In separate studies, pp120 was identified as a substrate for several receptor tyrosine kinases (Margolis *et al.*, 1988, and references therein). Thus insulin and growth factors stimulated its phosphorylation both in cell-free systems, in tissue cultures of intact hepatoma cells, and in the perfused liver, where the normal polarity of the plasma membrane is preserved. In the intact cells pp120 was phosphorylated on both serine and tyrosine residues (Phillips *et al.*, 1987). Identity between pp120 and HA4 was demonstrated by Margolis *et al.* (1988), and by comparison of the partial amino acid sequences of pp120/HA4 with the deduced amino acid sequence of the rat liver ecto-ATPase, it was concluded that they represent the same protein (Margolis *et al.*, 1990).

Another detailed biochemical analysis (Aurivillius *et al.*, 1990) established the identity between the deduced amino acid sequence of the rat liver ecto-ATPase and Cell-CAM 105 (C-CAM). C-CAM is the rat homolog of human biliary glycoprotein 1, and both contain four immunoglobulin-like loops, one V domain, and three C2 domains. C-CAM is a Ca^{2+}

-independent cell adhesion molecule with homophilic binding properties (Tingström *et al.,* 1990), and detailed biochemical characterization has been reported (Odin *et al.,* 1986). It mediates cell–cell adhesion of isolated rat hepatocytes (Ocklind and Öbrink, 1982) and is present in a variety of epithelia, vessel endothelia, granulocytes, and platelets (Odin *et al.,* 1988).

In rat liver ecto-ATPase/C-CAM/pp120/HA4 invariably appears as two structurally related, highly glycosylated peptide chains with apparent molecular masses of 105 and 110 kDa, and it was shown (Lin *et al.,* 1991; Culić *et al.,* 1992; Najjar *et al.,* 1993; Edlund *et al.,* 1993) that these two forms differ mainly in their cytoplasmic domains. The long form (C-CAM1) has a C-terminal, cytoplasmic domain of 71 amino acids, whereas that of the short form (C-CAM2) is only 10 amino acids long and lacks the putative sites for tyrosine- and serine-specific phosphorylation. Two variants (a and b) were identified in one of the four immunoglobulin-like loops, that is, so far there are four different protein-coding sequences of rat liver C-CAM: C-CAM1a and 1b, and C-CAM2a and 2b (Culić *et al.,* 1992; Edlund *et al.,* 1993.)

Lin and co-workers investigated the cell adhesion properties of selected C-CAM isoforms in a baculovirus expression system in which Sf9 cells expressed the full-length cDNA, the short isoform, or a deletion mutant (Cheung *et al.,* 1993,a,b,c). The ecto-ATPase activity of the transfected cells was however, not measured (see below).

In a detailed study of the immunohistochemical localization of C-CAM in rat tissues, no staining was found in the nervous system, myocardium, skeletal muscle, ductus deferens, larger arteries, or veins (Odin *et al.,* 1988) and C-CAM RNA signals were not detectable in kidney, heart, brain, adrenal glands, and pancreas (Cheung *et al.,* 1993a). All of these tissues have ecto-ATPases and/or E-type ATPase activity, which was obviously not catalyzed by C-CAM/pp120/HA4. But other cell adhesion molecules might possibly be ecto-ATPases and, supporting this hypothesis, Cunningham *et al.* (1993) published evidence indicating that the cell adhesion molecule T-cadherin was closely associated with, if not identical to, the ecto-ATPase of the skeletal muscle T-tubule system.

T-cadherin belongs to the cadherin superfamily of adhesion molecules. Members of this superfamily exhibit Ca^{2+}-dependent homophilic binding, and their function depends on both an HisAlaVal sequence in the first extracellular domain and the interaction of a conserved cytoplasmic region with intracellular proteins. T-cadherin is an unusual member of the cadherin superfamily that lacks the HisAlaVal sequence; furthermore, it is anchored to the membrane through a glycosyl phosphatidylinositol (GPI) moiety (Ranscht and Dours-Zimmermann, 1991). The mechanism of T-cadherin-induced adhesion is distinct from that of classic cadherins (Vestal and Ranscht, 1992).

Further support of the hypothesis that CAMs other than C-CAM may exhibit ecto-ATPase activity was presented by Dzhandzhugazyan and Bock (1993), who demonstrated E-type ATPase activity of immunoprecipitated neural cell adhesion molecule, NCAM. They solubilized brain microsomes under conditions that did not inactivate the E-type ATPase activity and, using different types of solid-phase immunoadsorption, it was shown that NCAM antibodies isolated the ATPase activity. Furthermore, agarose gel immunoelectrophoresis of solubilized brain microsomes followed by ATPase assay in the gel revealed ATPase activity associated with the NCAM immunoprecipitate. The specific activity was determined in excised NCAM precipitates, and found to be 4.5 U/mg of NCAM protein.

The neural cell adhesion and recognition molecule, NCAM, operates via both homophilic and heterophilic binding mechanisms. It belongs to the immunoglobulin superfamily and was shown to be phosphorylated by ectoprotein kinase on the external surface of neural cells (Ehrlich *et al.*, 1986). Three main classes of isoforms may be distinguished on the basis of their membrane association: two transmembrane forms of 180 and 140 kDa are named A and B, respectively, whereas NCAM-C of 120 kDa is linked to the membrane via a GPI anchor. Intact NCAMs of all three classes of isoforms were shed in soluble form in rat brain and cerebrospinal fluid (Olsen *et al.*, 1993). Adenylate cyclase activity has been suggested to be an intrinsic function of NCAM or of a tightly associated protein (Lipkin *et al.*, 1989, 1992); however, the E-type ATPase activity in the study of Dzhandzhugazyan and Bock (1993) was clearly distinguished from adenylate cyclase activity.

If ecto-ATPase activity is catalyzed by various cell adhesion molecules, then the putative bifunctional role of these proteins is in accordance with the initial idea of a mechanochemical function; as expressed by Aurivillius *et al.* (1990): "interactions between Ig-domains of C-CAM might be the structural basis for specific homophilic binding, and hydrolysis of ATP could modify this binding in a manner similar to the ATP-regulated binding between actin and moyosin."

Unfortunately, at present it is not possible to make conclusions with regard to the function of ecto-ATPases.

10. From rat liver canalicular vesicles a 110-kDa glycoprotein had been purified and identified as the canalicular **bile acid transport** protein. Internal amino acid sequence and chemical and immunochemical characteristics of the protein (Sippel *et al.*, 1990) were found to be identical to rat liver ecto-ATPase as determined by Lin and Guidotti (1989) and Lin *et al.* (1991). The cDNA for the liver ecto-ATPase/C-CAM/pp120/HA4 was obtained from the latter group and used to transfect COS cells (Sippel *et al.*, 1993). Transfection conferred *de novo* synthesis of a 110-kDa polypeptide, as immunoprecipitated by antibodies against the canalicular bile

acid transport protein and conferred on the COS cells the capacity to pump out taurocholate with efflux characteristics comparable to those previously determined in canalicular membrane vesicles (V_{max} = 0.6 nmol/min per milligram of protein). A truncated ecto-ATPase/C-CAM/pp120/HA4 cDNA, missing the cytoplasmic tail, did not confer bile acid transport activity on COS cells, and it was shown that phosphorylation within the cytoplasmic tail was essential for bile acid transport activity. An ATPase activity was measured in the cell lysates. In the cells transfected with the long form the ATPase activity of the cell lysates was equivalent to the taurocholate transport rate (0.75 mU/mg of protein), and the same ATPase activity was measured in the lysates of cells transfected with the short form, that is, the ATPase activity was not dependent on the cytoplasmic tail and its phosphorylation. The authors conclude that bile acid efflux and ecto-ATPase activity are two distinct properties of a single protein. In a later study (Sippel *et al.*, 1994) they show that ATP depletion of COS cells, tranfected with ecto-ATP ase/C-CAM/pp120/HA4 cDNA (Lin and Guidotti, 1989), resulted in abrogated bile acid transport, which was reconstituted on the addition of exogenous ATP. On the other hand, when mutations were introduced in the ATPase consensus sequence, the ATPase activity could be abrogated without any effect on bile acid transporter synthesis and function. It was concluded that bile acid efflux is dependent on ATP, but not on ATP hydrolysis by the transporter.

The ATPase activity in the lysate of the transfected COS cells was very low, but not lower than the ATPase activity (0.6 mU/mg of protein) measured in the lysate of the L cells that were transfected with the ecto-ATPase/C-CAM/pp120/HA4/bile acid transporter cDNA in the original work by Lin and Guidotti (1989). (For comparison, most cell homogenates exhibit an E-type activity that is 10 to 1000 times higher; see sections on individual tissues below.) In neither work was the ecto-ATPase activity measured and, as mentioned above, when various forms of the ecto-ATPase/C-CAM/pp120/HA4/bile acid transporter cDNA were expressed in insect cells to study the adhesive properties conferred by the transfection, neither ecto-ATPase nor cell lysate ATPase activity was measured (Cheung *et al.*, 1993a,b,c).

On the basis of published cDNA and the deduced amino acid sequence (Lin and Guidotti, 1989), a cDNA encoding for this sequence was recloned and used to transfect NIH-3T3 mouse fibroblasts to obtain stable transfectants (S. M. Najjar, personal communication). The transfected cells expressed pp120, but their ecto-ATPase activity was only 0.5 mU per mg of protein, that is, the same activity that was obtained in the transfected L cells and COS cells (Lin and Guidotti, 1989; Sippel *et al.*, 1993).

These studies may show that the ecto-ATPase/C-CAM/pp120/HA4/bile acid transporter protein needs modification to become hydrolytically active, that is, the transfected cells will not exhibit ecto-ATPase activity unless the modifying system is also expressed. On the other hand, the results may also show that the C-CAM/pp120/HA4/bile acid transporter cDNA does not code for rat liver ecto-ATPase. The latter interpretation is supported by the following observations:

1. Cell adhesion properties of isolated ecto-ATPase have not been demonstrated, and attempts to detect any ATPase activity in purified C-CAM/pp120/HA4 have so far been unsuccessful (Öbrink, 1991; S. M. Najjar, personal communication).
2. The amino acid sequence deduced by Lin and Guidotti (1989) has a blocked N terminal that exhibits no homology to the N-terminal sequences of rabbit T-tubule and chicken gizzard ecto-ATPase and no homology as well to the sequence of a peptide isolated from the latter enzyme (Treuheit *et al.*, 1992; Stout and Kirley, 1994a,b) (see Section IV,A).
3. An ecto-ATPase with properties similar to those of the ecto-ATPase from rat liver was expressed in large amounts in factor-treated human hepatoma cells (Gao and Knowles, 1993; see Section II,G). A cDNA library was obtained from these factor-treated cells and, probing with the rat liver ecto-ATPase/C-CAM/pp120/HA4/bile acid transporter cDNA of Lin and Guidotti (1989), a partial CEA cDNA was cloned instead of an ecto-ATPase gene (Knowles, 1994).

In summary, these observations indicate that the exact nature of the relationship between cell adhesion molecules, bile acid transporter, and ecto-ATPases remains to be established. Some cell adhesion molecules may be ecto-ATPases or, alternatively, ecto-ATPases tend to be intimately associated with some cell adhesion molecules. Meanwhile, most of the suggested functions described above, may still be valuable as working hypotheses. It is possible that different subfamilies of ecto-ATPases each have a different and specific tissue distribution, function, and protein structure. It is also possible that certain "functional complexes," each with a different and specific tissue distribution, function, and composition, may contain the same ecto-ATPase protein.

The question of ecto-ATPase identities and functions has not met with a definitive answer, and implying that this is caused mainly by the specific and substantial experimental problems presented by these enzymes, the introduction is concluded by summarizing the most important difficulties.

C. Specific Experimental Problems

1. Detergent Inhibition

With a few exceptions (given in Section III,A) ecto-ATPases are inactivated by those detergents normally used to solubilize membrane-bound proteins. As a result most work was done on either intact cells or isolated membranes, the latter often in vesiculated form. Quite a number of these papers conclude with the same line: "Further elucidation will await solubilization and purification of the enzyme," and typically the next publication by the authors was in a different field.

2. Lack of Specificity

The ecto-ATPase is supposed to be responsible for E-type ATPase activity in isolated preparations, but E-type ATPase activity was defined by a number of absent properties such as lack of specific divalent cation requirement, lack of inhibition by certain inhibitors, and lack of substrate specificity.

One example of the problems caused by the lack of specific divalent cation requirements was mentioned above: until a few years ago all Ca^{2+}-activated ATPase activity of plasma membranes was mistakenly attributed to the Ca^{2+} transporter. One example of the problems caused by the lack of substrate specificity is mentioned below (Section II,C): the E-type ATPases tend to hydrolyze the affinity probes that are normally useful for labeling and identification of ATPases. The lack of sensitivity to specific inhibitors of P-type, F-type, and V-type ATPases is a distinctive characteristic, but it would be useful if the ecto-ATPases had a specific inhibitor of their own. The search for such a specific inhibitor has been intensive throughout the years, and in this matter immense effort was also contributed by the large group of researchers who study the purinoceptors, as their work is constantly hampered by the ecto-ATPases hydrolyzing at a high rate one group of agonists to produce another with different (often opposite) effects.

3. Glycoprotein

Less specific, but still substantial, are the problems associated with the presence of carbohydrate in these highly glycosylated membrane proteins. Glycoproteins tend to give broad, poorly defined bands by sodium dodecyl sulfate (SDS)-gel electrophoresis or gel-permeation chromatography and multiple peaks by ion-exchange chromatography or isoelectric focusing. Glycoproteins are often resistant to proteases; moreover, heterogeneity of the resulting glycopeptides on digestion is another serious obstacle with respect to sequence analysis. Antibodies against a glycoprotein often cross-react with other glycoproteins, and in this respect apparently the CEA gene family members are unsurpassed. Some ecto-ATPases are either identical to, or intimately associated with, a CEA gene family member (see Section I,B).

4. Difficulty in Unfolding

Evidence (Stout and Kirley, 1994a) indicates that the ecto-ATPase protein may be unusually hard to unfold: antibodies would recognize the catalytic subunit only under specific ionic conditions, and changing the ionic conditions of samples for SDS-electrophoresis would change the apparent molecular weight obtained on the gel.

5. Copurification

The chicken oviduct and liver enzymes were purified by immunoaffinity chromatography and high specific activities were obtained (Strobel and Rosenberg, 1992; Strobel, 1993) (see the next section). Yet the authors describe ''microheterogeneity'' in the liver preparation and the presence of two very minor bands in the purified oviduct enzyme (Strobel, 1993). When a preparation of rat hepatocyte C-CAM was submitted to immunoaffinity purification, the coisolation of three immunologically unrelated proteins was reported (Lim *et al.,* 1993). So far these are the only published immunoaffinity purification results, but in our laboratory immunoaffinity purification of E-type ATPase from rat ductus deferens was performed and resulted in coisolation of several immunologically unrelated proteins (Plesner, 1994). These examples of copurification, even through immunoisolation, may illustrate yet another ''specific experimental problem'' presented by these enzymes, but they may also possibly indicate that the ecto-ATPases tend to be part of (functional?) complexes. Under nondenaturing conditions, in which the E-type ATPase activity is readily demonstrated on the gel, the proteins do not separate. On the other hand, under conditions in which separation is obtained, no activity remains, and the identification of the individual bands must rely on immunostaining of Western blots with the uncertainty inherent in this method.

6. Low Abundance

In most tissues the E-type ATPase activity is more than two orders of magnitude higher than the ATPase activity of the plasma membrane Ca^{2+} pump (e.g., Minami and Penniston, 1987; Birch-Machin and Dawson, 1988; Kelley *et al.,* 1990) and often more than 30-fold higher than the ATPase activity of the Na^{+} pump (Bonting, 1970; Juul *et al.,* 1991b). Examples of ecto-ATPase-rich tissues are rat pancreas with 100 U/10^9 cells (Hamlyn and Senior, 1983), cultured chromaffin cells with 170 U/10^9 cells (Torres *et al.,* 1990), and chicken gizzard smooth muscle cells with 230 U/10^9 cells (measured in the presence of concanavalin A) (Cunningham *et al.,* 1993). Unfortunately, these high activities do not necessarily reflect similar high quantities of enzyme protein in the tissues. That instead the enzyme may have an extremely high turnover number was already evident from studies on the ecto-ATPase of certain viruses (Banerjee and Racker, 1977; Banerjee, 1978, 1979). Specific activities of more than 30 U/mg of virus protein were repeatedly measured and the same value was determined in membrane preparations from rat pancreas (Martin and Senior, 1980), that is, specific activities equivalent to the highest ever published for the purified Na^+ pump. Immunoaffinity purification of the enzyme from chicken oviduct and liver were reported to yield specific activities of 806 and 1242 U/mg of protein, respectively (Strobel and Rosenberg, 1992; Strobel, 1993), and a specific activity of 1700 U/mg of protein was measured in the presence of concanavalin A in E-type ATPase purified from chicken gizzard (Stout and Kirley, 1994b). The most extensive purification of an E-type ATPase was performed with the T-tubule enzyme (Kirley, 1988, 1991, Treuheit *et al.*, 1992) and a specific activity of close to 7000 U/mg of protein was obtained. Using this figure and a molecular mass of 70 kDa (Treuheit *et al.*, 1992), a turnover number of 500,000 min^{-1} may be calculated; this is only 3 times lower than the value for acetylcholinesterase and more than 50 times higher than the value of the Na^+,K^+-ATPase. If all ecto-ATPases have such a high turnover number, then a preparation that is purified to a specific activity of, say, 70 U/mg of protein would contain only 1% of enzyme protein, and this fraction probably would not show up as a visible band in electrophoretic gels.

II. Occurrence and Localization

A table showing the specific activity of ecto-ATPases in different cells and tissues from different animals would be extremely useful for a series of purposes, but unfortunately it is not possible to supply such a table. Whereas "K_m" values have been extensively reported and compared in the ecto-ATPase literature, there has been only limited interest in reporting figures that would allow a comparison of the specific ecto-ATPase or E-type ATPase activity (i.e., units per gram of tissue). To the extent that it was possible to calculate this figure from the reported experimental results, it is given in the sections on the individual tissues below. Please note, however, that these calculated values should be accepted with caution. To ease comparison only three measures of specific activity were used: "units per milligram of protein" when the activity was determined in isolated membranes or purified enzyme preparations, and "units per 10^9 cells" or "units per gram of tissue" when either the ecto-ATPase activity was measured in intact cells or an E-type ATPase activity was determined in crude preparations. "Units per gram of tissue" was in the latter case calculated assuming protein to be 10% of tissue mass. It is hoped that in the future there will be more focus on the quantitative measures of specific activities, for comparison of results, for planning of experiments, for identification of possible species differences, and for qualified speculations on possible functions in different tissues.

A. Vascular Tissues

ATP and its hydrolysis products at low micromolar concentrations strongly influence vascular tone, cardiac function, platelet aggregation, and the function of lymphocytes and granulocytes (cf. the reviews listed in Section I,A).

Platelets secrete ATP and ADP when aggregating, and so do endothelial and vascular smooth muscle cells (Pearson and Gordon, 1979) responding to certain vasoactive mediators. The nucleotides are further released into the circulation following tissue damage and hypoxia. The endothelial cells and the vascular smooth muscle cells play a dominant role in the hydrolysis of extracellular nucleotides in the vascular system. Organ perfusion and tracer dilution studies have been demonstrated that the *in vivo* half-life of ATP or ADP injected into the blood stream is less than the time needed to pass one capillary bed, (i.e., a few seconds) whereas the half-life of ATP in cell-free plasma or whole blood *in vitro* is several minutes (Binet and Burstein, 1950; Brashear and Ross, 1969; Smith and Ryan, 1970; Chelliah and Bakhle, 1983; Fleetwood *et al.*, 1989). The maximum degradation rate for ATP in isolated whole blood was determined to be 2–4 U/liter (Coade and Pearson, 1989; Beukers *et al.*, 1993) at ATP concentrations between 50 and 100 μM.

The role of the ecto-ATP ase in the control of nucleotide concentration in blood vessels has been reviewed (Côté *et al.*, 1992c).

1. Lymphocytes

The ecto-ATPase activity of lymphocytes from guinea pig leukocyte exudate was 0.24 U/10^9 cells (DePierre and Karnovsky, 1974b). The activities measued in lymphocytes from human peripheral blood were 2.6 U/10^9 cells (Medzihradsky *et al.*, 1980, 0.3 U/10_9 cells (Gutmann *et al.*, 1983), and 0.14 U/10^9 cells (Beukers *et al.*, 1993). ATP concentrations were 0.5–2 mM.

In mouse thymocytes and mouse spleen lymphocytes the ecto-ATPase activity was 2.7 and 12 U/10^9 cells, respectively, at 2 mM ATP. Lymphocytes were stimulated by concanavalin A to become active blast cells. As this lectin strongly activated the ecto-ATPase a possible role for the enzyme in lymphocyte blast transformation was suggested (Pommier *et al.*, 1975).

The ecto-ATPase activity of human B cells was higher than that of T cells (Kragballe and Ellegaard, 1978; Gutmann et al., 1983). Barankiewicz and co-workers (1988) found that the ecto-ATPase activity of the T cell was virtually 0, and as adenosine receptors are present on the T lymphocytes it was suggested that adenosine may serve as a means of communication between human B and T cells in lymphoid organs, B lymphocytes being the sole producers of adenosine and T lymphocytes being the recipients of this signal (Barankiewicz *et al.*, 1988).

The ecto-ATPase activity of cytolytic T lymphocytes was 7 U/10^9 cells when assayed at 3 m*M* ATP (Filippini *et al.*, 1990). Extracellular ATP was considered an intermediate in lysis of tumor target cells by cytotoxic T lymphocytes, which perform accumulation of extracellular ATP after triggering via their target cell receptors. Possibly ATP does not act as a "hit" molecule, but merely as a substrate for ectoprotein kinases. Ecto-ATPase and ectoprotein phosphatases would regulate this system of extracellular phosphorylation (Apasov *et al.*, 1993; Redegeld *et al.*, 1993).

The ecto-ATPase activity of human natural killer cells was studied by Bajpai and Brahmi (1993) and Dombrowski *et al.* (1993). An activity of about 4 U/10^9 cells was reported by the latter authors, who used 0.3 m*M* ATP in the assay medium.

2. Polymorphonuclear Leukocytes

The ecto-ATPase activity of intact leukocytes (from humans and rabbits) was first measured by Tenney and Rafter (1968).

DePierre and Karnovsky (1974a,b) established criteria for the identification of "ecto-enzyme-activity" by studying guinea pig exudate leukocytes (Section V,A,1). They measured the ecto-ATPase activity at 1 m*M* ATP in Krebs–Ringer phosphate buffer, and the phosphate (15 m*M*) was shown not to inhibit the activity. The following values were obtained: 0.24 U/10^9 polymorphonuclear leukocytes; 7 times higher for monocytes and 70 times higher for eosinophil granulocytes.

At 1–2 m*M* Mg^{2+} and ATP the ecto-ATPase activity of human polymorphonuclear leukocytes was 0.8–1.4 U/10^9 cells (Harlan *et al.*, 1977; Smolen and Weissmann, 1978; Medzihradsky *et al.*, 1980), whereas the activity on exudate granulocytes from mice was 3 U/10^9 cells at 0.25 m*M* ATP (Weiss and Sachs, 1977).

Suramin was found to inhibit (by 65%) the ecto-ATPase of human polymorphonuclear leukocytes. As suramin inhibition of the ecto-ATPase had no effect on chemotaxis, superoxide anion generation, or phagocytosis, it was considered unlikely that the enzyme plays a major role in these functions (Smolen and Weissmann, 1978).

3. Platelets

The ability of ADP to activate platelet aggregation, secretion, and thrombus formation has been recognized since the early 1960s (Gaarder *et al.*, 1961; Born, 1962; Born and Cross, 1963), and represents one of the most intensively studied physiological responses to extracellular nucleotides. Because activated platelets also release ATP and ADP from granule stores, ADP constitutes a powerful paracrine signal for the rapid "feed forward" recruitment of new platelets to the developing thrombus. ATP and adenosine antagonize the actions of ADP; ATP competes with ADP for the P_{2t} purinoceptor, whereas adenosine antagonizes the ADP effect by binding to P_1 purinoceptors on the platelets (for references, see Olsson and Pearson, 1990; Dubyak and El-Moatassim, 1993).

ATPase reaction product P_i, was localized on the surface membranes of human platelets by histochemical and electron microscopy methods (White and Krivit, 1965), and several mechanisms involving the ecto-ATPase activity were suggested to explain the initiation of platelet aggregation by ADP (Robinson *et al.*, 1965; P. C. T. Jones, 1966; B. M. Jones, 1966; Salzman *et al.*, 1966; Chambers *et al.*, 1967; Mason and Saba, 1969; Saba *et al.*, 1969; Jones and Kemp, 1970a,b; Wang *et al.*, 1977).

Salzman *et al.* (1966) proposed that ADP-induced aggregation was due to ADP inhibition of the ecto-ATPase, "the function of which is customarily to keep the platelet in an unsticky state and to maintain the cell size and shape." Ca^{2+} or Mg^{2+} activated the ecto-ATPase activity of human platelets that also hydrolyzed other nucleoside triphosphates and ADP (Chambers *et al.*, 1967). Antisera against the contractile protein of the cell surface inhibited the ecto-ATPase activity (Chambers *et al.*, 1967), and aggregation was inhibited by antibodies against actomyosin (Jones and Kemp, 1970b). Against this background the possible identity between the ecto-ATPase and the contractile protein of the cell surface was discussed by the quoted authors.

Cumulative addition of ADP to platelets in whole blood *in vitro* resulted in only a modest amount of aggregation, and on the basis of this observation the rate of degradation of ATP to ADP in whole blood was found to be too slow to account for a significant role in platelet aggregation (Beukers *et al.*, 1993). This finding is not surprising because the ecto-ATPase activity of endothelial and vascular smooth muscle is the dominant contributor to the ATP degradation rate in the vascular system (Côté *et al.*, 1992a).

4. Erythrocytes

The ecto-ATPase activity of guinea pig erthrocytes was 0.0025 U/10^9 cells (DePierre and Karnovsky, 1974b). The ecto-ATPase activity of human erythrocytes was demonstrated by histochemical and electron microscopy methods (White and Krivit, 1965) and the ecto-ADPase activity of these cells was studied more recently by Lüthje *et al.* (1988).

5. Cultured Endothelial Cells, Vascular Smooth Muscle Cells, and Cardiac Cells

In a series of papers the hydrolysis of extracellular ATP, ADP, and AMP by cultured vascular cells was studied (Pearson *et al.*, 1980, 1985; Cusack *et al.*, 1983; Pearson and Gordon, 1985; Gordon *et al.*, 1986, 1989; Pearson, 1986; Slakey *et al.*, 1986; Coade and Pearson, 1989; Fleetwood *et al.*, 1989; Meghji *et al.*, 1992).

The kinetic properties of the ATP-, ADP-, and AMP-hydrolyzing enzymes were investigated by recirculating exogenous substrates over flowthrough columns packed with endothelial cells from pig aorta cultured on microcarrier beads (Gordon *et al.*, 1986). This approach permitted the study of the reactions at volume-to-cell surface ratios approaching those of small blood vessels (Gordon *et al.*, 1986). The rate of adenosine formation was found to decrease as the initial ATP or ADP concentration increased, which was explained by ADP inhibiting the 5′-nucleotidase. It was suggested that this feed-forward inhibition would create, at sites of platelet degranulation, a time gap between ADP release (a proaggregatory milieu) and the appearance of adenosine (an antiaggregatory milieu). Kinetic constants for each of the reactions were obtained, assuming that the ATPase and ADPase obeyed simple Michaelis–Menten kinetics and fitting simulated reaction curves to observed time courses.

A similar approach was used to monitor the time course for the ATP, ADP, and AMP hydrolysis by pig vascular smooth muscle cells attached to polystyrene beads (Gordon *et al.*, 1989). The observations were in marked contrast to the findings with endothelial cells: adenosine was produced rapidly in spite of the inhibition of 5′-nucletidase by ADP. This was explained by assuming local concentrations of substrate (ADP and AMP) supplied from the preceding reactions to be much higher than those in the bulk phase. The enhancement of efficiency by preferential delivery of products to become substrates was sufficient to overcome the effect of the feed-forward inhibition.

By analyzing the time courses of ATP, ADP, and AMP hydrolysis by adult rat ventricular myocytes, monitored in petri dishes with attached myocytes (Meghji *et al.*, 1992), it was concluded that the supply of ATP and

ADP as initial substrates was rate limiting, and that preferential delivery of intermediates to become substrates for downstream reactions played a significant role in regulation of adenosine production. As adenosine was taken up inefficiently relative to the rate of its production, the milieu at the cardiac myocyte surface would be nucleotide poor and adenosine rich, even with a supply of extracellular nucleotides. Preferential delivery has more impact on the delivery of ADP from the ATPase to the ADPase than on delivery of AMP from the ADPase to the 5′-nucleotidase; nonetheless, on the basis of other evidence, the authors throughout the quoted papers assume that ATP and ADP were hydrolyzed by different enzymes. This is discussed in Section V,B,1.

The maximal rates calculated for ATP hydrolysis were 22 U/10^9 endothelial cells and 14 U/10^9 vascular smooth muscle cells. ADP hydrolysis rates were about 50 and 80%, respectively. Whereas endothelial and vascular smooth muscle cells were isolated from pig aorta, the ventricular myocytes were from rat. With the latter tissue the kinetic parameters derived were strikingly different from those obtained with the pig tissues (see Section V,B,5).

Cusack *et al.* (1983) measured the ecto-ATPase activity of cultured endothelial cells from pigs and found 67 U/10^9 cells and the ADP hydrolysis rate amounting to 14% of this value. Cultured human endothelial cells also hydrolyzed ATP and ADP (Dosne *et al.*, 1978). When the time course of ATP hydrolysis by cultured bovine endothelial and vascular smooth muscle cells was measured, no transient accumulation of ADP was observed (Yagi *et al.*, 1991). The discrepancy with the results of Gordon et al., (1986, 1989) may probably be explained by the volume-to-cell surface ratio being much smaller in the work by the latter authors.

6. Blood Vessels

ADP-induced platelet aggregation could be released by the addition of isolated bovine aorta microsomes, and the microsome fraction exhibited. ADPase activity (Miura *et al.*, 1987). The enzyme, which was solubilized and partly purified, was an E-type ATPase that hydrolyzed ATP and ADP almost equally well. In a subsequent work by the same authors the enzyme was purified to a specific activity of 58 U per mg of protein (Yagi *et al.*, 1989). The ectoorientation of the enzyme responsible for the E-type ATPase activity was demonstrated in bovine aorta endothelial and vascular smooth muscle cells (Yagi *et al.*, 1991). From human umbilical vessels E-type ATPase was characterized and purified to a specific activity of 37 U/mg of protein (Yagi *et al.*, 1992). The ATP- and ADP-hydrolyzing activity copurified, indicating that the same catalytic site was responsible for both activities.

Côté *et al.* (1991, 1992b) compared the ATP- and the ADP-hydrolyzing activity of bovine aorta with the E-type ATPase that they had characterized previously in the pancreas (see Section II,F,3). On the basis of pH dependency profiles, heat denaturation curves, ^{60}Co irradiation–inactivation curves, and enzyme activity localization on a native gel it was concluded that the two enzymes has different properties. But with both enzymes ATP and ADP hydrolysis varied in parallel in each of these experiments, indicating that the same catalytic site was responsible for ATP and ADP hydrolysis. From the inner layer of bovine aorta two zones, the intima and the media, were separated and shown to contain ATPase activities with virtually identical characteristics and potency to inhibit ADP-induced platelet aggregation. When both ATP and ATPase were added to platelet-rich plasma, there was an immediate dose-dependent aggregation of platelets followed by a slowly developing disaggregation. The enzyme concentrations used in these experiments (Côté *et al.*, 1992a) were up to 30 times higher than quoted above for the enzyme concentration in blood [2–3 U/liter (Coade and Pearson, 1989; Beukers *et al.*, 1993)]. The specific activity was 8 and 10 U/g of intima and media, respectively (Côté *et al.*, 1992a).

E-type ATPase activity was also found in plasma membrane fractions from porcine aortic and coronary artery smooth muscle (Sun *et al.*, 1990).

The effects of exogenous nucleoside phosphates on tone, membrane potential, and intracellular Ca^{2+} were measured in intact small mesenteric arteries from rat (Juul *et al.*, 1991a, 1992, 1993) and at the same time the hydrolysis of the nucleoside phosphates catalyzed by the vessels was measured (Juul *et al.*, 1991b). Time courses were nonlinear, and evidence was presented for a concentration- and time-dependent modification of the enzyme by the substrates. The modified enzyme had a much reduced affinity for the substrates, but the maximal rate of hydrolysis (14 U/g of tissue) was not changed. A membrane-bound E-type ATPase, isolated from the vessels, hydrolyzed the substrates with high affinity and time courses were linear (Plesner *et al.*, 1991) (see Section V,A and V,B,5).

The ecto-ATPase of the endothelial and vascular smooth muscle cells has been used histochemically to visualize the vasculature in many tissues, for example, retina (Lutty and McLeod, 1992) and organs from hypertensive rats (Herrmann *et al.*, 1992).

B. Muscle

1. Skeletal Muscle

In striated muscle cells the contraction–relaxation mechanism is controlled by a sarcotubular membrane system consisting of the transverse tubules (T-tubules) and the sarcoplasmic reticulum. The T-tubule membranes are continuous invaginations of the plasma membrane, and constitue over 80% of the surface membrane of striated muscle. Following nerve excitation, an action potential generated at the neuromuscular junction of the plasma membrane is propagated into the interior of the muscle fiber through the transverse tubules. Depolarization of the T-tubule membrane facilitates the release of Ca^{2+} from the terminal cisternae of the sarcoplasmic reticulum.

The most prominent enzymatic activity found in highly purified T-tubule membranes is the T-tubule Mg^{2+}-ATPase. In 1991 evidence for the extracellular orientation of the active sites of the chicken and rabbit enzymes was published by Saborido *et al.* (1991a), and the finding was confirmed by Treuheit *et al.* (1992) and Cunningham *et al.* (1993), who worked with rabbit and chicken enzymes, respectively.

In their paper, Sabbadini and Dahms (1989) give a comprehensive and excellent review of this enzyme, the T-tubule Mg^{2+}-ATPase. It has E-type ATPase activty (Hidalgo *et al.*, 1983; Beeler *et al.*, 1983) and it is a glycoprotein modulated by lectins, strongly inhibited by detergents, and resistant to protein hydrolysis. Methods have been devised for obtaining highly purified T-tubule membranes that are virtually devoid of sarcoplasmic reticulum (SR) and attendant SR Ca^{2+}-ATPase activity (for references, see Sabbadini and Dahms, 1989).

Some major points are summarized in the following, and more recent results are described.

Two different patterns of substrate inactivation were observed in rat and chicken T-tubule vesicles (see Section V,B,5). In both preparations the inactivation could be prevented by preincubation with concanavalin A (Beeler *et al.*, 1983; Moulton *et al.*, 1986). The E-type ATPase activity of chicken T-tubules exhibits an unusual bell-shaped temperature dependence, with an optimum at 27°C and only 50% of maximum activity at 10 and 37°C. Following preincubation with concanavalin A the temperature dependence of the enzyme activity becomes completely normal, the lectin giving rise to increasing activation at temperatures above 27°C. At 37°C the activation is four- to eightfold (Sabbadini and Okamoto, 1983; Moulton *et al.*, 1986).

In contrast, the E-type ATPase activity of rabbit T-tubule vesicles was not modified by lectins, and saponin treatment increased the E-type ATPase activity. These results were interpreted to show that the orientation of the T-tubule vesicles was inside out, that is, opposite to the orientation of the chicken vesicles (Hidalgo *et al.*, 1986; Horgan and Kuypers, 1988). But the fact that high ATPase activity was found in vesicles from both animals led Saborido *et al.* (1991a) to compare the leakiness and the sidedness of T-tubule vesicles from chicken and rabbit; and they found that the percentage of sealed vesicles averaged 88 and 78%, respectively, whereas the sidedness of the sealed vesicles was predominantly right side out in both (69–76 and 62–70%, respectively). Thus, the absence of significant effects of concanavalin A on the activity of rabbit T-tubule vesicles was not due to inside orientation of the glycosylation sites (see Section V,B,5), and the activation by saponin was also not due to exposure of inward-facing substrate sites (see Section V,B,3).

The possibility that the T-tubule Mg^{2+}-ATpase might be an ecto-ATPase was discussed first by Beeler *et al.* (1983), on the basis of the observation that lectins activate on the surface of tight T-tubule vesicles where ATP is also hydrolyzed, and the question was further debated by Sabbadini and Dahms (1989), who also underlined the E-type characteristics of the T-tubule enzyme. But cytochemical localization of Mg^{2+}- or Ca^{2+}-activated ("basic") ATPase in chicken skeletal muscle had shown that the lead deposits, formed enzymatically by ATP hydrolysis, were on the cytoplasmic side of the plasmalemma and T-system membranes (Malouf and Meissner, 1979). Although the authors did not exclude the possibility that the inner location of the lead deposits could be caused by an artifact of the cytochemical technique, it was not until the work of Saborido *et al.* (1991a) that it was accepted that an ecto-ATPase is responsible for the E-type ATPase activity of the T-tubule.

Kinetic results reported on the T-tubule enzyme are discussed in Section V, whereas the effects of free fatty acids, fatty-acyl-CoAs, and lipid second messengers on the T-tubule system (Okamoto *et al.*, 1985; Moulton *et al.*, 1986; Kang *et al.*, 1991) are discussed in Section VI.

Specific activities of isolated T-tubule membrane vesicles ranged between 1 and 25 U/mg of protein (Beeler et al., 1983; Hidalgo *et al.*, 1983; Moulton *et al.*, 1986; Rosemblatt and Scales, 1989; Valente *et al.*, 1990; Saborido *et al.*, 1991a). Additional values are tabulated in Hidalgo *et al.* (1986).

As a result of a persistent effort (Kirley, 1988, 1991; Treuheit *et al.*, 1992) the rabbit T-tubule enzyme was extensively purified, and a specific activity of close to 7000 U/mg of protein was obtained. The implication of this high specific activity was discussed in Section I,C. The N-terminal sequence determined (Treuheit *et al.*, 1992) is given in Section IV,A, and the procedures used by this group and by groups that partly purified the T-tubule enzyme are described in Section III.

The occurrence in rat heart, spleen, lung, kidney, liver, brain, and adipose tissue of an enzyme activity with properties similar to those of the T-tubule enzyme was reported by Beeler *et al.* (1983) and described in detail (Beeler *et al.*, 1985) (see Section V,B,5). In chicken gizzard smooth muscle, brain, heart, spleen, and lung polypeptides reacted with inhibitory polyclonal antibodies raised against the chicken T-tubule enzyme (Cunningham *et al.*, 1993). Moreover, an ATPase activity in chicken gizzard and brain had properties indistinguishable from those of the T-tubule enzyme and the activities were found to copurify with the antibody reactive protein (see Section IV,B).

2. Heart

"Basic" ATPase was localized by ultrastructural cytochemistry in paraformaldehyde-fixed canine heart under conditions that inhibited Na^+,K^+-ATPase and sarcoplasmic reticulum Ca^{2+}-ATPase activities (Malouf and Meissner, 1980). The enzyme was present at plasmalemma, T-tubule system, and intercalated disk membranes, with the exception of the nexus. (Lead deposits were found to be on the cytoplasmic side of the membranes, see Section II,B,1.)

Isolating low-density vesicles from a series of rat organs and comparing their E-type ATPase activities, the specific activity obtained in the heart preparation was among the highest measured, 11 U/mg of protein (Beeler *et al.*, 1985).

An E-type ATPase activity was characterized in rat heart sarcolemma (Tuana and Dhalla, 1988). The rate of ATP hydrolysis decayed exponentially (Zhao and Dhalla, 1988), and the inactivation depended on the presence of ATP or other high-energy nucleotides (see Section V,B,5). Following solubilization and purification (Zhao and Dhalla, 1991) the specific activity obtained was 80 U/mg of protein (Section III). In the purified enzyme the phospholipid content was 60 times higher than the content of the plasma membrane, whereas the cholesterol and polysaccharide content was increased by a factor of 8 (Zhao *et al.*, 1991)

The ecto-ATPase activity of cultured rat ventricular myocytes was measured by Meghji *et al.* (1992) (see Section II,A,5).

3. Smooth Muscle

The vascular smooth muscle ecto-ATPase was described above (Sections II,A,5, and 6).

Magocsi and Penniston (1991) compared the ecto-ATPase activity of intact rat myometrial cells to the E-type ATPase activity of isolated plasma membranes. Nonlinear progress curves were interpreted to indicate that two different ecto-ATPases were contributing to the ATP hydrolysis, and a labile as well as a stable E-type ATPase activity was characterized (see Section V,B,5). The specific activity of the stable ecto-ATPase was about 50 U/g of tissue.

The chicken gizzard was shown to be a rich (and easily available) source of E-type ATPase, reacting with and inhibited by antibodies against the chicken T-tubule ecto-ATPase (Yazaki *et al.*, 1992). The smooth muscle enzyme exhibited properties indistinguishable from the skeletal muscle enzyme with regard to the effects of several activators and inhibitors. The ectoorientation of the enzyme was demonstrated in isolated gizzard smooth muscle cells, where the activity was 70 U/10^9 cells, increasing to 230 U/10^9 cells in the presence of concanavalin A (Cunningham *et al.*, 1993).

The gizzard enzyme was purified to a specific activity of 90 U/mg of protein, increasing to 1700 U/mg of protein in the presence of concanavalin A (Stout and Kirley, 1994b) (see Section III). The chromotographic and electrophoretic characteristics of the chicken gizzard enzyme were found to be similar to those reported for the rabbit skeletal muscle enzyme (Treuheit *et al.*, 1992), and the N-terminal amino acid sequences exhibited pronounced homology (see Section IV,A). Using an anti-peptide antibody raised against an internal protein sequence of the purified gizzard ecto-ATPase, the tissue distribution of the enzyme in adult and embryonic chicken was investigated (Stout and Kirley, 1994a). Antibody-reactive material was found in virtually all tissue types examined, but was most abundant in excitable tissues. The immunological tissue distribution results were confirmed in adult chicken tissues, in which the ATPase activity and the influence of inhibitors and activators were measured (Stout and Kirley, 1994a).

C. Nervous Tissue

ATP is copackaged in both adrenergic and cholinergic neurotransmitter granules and released during neurotransmission into synaptic spaces. There is considerable evidence for ATP acting as transmitter or cotransmitter at neuroeffector junctions in the periphery (for review, see White, 1988; Bean, 1992), and three reports provide direct evidence that ATP acts as a fast transmitter between neurons of the central nervous system (CNS) (Edwards *et al.*, 1992; Silinsky *et al.*, 1992; Evans *et al.*, 1992). Furthermore in the mammalian CNS adenosine, the end product of ATP hydrolysis, has been shown to be an inhibitory neuromodulator both pre- and postsynaptically (Richardson *et al.*, 1987).

As mentioned above (Section II,B,1), in low-density vesicles, isolated from a series of tissues, including brain, an ATPase activity was found that had properties similar to the T-tubule enzyme (Beeler *et al.*, 1983, 1985), and an ATPase activity in chicken brain has properties (including immunological properties) indistinguishable from those of the chicken T-tubule enzyme (Cunningham *et al.*, 1993) and the chicken gizzard enzyme (Stout and Kirley, 1994a). It was also mentioned above (Section I,B) that

E-type ATPase activity was demonstrated in immunoprecipitated N-CAM from rat (Dzhandzhugazyan and Bock, 1993).

Ecto-ATPase activity was reported for the first time in brain tissue, when rabbit glia cells were shown to hydrolyze extracellular ATP whereas intact neurons did not (Cummins and Hydén, 1962). The presence of ecto-ATPase was reported on intact human glia cells (Ågren *et al.*, 1971) and on various neuronal and glial cell lines (Stefanovic *et al.*, 1976; 1977). Trams and Lauter (1978a) compared the properties to the E-type ATPase activity of nerve cells and brains of various species and suggested that the major part of the measured ATPase activity was due to plasma membrane ecto-ATPase.

The search for the ecto-ATPase activity responsible for hydrolysis of ATP in the synaptic cleft had been hampered by leakage of intracellular enzymes, breaking of nerve endings during isolation, and contamination by other subcellular particles. In 1983, however, Nagy *et al.* reported an ATP-hydrolyzing activity on the external surface of intact synaptosomes from chicken forebrain. The procedure for synaptosome isolation was further developed (Nagy and Delgado-Escueta, 1984) and the preparation further characterized (Nagy *et al.*, 1986; Nagy, 1986), comparing also synaptosomes from different parts of rat, mouse, gerbil, and human brain. Mg^{2+} and Ca^{2+} were found to activate the ecto-ATPase activity equally well; GTP, UTP, and ITP were also hydrolyzed to a similar extent as ATP, and ADP was hydrolyzed at 30% of the rate for ATP. The specific activity of the synaptosome preparations was around 0.1 U/mg of protein.

Similar characteristics were demonstrated for the ecto-ATPase activity of synaptosomes isolated from *Torpedo mamorata* electric organ (Grondal and Zimmermann, 1986; Sarkis and Salto, 1991) and rat brain (Gandhi and Ross, 1988; Battastini *et al.*, 1991).

In immunoisolated rat cholinergic nerve terminals from the striatum, ectoenzymes were shown to convert released ATP to adenosine, which inhibited acetylcholine release (Richardson *et al.*, 1987). However, this inhibitory effect was not seen in cortical cholinergic terminals lacking the enzymes necessary for the degradation of ATP to adenosine. It was therefore concluded that the differing adenosine-mediated modulation in different brain areas is controlled by the presence and activity of synaptic ecto-ATPase and 5′-nucleotidase.

In a study on the ATPase activities of lysed synaptosomal plasma membranes and synaptic vesicles, the E-type ATPase activity was found to be three to four times higher in the latter (S. C. Lin and Way, 1984).

Solubilization and partial purification of ecto-ATPase from bovine brain synaptic membranes has been reported (Hohmann *et al.*, 1993) (see Sections III and IV,B).

8-Azido-ATP and 8-azido-ADP were found to be excellent substrates of ecto-ATPase in cultured bovine chromaffin cells (Rodriquez-Pascual *et al.*, 1993). The affinity for these substrates and the maximal hydrolysis rates were similar to what was obtained with ATP and ADP in the same culture and assay conditions. Ultraviolet (UV) photoactivation of the 8-azido-nucleotides (100 μM) irreversibly inhibited the enzyme activity by about 50%, and the presence of ATP or ADP in the same concentration range effectively protected against the inhibition. To obtain photolabeling to a significant extent high concentrations of azido derivatives were necessary, see Section IV,B.

Cultured bovine chromaffin cells hydrolyzed extracellular ATP and ADP at a high rat (170 and 125 U/10^9 cells, respectively) (Torres et al., 1990). The half-lives of the ATP- and ADP-hydrolyzing activites were studied in chromaffin cells cultured in the presence of cycloheximide (Rodriguez-Pascual *et al.*, 1992). Similar values were obtained for the two activities (22.7 and 25.2 hr, respectively), and the results thus indicate that one enzyme may catalyze both activities.

Cytochemical demonstration of ecto-ATPase reaction product on the cells surface was reported in rat cerebellum glia cells, neuronal cell processes (Mughal *et al.*, 1989), and cultured chromaffin cells (Kriho *et al.*, 1990).

D. Lung

As shown by perfusion studies, ATP is rapidly metabolized into adenosine on a single passage through the pulmonary capillary bed (Binet and Burstein, 1950; Chelliah and Bakhle, 1983; Hellewell and Pearson, 1987). From the measured E-type ATPase activity in bovine lung homogenate (Picher *et al.*, 1993) the specific activity may be around 12 U/of lung tissue.

It is not known if this activity is derived solely from vascular endothelial and smooth muscle cells. Extracellular nucleotides induced chloride secretion through apical membrane purinergic receptors and the potential therapeutic role of aerosolized nucleotides in the treatment of the lung disease of cystic fibrosis has been discussed (Knowles et al., 1991; Merten *et al.*, 1993); but apparently ATP is rapidly broken down in human airway epithelia (Knowles *et al.*, 1991).

E. Reproductive System

1. Oviduct and Sperm

Rosenberg and co-workers described the oviduct ecto-ATPase in a series of studies. They found large quantities of ATPase to be present in oviductal secretions from chicken (Anderson *et al.*, 1974), mice, and humans (Rhea *et al.*, 1974). An intensive ATPase activity was also demonstrated histochemically on the free lumenal surfaces of the secretory cells of the proximal portions of the oviduct from chicken, mice, and humans (Rhea *et al.*, 1974). From the histochemical investigation of the human oviducts it seemed that the amount of ATPase on the secretory cells could be related to the hormonal status, and in chicken it was shown that oviductal ATPase was three times higher in the preovulatory state compared to the postovulatory state of the cycle (Anderson and Rosenberg, 1976). Within minutes, as the ovulated oocyte passes through the proximal part of the oviduct, ATPase is transferred from the outer surface of the secretory cells to be bound by strong bonds to the surface coat of the ovum, the fertilized oocyte picking up three times the amount of ATPase compared to the unfertilized oocyte (Haaland and Rosenberg, 1969; Haaland *et al.*, 1971; Etheredge *et al.*, 1971; Rhea and Rosenberg, 1971; Anderson *et al.*, Rhea *et al.*, 1974). The ATPase was secreted and transferred in the form of enzymatically active microvesicles, roughly 150 nm in diameter and consisting of half-protein, half-lipid (Rosenberg *et al.*, 1977). The enzyme had E-type ATPase activity, and the name *exo-ATPase* was used (Rosenberg *et al.*, 1977, 1980) because it was secreted and transferred (see Section I,B). Contrary to other E-type ATPases the activity of the microvesicle-associated oviduct enzyme was highly labile. In spite of this, extensive purification by immunoaffinity chromatography was reported (see Sections I,B, III,A, III,B, and IV,B). The specific activity obtained for the purified chicken oviduct enzyme was 806 U/mg of protein (Strobel, 1993).

To fertilize an egg, spermatozoa need to undergo a complex chain of biochemical and morphological events collectively known as capacitation and acrosome reaction. These activating reactions usually take place in the proximal portion of the oviduct just before the spermatozoa approach the egg. In one report it was shown that extracellular ATP effectively and rapidly induced the acrosome reaction in human spermatozoa; furthermore, ATP-treated sperm had a high success rate in the hamster egg fertilization test (Foresta *et al.*, 1992). Majumder and Biswas (1979) characterized the ecto-ATPase activity of rat epididymal spermatozoa. One-third of the activity of the most immature spermatozoa was sensitive to vanadate (0.25 mM), and while the vanadate-sensitive ecto-ATPase activity stayed constant during maturation in the epididymis, the vanadate-insensitive ecto-ATPase activity declined and approached zero in the mature cell (Majumder, 1981).

Prostasomes are intact organelles secreted by the acinar epithelial cells of the human prostate gland, see Section I,B. They are located within storage vesicles that are translocated *in toto* from the cell interior into the acinar lumen through the plasma membrane, or alternatively they deliver their content into the glandular lumen by exocytosis. The prostasomes are 125-nm microvesicles that exhibit E-type ATPase activity and their function is not shown (Ronquist and Brody, 1985).

2. Placenta

Kelley and Smith (1987) studied the Ca^{2+}-transporting ATPase in human placental trophoblast basal plasma membrane and suggested the addition of 5 mM GTP to reduce the background release of $^{32}P_i$ in the assay of the Ca^{2+}-transporting enzyme. The "background release" was later shown to be caused by an E-type ATPase activity that was much higher than the ATPase activity of the Ca^{2+} pump. Lacking substrate specificity, it hydrolyzed the abundant GTP, leaving the [^{32}P]ATP (100 μM) to be hydrolyzed by the substrate-specific Ca^{2+} transporter (Kelley *et al.*, 1990). The E-type ATPase activity hydrolyzed ADP at a higher rate and several nucleoside triphosphates at the same rate as ATP.

In contrast, Papamarcaki and Tsolas (1990), also working with human placenta, found extremely low hydrolysis rates for nucleoside triphosphates other than ATP, but in accordance with the results of Kelley *et al.* (1990) P_i was produced at a higher rate when ADP rather than ATP was substrate (see Section V,B,1).

F. Exocrine Glands

1. Liver

The ecto-ATPase activity was found to be suitable for the reproducible demonstration of biliary canaliculi in all species examined, including humans (Wachstein and Meisel, 1957). It was shown that on ligation of the bile duct in the rat and in obstructive jaundice in humans, the microvilli of the dilated canaliculi were reduced and the ATPase activity lost (Novikoff and Essner, 1960). On surgical removal of the obstruction, the ATPase activity reappeared.

The occurrence of Mg^{2+}-activated ATPase in rat liver microsomes was first demonstrated by Novikoff *et al.* (1952). Further studies on rat microsome preparations revealed an apparent lack of substrate specificity and activation by either Mg^{2+} or Ca^{2+} (Ernster and Jones, 1962). Ouabain and oligomycin did not inhibit the ATPase activity, which was highly sensitive to detergents (Emmelot and Bos, 1966). Activation by lectins of the rat liver enzyme was described (Riordan *et al.*, 1977), and the quantity of this E-type ATPase activity was determined to be 12–15 U/g of rat liver (Luu-The *et al.*, 1987). A similar quantity seems to be present in chicken liver, where 13 U was recovered in the crude membranes per gram of tissue homogenized (Strobel, 1993). Lin and Russell (1988) determined the ecto-ATPase activity to be 23 U/10^9 intact hepatocytes in primary culture.

The rat liver enzyme was solubilized and purified to a specific activity of 20 U/mg of protein (Lin and Fain, 1984), and two subsequent studies presented evidence that the purified enzyme was distinct from the Ca^{2+} pump (Lin, 1985a,b). The same conclusion was reached by Birch-Machin and Dawson (1988), who also worked with the rat liver enzyme. By measuring the ecto-ATPase activity of rat hepatocytes in primary culture and comparing its properties with those of the purified E-type ATPase activity, Lin and Russell (1988) were able to conclude that they were catalyzed by the same enzyme. Two polyclonal antibodies were raised against the purified rat liver ecto-ATPase and used to determine the surface distribution of the enzyme by immunofluorescent staining. In addition, canalicular plasma membranes were separated from sinusoidal plasma membranes for determination of E-type activity in either. By both methods the ecto-ATPase was shown to be concentrated on the canalicular surface of the hepatocyte (Lin, 1989). The amino acid sequence of the rat liver enzyme, deduced from cDNA cloning, was published in an accompanying paper (Lin and Guidotti, 1989). This work and the more recent papers by Lin and co-workers were described in Section I,B.

2. Kidney

It is a problem to obtain membrane preparations of well-defined cellular origin from the kidney; moreover, an active, Mg^{2+}-stimulated V-type proton pump ATPase is difficult to separate from the ecto-ATPase in the brush border membranes where both enzymes are found.

Busse *et al.* (1980) demonstrated an Mg^{2+}-dependent enzyme that catalyzed at equal rates the hydrolysis of several nucleoside triphosphates in a brush border preparation from rabbit kidney. The enzyme activity was insensitive to vanadate and azide. In a subsequent publication (Mörtl *et al.*, 1984) the E-type ATPase was solubilized from the brush border preparation and partly purified. Whereas the plasma membrane Ca^{2+} pump ATPase, in analogy to the Na^+,K^+-ATPase, was identified solely in the basolateral membrane of the tubule cell (Kinne-Saffran and Kinne, 1974) the E-type ATPase activity seemed to be abundantly present not only in brush border membranes but also in basolateral membrane preparations (Gmaj *et al.*, 1982; Ghijsen *et al.*, 1984; Ilsbroux *et al.*, 1985; Turrini *et al.*, 1989). Following improved separation, however, the E-type ATPase was found only in the brush border fraction (Mörtl *et al.*, 1984; Van Erum *et al.*, 1988).

Turrini *et al.* (1989) prepared sealed rat brush border membrane vesicles and found that they hydrolyzed several externally added nucleotides in the presence of inhibitors of P-type and F-type ATPases. The primary goal of this study was to demonstrate *N*-ethylmaleimide (NEM)-sensitive ATP-drive H^+ secretion by the brush border vesicles, but the ecto-ATPase of such vesicles from rat kidney was characterized by Culić *et al.* (1990). ATP and ADP hydrolysis rates varied in parallel under a series of assay conditions, indicating that the activities possibly reside on the same protein. The specific activity of brush border membranes from rabbit and rat was 0.12 U/mg of protein (Mörtl *et al.,* 1984) and 0.4 U/mg of protein (Culić *et al.,* 1990), respectively.

Sabolić *et al.* (1992) examined the localization of ecto-ATPase/C-CAM in rat kidney and isolated renal cortical membranes, using antibodies against rat liver ecto-ATPase/C-CAM and found that brush border in contrast to basolateral membranes were antibody positive. They also found that preparations of basolateral as well as brush border membranes were contaminated with endothelial cell plasma membranes that are known to contain an active ecto-ATPase. Finally, using polyclonal antipeptide antibodies against the "long" form (C-CAM1) of liver ecto-ATPase/C-CAM, no specific staining in the kidney was obtained.

ADPase activity was demonstrated in the glomerular basement membrane of rat kidney (Bakker *et al.,* 1987; Poelstra *et al.,* 1991).

3. Pancreas and Parotid Glands

Ca^{2+} movements across the plasma membrane were related to the secretory mechanism, which led Lambert and Christophe (1978) to look for a Ca^{2+}-stimulated ATPase in exocrine rat pancreas membranes. The dominant ATPase activity of their preparation was typical E-type, and the same result was obtained by Martin and Senior (1980), who were in search of an ATPase playing a possible role in the bicarbonate secretion in the same tissue. The E-type ATPase activity in the plasma membranes prepared by the latter authors was remarkably high (30 U/mg of protein). They noticed that an ATPase was also present in the light membrane fractions derived from zymogen granule membranes.

The E-type ATPase activity of rat zymogen granule membranes had been characterized by Harper *et al.* (1978), who reached the conclusion that a single protein catalyzed the hydrolysis of nucleoside di- and triphosphates. They found evidence that the active site was localized on the internal face of the zymogen granule, that is, the face that becomes extracellular on exocytosis.

The ectoorientation of the enzyme was shown in a preparation of enzymatically dispersed rat pancreatic cells that hydrolyzed externally added nucleoside triphosphates and diphosphates at high rates; 100 U per 10^9 of cells was measured with 5 m*M* ATP as substrate (Hamlyn and Senior, 1983). The ecto-ATPase activity was compared to the high-activity membrane preparation (Martin and Senior, 1980). Both were activated by more than 100% by concanavalin A.

In contrast to these studies, all of which were done on rat tissue, Beaudoin and co-workers used pig pancreas for characterization and purification of the enzyme (LeBel *et al.*, 1980; Laliberté *et al.*, 1982; Laliberté and Beaudoin, 1983). The E-type ATPase was solubilized with no considerable loss of activity (LeBel *et al.*, 1980; Laliberté *et al.*, 1982) (see Section III). In the latter work the kinetics of divalent cation activation was described, and a subsequent publication analyzed the kinetics of the sequential hydrolysis by the enzyme of the γ- and β-phosphate groups of ATP (Laliberté and Beaudoin, 1983) (see Section V).

In plasma membranes from rat parotid gland an ATPase was found that had properties distinctly different from those of the Ca^{2+} pump ATPase (Teo *et al.*, 1988), and in a plasma membrane-rich fraction from bovine parotid gland a Ca^{2+}-dependent ATPase activity was described that was insensitive to P-type ATPase inhibitors and hydrolyzed nucleoside tri- and diphosphates at comparable rates. However, activity could be measured only in the presence of Ca^{2+}; the enzyme was not activated by Mg^{2+} (Matsukawa, 1990).

G. Cancer

Karasaki (1972) studied the distribution of ecto-ATPase activity in the precursor lesions and tumor cells of malignant hepatoma produced in rat liver following hepatocarcinogen feeding. The ecto-ATPase activity was demonstrated histochemically. Whereas the ecto-ATPase activity of the normal liver cell was present only at the surface membranes limiting the bile canaliculi, the cells in the precursor lesions, which represent sites of extensive dedifferentiation, exhibited activity over the entire surface, and it was suggested that the changes in distribution of surface ATPase activity were associated with early events of neoplastic development. In a subsequent study, it was found that the ecto-ATPase activity in epithelial cells derived from normal rat liver, carcinogen-induced rat hepatoma, and spontaneous transformants of cultured liver cells, could be correlated to their tumorigenicity in neonatal Wistar rats. It was concluded that the ecto-ATPase activity may provide a reliable index of oncogenic potential of different liver epithelial cells (Karasaki and Okigaki, 1976). Furthermore, whereas preoncogenic cultures were devoid of ectoenzyme activity and had smooth surfaces, the transformed cells showed complexities of surface membranes in the form of ruffles and microvilli, and combined cytochemistry and transmission electron microscopy indicated that the ecto-ATPase was concentrated on ruffles and microvilli (Karasaki *et al.*, 1977).

Working with primary cultured hepatoma cells, Ohnishi and Kimura (1976) noticed that the ecto-ATPase activity per cell increased as the density of the cell increased, and histochemical investigation revealed that the ecto-ATPase activity was located intensively and almost solely on the surface of cell contact. Cultured hepatic parenchymal cells in monolayer (and less malignant hepatoma cells) had little or no ecto-ATPase activity at their surface immediately after contact or after growth to a confluent state (Yamaguchi and Ohnishi, 1977), and when the malignant hepatocytes were treated with dibutryl cyclic AMP and theophyllin they no longer developed ecto-ATPase activity at cell contacts; the enzyme was uniformly distributed over the cell surface, and growth became contact inhibited. The authors thus suggested that the ecto-ATPase activity developed by hepatoma cells at cell contacts is important for the ability of the cell to overcome contact inhibition.

In contrast to these studies San *et al.* (1979) found no correlation between ecto-ATPase activity during transformation and tumorigenicity in adult rat liver epithelial-like cells. Furthermore, the ecto-ATPase of normal glia-like cells was high compared to transformed glia cells and lines from gliomas and sarcoma cells (Ågren *et al.*, 1971). Weiss and Sachs (1977) likewise found that the ecto-ATPase activity of early myeloid cells was similar to that of normal mature granulocytes and higher than the activity of any cells from nine clones of myeloid leukemic cells.

It is possible that the apparent conflict between the results of San, Ågren, Weiss, Sachs, and co-workers on the one hand and Karasaki, Ohnishi, and co-workers on the other may be explained by the presence of two separate ecto-ATPases or two different forms of the enzyme in the reported experiments. The existence of two distinct ecto-ATPases in liver epithelial cultures was suggested by the results of Ohnishi and Yamaguchi (1978) and Karasaki *et al.* (1980). One was characterized by a high substrate affinity and present in nononcogenic cell lines and in sparse cells of hepatoma cultures. In contrast, the ecto-ATPase at cell–cell contacts in transformed cells had a low substrate affinity, and as the transformed cells develop microvilli and/or extensive membrane infoldings at cellular boundaries where ecto-ATPase is concentrated, the hydrolytic activity per cell was high. By studying the kinetic properties of the ecto-ATPase of B cells isolated from patients with chronic lymphatic leukemia and comparing them to the properties of normal B cells, Gutmann *et al.* (1983) also found a low substrate affinity combined with a high hydrolysis rate per cell in the leukemic B cells.

It therefore may be that San, Ågren, Weiss, Sachs, and co-workers found malignant transformation to be associated with a block in ecto-

ATPase activity because they used lower ATP concentrations (8–250 μM) than did Ohnishi, Karasaki, Gutman, and co-workers (0.5–3 mM), that is, a substrate concentration that may have been insufficient with the low-affinity ecto-ATPase of the transformed cells.

More recently, ecto-ATPase activity of a human hepatoma cell line (Li-7A) was investigated. Untreated Li-7A cells exhibited a mercurial sensitive (MS) ecto-ATPase activity. On supplementation of the culture medium with epidermal growth factor (EGF), hydrocortisone, and choleratoxin, an ecto-ATPase was expressed that was insensitive to mercurials (MI ecto-ATPase) (Knowles *et al.*, 1985; Knowles, 1988a). The induction was a consequence of the synergistic actions of EGF and cyclic AMP, and the tyrosine kinase activity of the EGF receptor was essential, because enzyme induction was abolished by the tyrosine kinase inhibitor, genistein. Cycloheximide and actinomycin were also inhibitory, indicating that enhancement of MI ecto-ATPase activity was not due to posttranslational modification (Knowles, 1988b, 1990). In contrast, when human hepatoma cells were supplemented with butyrate an ecto-ATPase highly sensitive to mercurials was expressed. Maximal induction of MS ecto-ATPase by butyrate required 48 hr; it was dependent on the butyrate concentration, but independent of EGF and cyclic AMP-elevating agents. Expression was prevented by actinomycin D and cycloheximide, indicating that both transcription and translation were necessary for the butyrate-induced production of MS enzyme (Murray and Knowles, 1990; Knowles and Murray, 1990). MS ecto-ATPase was characterized and partly purified from human carcinoma cells (Shi and Knowles, 1994; Knowles and Leng, 1984). Also, MI ecto-ATPase was partly purified (from human hepatoma cells treated with EGF and cyclic AMP-elevating agents) (Gao and Knowles, 1993). It was similar to the rat liver ecto-ATPase with respect to divalent ion activation and response to sulfhydryl reagents and cross-reacted with an antibody to the rat liver ecto-ATPase/C-CAM. Whereas the MS ecto-ATPase preferred Mg^{2+} for activity, MI ecto-ATPase was more active with Ca^{2+}.

Whether MS ecto-ATPase is identical to the low-affinity ecto-ATPase described by Ohnishi, Karasaki, and Gutmann remains to be seen. It is certainly possible because MS ecto-ATPase has low substrate affinity (Knowles, 1988a; Murray and Knowles, 1990), and low-affinity ecto-ATPase was inhibited by mercurials (Karasaki, 1972; Yamaguchi and Ohnishi, 1977).

An ecto-ATPase with decreased substrate affinity was also identified in neuroblastoma cells on coculture with glia cells (Stafanović *et al.*, 1977). The originally low ecto-ATPase activity in both cell lines increased manyfold when they were cocultured, and when the neuroblastoma cells were separated from coculture the higher ecto-ATPase activity was permanent, that is, it continued for more than 20 replications (5 months). The substrate affinity exhibited by the ecto-ATPase of the reisolated cells was decreased (more than twofold). Cells cocultured for longer periods had higher enzymatic activity, corresponding roughly to the period of cocultivation. An increase in neuroblastoma ecto-ATPase activity could also be brought about by treatment with the medium from glia cells, but this increase disappeared when the supplementation with glia cells medium was discontinued. Following animal passage no change in the ecto-ATPase activity could be observed when the enzyme properties were compared to those of the original neuroblastoma cell line (Stefanović *et al.,* 1977).

It should be mentioned that extracellular ATP was shown to function as a mitogen in a number of mammalian cell lines and to act synergistically with other growth factors (Huang *et al.,* 1992, and references therein). This is another reason why ecto-ATPases would be involved in cell growth control and malignant transformation.

The quantities of ecto-ATPase in the cell lines used in the quoted work were about 2 U/10^9 glia-like and glioma cells (Ågren *et al.,* 1971), 2 U/10^9 myeloid leukemic cells (Weiss and Sachs, 1977), 10 and 20 U/10^9 AH66 cells in sparse and dense culture, respectively (Ohnishi and Yamaguchi, 1978), and 3–20 U/10^9 Li-7A cells, increasing by a factor of 7–9 on factor treatment (MI ecto-ATPase) (Knowles, 1990) and by a factor of 5–7 on treatment with butyrate (MS ecto-ATPase) (Murray and Knowles, 1990). An activity of 85 U/10^9 mammary adenocarcinoma ascites cells was determined by Carraway *et al.* (1980). The ecto-ATPase activity of neuroblastoma cells increased from 2 to 10 U/g of cells when reisolated after 2 months of coculture (Stefanović *et al.,* 1977).

H. Viruses, Insects, and Plants

In the introduction (Sections I,B and C) it was mentioned that the ecto-ATPase activity of avian myeloblastosis virus was 30 U/mg of virus protein. The enzyme was solubilized and purified and the E-type ATPase activity characterized (Banerjee and Racker, 1977). The ecto-ATPase of the host cell, the myeloblast, had molecular, catalytic, and immunological properties similar to those of the virus (Banerjee, 1979). Host cell ecto-ATPase was found in the envelope of several viruses (Novikoff *et al.,* 1962b; Epstein and Holt, 1963b), but the specific activity was significantly higher in the avian myeloblastosis virus than in other viruses, for example, Rous and reoviruses (Banerjee, 1978).

An enzyme that hydrolyzed ATP and ADP was described in the saliva of blood-sucking bugs (Ribeiro and Garcia, 1980; Smith *et al.,* 1980; Sarkis *et al.,* 1986), tse-tse flies (Mant and Parker, 1981), mosquitoes (Ribeiro *et al.*, 1984), and ticks (Riberio *et al.*, 1985), where it supposedly accounts, at least in part, for the anti-platelet-aggregating properties of the saliva of these unrelated insects. The activity of the enzyme presumably prevents blood clotting during feeding, and Sarkis *et al.* (1986) suggest that it is "a normal housekeeping enzyme" in nonhematophagous insects that evolved into an abundant secretory product on adoption of blood feeding. The quantities are surprising: 10 U in 1 μl of bug saliva with a specific activity of 10–40 U/mg of protein (Smith *et al.*, 1980). Sarkis *et al.* (1986) found similar values and purified the enzyme to a specific activity of almost 600 U/mg of protein.

These enzymes are probably E-type ATPases: the latter authors found that basic kinetic characteristics were shared by the isolated enzyme and vertebrate E-type-ATPases described by LeBel *et al.* (1980) and Knowles *et al.* (1983). Thus the insect enzyme(s) are activated by Ca^{2+} or Mg^{2+} (Ribeiro *et al.*, 1984) and lack substrate specificity (Sarkis *et al.*, 1986). Furthermore, it is interesting that the enzyme activity was found on a native gel at a molecular mass greater than 10^3 kDa, where no protein band could be seen, suggesting a high-activity low-abundance complex (Smith *et al.*, 1980).

A highly active ecto-ATPase was localized on the external surface of the tegument of *Schistosoma mansoni*, and E-type ATPase characteristics were demonstrated in analyzing the isolated enzyme (Vasconcelos *et al.*, 1993). The authors suggest that the enzyme may be involved in the mechanisms that enable schistosomes to escape platelet-mediated responses from the host.

An enzyme that hydrolyzed ATP and ADP was isolated from plants. Traverso-Cori *et al.* (1965) purified the enzyme from potato and obtained a specific activity of 300 U/mg of protein, whereas Vara and Serrano (1981), working with chick-pea roots, obtained a specific activity of 1800 U/mg of protein. The enzyme needed Ca^{2+} or Mg^{2+} for activity and hydrolyzed other nucleotides as well (Vara and Serrano, 1981). The function of this plant enzyme, which may be E-type, is not known.

III. Solubilization and Purification

A. Solubilization

Some of the specific experimental problems presented by the ecto-ATPases during isolation are described above (Section I,C). The difficulty of solubilizing the enzyme in an active state was encountered in most studies, and in 1981, according to Banerjee (1981), only one method had been published for solubilization of the ecto-ATPase in an active state.

The procedure involved treatment (of virus particles) with 98% ethanol followed by sonication of the protein pellet at an alkaline pH, and it decreased the specific activity by 80% (Banerjee and Racker, 1977).

From the following years there are, however, five reports of solubilization without considerable loss of activity: (1) LeBel *et al.* (1980) and Laliberté *et al.* (1982) used Triton X-100, 2 mg/ml, pH 8 and (in the latter work) incubation overnight at 4°C under gentle agitation. Recovery of pig pancreas zymogen granule membrane E-type ATPase activity was 104% by this solubilization procedure; (2) Lin and Fain (1984) solubilized 60% of the E-type ATPase activity from rat liver membranes by incubation with polidocanol ($C_{12}E_9$) at a detergent-to-protein ratio of 3 to 10 for 15 min at 3°C and pH 7.4; (3) more than 75% of the chicken oviduct exo-ATPase was solubilized without significant loss of activity in several nonionic, anionic, and zwitterionic detergents, and solubilization of chicken liver E-type ATPase without loss of enzyme activity was obtained through incubation with Nonidet P-40 (NP-40), 50 mg/ml, at pH 7.4 and stirring overnight at 4°C (Strobel, 1993); (4) 80% of the E-type ATPase activity was solubilized from bovine brain synaptic membranes on addition of polidocanol, 20 mg/ml, at pH 7.4 and unspecified temperature. Incubation time was 40 min, including 30 min of agitation (Hohmann *et al.*, 1993); (5) from rat brain microsomes 75% of the E-type ATPase activity was solubilized by incubation for 30 min at 0°C and pH 7.5 in CHAPS (5 mg/ml) and in the presence of 25% glycerol (Dzhandzhugazyan and Bock, 1993).

It is not clear why in these studies the solubilization did not present problems; for example, Lambert and Christophe (1978) also worked with pancreas and, like LeBel *et al.* (1980) and Laliberté *et al.* (1982), they used Triton X-100 for solubilization. But at 0.2 mg/ml they lost half of the activity, and at 1 and 3 mg/ml only 65 and 9%, respectively, of the remaining activity was solubilized. It may be a matter of species differences, the rat enzyme (Lambert and Christophe, 1978) being particularly difficult to solubilize. Other examples may support this explanation; whereas Hohmann *et al.* (1993) solubilized 80% of the bovine brain enzyme using polidocanol (20 mg/ml), Dzhandzhugazyan and Bock (1993) found that this detergent at 10 mg/ml decreased the activity of rat brain enzyme by 75%; and while the liver enzyme from chicken could be solubilized without loss of activity (Strobel, 1993), 40% was lost solubilizing the liver enzyme from rat (Lin and Fain, 1984).

Several investigators reported the inhibition of T-tubule ecto-ATPase by low concentrations of detergents (Malouf and Meissner, 1979; Beeler *et al.*, 1983, 1985; Hidalgo *et al.*, 1983; Okamoto *et al.*, 1985; Saborido *et al.*, 1991a), but in 1983 Hidalgo and co-workers found that egg yolk lysophosphatidylcholine (LPC) did not inhibit the activity, and solubilization of 50% of rabbit T-tubule E-type ATPase activity was obtained by extracting a membrane preparation twice with LPC, 0.05 mg/ml, for 5 min at room temperature and pH 7. This method was adopted by Horgan and Kuypers (1988) with identical results. Kirley (1988, 1991) and Treuheit *et al.* (1992) modified the procedure slightly by using higher concentrations of LPC or digitonin, 10 mg/ml, whereas Valente *et al.* (1990) extracted "most of the total activity" with a combination of LPC (1.8 mg/ml) plus digitonin (5 mg/ml), incubating for 30 min at room temperature, and pH 7.5. Like Hidalgo *et al.* (1983), the latter three groups worked with rabbit T-tubule enzyme.

Tuana and Dhalla (1988) solubilized 25% of the E-type ATPase activity present in rat heart sarcolemma by sonication at room temperature in the presence of 1% Triton X-100 at pH 7.4. Prior treatment of the membranes with trypsin was found to be essential even for this low degree of solubilization of the enzyme. In a later study (Zhao and Dhalla, 1991) the combination of LPC (5 mg/ml), CHAPS (5 mg/ml), and the chaotropic salt NaI (0.6 *M*) was used to extract 40% of rat heart plasma membrane E-type ATPase activity in 10 min at room temperature.

Yagi *et al.* (1989) solubilized 40% of the E-type ATPase activity present in bovine aorta microsomes by a procedure involving incubation at pH 10 with Triton X-100 (0.25 mg/ml) for 2 hr at 4°C, followed by centrifugation and extraction of the pellet with Triton X-100 (5 mg/ml) and 10% ethylene glycol at pH 7.4, 4°C and gentle agitation overnight. From human umbilical vessels 20% of the E-type ATPase activity was solubilized from microsomes in a solution consisting of Tween 20 (10 mg/ml) and 10% glycerol by stirring for 24 hr at pH 9 (temperature not specified) Yagi *et al.*, 1992).

By incubating factor-treated human hepatoma Li-7A cells for 30 min at 0°C and pH 7 in the presence of *n*-dodecylmaltoside (5 mg/ml), Gao and Knowles (1993) solubilized 24% of the mercurial-insensitive (see Section II,G) ecto-ATPase activity. Digitonin was found to be the only detergent that did not inactivate the mercurial-sensitive (see Section II,G) ecto-ATPase activity of human small cell lung carcinoma and approximately 20% of the plasma membrane ecto-ATPase was extracted with this detergent (Shi and Knowles, 1994).

Saponin and digitonin both increase the enzyme activity (see Section V,B,5).

B. Purification

The following procedures have been used: density gradient centrifugation (Hidalgo *et al.*, 1983; Zhao and Dhalla, 1991; Hohmann *et al.*, 1993), anion-exchange chromatography (Lin and Fain, 1984; Miura et al., 1987; Yagi *et al.,* 1989, 1992; Valente *et al.,* 1990; Hohmann *et al.,* 1993), concanavalin A (Con A)–Sepharose affinity chromatography (Lin and Fain, 1984; Yagi *et al.,* 1992), 5′-AMP–Sepharose 4B (Laliberté *et al.,* 1982; Yagi *et al.,* 1989, 1992), Affi-Gel blue (LeBel *et al.,* 1980; Zhao and Dhalla, 1991), blue dextran-conjugated Sepharose 4B (Miura *et al.,* 1987), and Sephacryl S-300 (Hohmann *et al.,* 1993).

The specific activity obtained by the published purifications are given in units per milligram of protein, followed by the tissue and the reference in parentheses: 60 (avian myeloblastosis virus, Banerjee and Racker, 1977), 13 (pig pancreas, LeBel *et al.,* 1980), 70 (pig pancreas, Laliberté *et al.,* 1982), 17 (rabbit skeletal muscle, Hidalgo *et al.,* 1983), 20 (rat liver, Lin and Fain, 1984), 45 (bovine aorta, Miura *et al.,* 1987), 6 (rat heart, Tuana and Dhalla, 1988), 58 (bovine aorta, Yagi *et al.,* 1989), 80 (rat heart, Zhao and Dhalla, 1991), 37 (human umbilical vessels, Yagi *et al.,* 1992), and 10 (bovine brain, Hohmann *et al.,* 1993).

The most successful purification of any E-type or ecto-ATPase was performed on the rabbit T-tubule enzyme: in a first attempt Kirley (1988) used a combination of selective solubilization with lysolecithin and digitonin followed by lectin affinity chromatography, ion-exchange chromatography, and native gel electrophoresis, and a specific activity of 100 U/mg of protein was obtained. Changing the solubilization procedure (Kirley, 1991) and replacing the final step of native gel electrophoresis by high-performance liquid chromatography (HPLC) size-exclusion chromatography resulted in a further 10-fold purification. Gel electrophoresis revealed the presence of four bands with apparent molecular masses of 160, 100, 70, and 43 kDa. The 160-kDa band was identified as angiotensin-converting enzyme. The 70- and 43-kDa bands were broad and difficult to detect without deglycosylation (Kirley, 1991). In a third study (Treuheit *et al.,* 1992) the enzyme was purified to a specific activity of close to 7000 U/mg of protein: following solubilization with either lysolecithin or digitonin, lectin affinity and ion-exchange chromatography at pH 7.1 were performed. The latter step was repeated at pH 9.1, and repeated again at pH 9.1 after treatment with neuraminidase. As most of the net charge was due to the presence of sialic acid on the associated glycan chains, this treatment changed the binding of the enzyme dramatically, allowing another fivefold purification, and in this step a specific activity of close to 7000 U/mg was achieved. The preparation was concentrated and run on a native gel where the band with enzyme activity was located, electroeluted, and submitted to SDS gradient gel electrophoresis. This showed a single diffuse band, typical of glycosylated proteins, at approximately 67 kDa, which was reduced to a single tight band at 52 kDa on further deglycosylation.

The purification of chicken gizzard E-type ATPase by a simple procedure has been reported (Stout and Kirley, 1994b). The purification involves three anion-exchange and two lectin affinity chromatography steps. The specific activity of the final preparation was 90 and 1700 U/mg of protein, measured in the absence and in the presence of concanavalin A, respectively.

Strobel (1993) obtained comparable high specific activities by immunoaffinity purification of E-type ATPase from chicken liver (1242 U/mg of protein) and oviduct (806 U/mg of protein); see Section I,C.

IV. Protein Structure

A. Primary Structure

The amino acid sequence of rat liver ecto-ATPase/pp120/HA4/C-CAM/bile acid transporter was deduced from a cDNA clone that contained the sequence of a tryptic peptide isolated from a rat liver ecto-ATPase separation (Lin and Guidotti, 1989). The amino acid sequence of NCAM is also known (Lipkin *et al.*, 1989, 1992; Premont, 1991). The possible ecto-ATPase activity of these (and other) cell adhesion molecules is discussed in Section I,B, where the as yet known isoforms are mentioned as well.

From purified ecto-ATPases the following sequences were published: rabbit T-tubule ecto-ATPase N terminal (Treuheit *et al.*, 1992) and chicken gizzard ecto-ATPase N terminal (Stout and Kirley, 1994b):

NH_2-Ala-Lys-Lys-Val-Leu-Pro-Leu-Leu-Leu-Pro- Pro- Leu-Val-Pro-Ala- (Treuheit *et al.*, 1992)

NH_2-Ala-Arg-Arg-Ala-Ala- Ala-Val- Leu-Leu-Leu-Leu-Ala- (Stout and Kirley, 1994b).

The homology between these two sequences indicates that they may be isoforms of the same protein. In contrast, the amino acid sequence of rat liver ecto-ATPase/pp120/HA4/C-CAM/bile acid transporter had a blocked N terminal (Lin and Guidotti, 1989), and it shared no homology with the N terminal or a peptide obtained from the chicken gizzard enzyme by proteolysis (Stout and Kirley, 1994a). The purified peptide had the following sequence: NH_2-(Lys)-Ile-Leu-Ser-Gly-Glu-Glu-Glu-Gly-Val-Phe-Gly-X-Ile-Thr-Ala-Asn-Tyr-Leu-Leu-Glu-Asn-X-X-(Lys) [(Lys) inferred from the known specificity of the protease) (Stout and Kirley, 1994a). (This sequence was used to synthesize a multiple antigenic peptide and generate antisera in rabbits; see the next section IV,B).

Yazaki *et al.* (1992) showed that Fab fragments, prepared from polyclonal antibodies raised against the partly purified chicken skeletal muscle ecto-ATPase, specifically inhibited the enzyme activity, and Cunningham *et al.* (1993) used this antibody to identify an 85-kDa protein in chicken skeletal muscle, gizzard, and brain (see the next section). The N-terminal amino acid sequence of protein eluted from the 85-kDa bands of the three tissues were identical: Ala-Ile-Leu-Ala-Thr-Pro-Ile-Leu-Ile-Pro-Glu-Asn-Gln-Arg-Pro-. This sequence shares 100% homology with the cell adhesion molecule, T-cadherin (Cunningham *et al.,* 1993), but the N-terminal sequence determined by Stout and Kirley (1994b) for the chicken gizzard enzyme is different (see above). The latter authors presented evidence against identity between T-cadherin and the gizzard ecto-ATPase (Stout and Kirley, 1994a,b) (see the next section IV,B).

The nature of the sugar moiety has been studied using different lectins (Beeler *et al.*, 1983; Moulton et al., 1986; Yazaki *et al.*, 1992; Cunningham *et al.*, 1993) and different deglycosylation procedures (Kirley, 1991; Yazaki *et al.*, 1992).

The possibility that the ecto-ATPase might be GPI anchored was investigated in several studies, and evidence was presented against this hypothesis (Torres *et al.*, 1990; Strobel, 1993; Hohmann *et al.*, 1993).

B. Molecular Weight

The molecular weight of the "native complex" was estimated to be high by various methods (Banerjee and Racker, 1977; Rosenberg *et al.*, 1977; Valente *et al.*, 1990; Yazaki *et al.*, 1992; Cunningham *et al.*, 1993; Smith *et al.*, 1980). The nucleoside phosphate-hydrolyzing activity of the native complex is readily demonstrated following electrophoresis of a solubilized and purified preparation under nondenaturing conditions. But when the individual constituents of the native complex are separated from each other on a SDS gel, the ATPase activity is lost.

The problem of identification of the catalytic protein was approached using affinity labeling. Martin and Senior (1980) used ^{14}C-labeled *p*-fluorosulfonylbenzoyl-5′-adenosine (FSBA) binding (see Section V,B,2) to identify a 35-kDa band in a rat pancreas membrane preparation with a specific activity of 30 U/mg of protein. Photoaffinity labeling of intact NK3.3 cells with 8-azido-[α-^{32}P]ATP demonstrated an ATP-binding protein of 68–80 kDa (Dombrowski *et al.,* 1993), and on the basis of both photoaffinity labeling with 8-azido-[^{32}P]ATP or -GTP and immunostaining, a 50-kDa band was suggested to be the catalytic subunit of synaptosomes from bovine brain (Hohmann *et al.*, 1993). On incubation of the intact synaptozomes with diazotized sulfanilic acid, a shift in the apparent molecular mass of this band was observed, further suggesting that the band represents the catalytic subunit of the synaptosomal ecto-ATPase. As mentioned above (Section II,C) the 8-azido derivatives of ATP and ADP were excellent substrates for the ecto-ATPase of cultured chromaffin cells (Rodriguez-Pascual *et al.*, 1993), and precautions are necessary using this probe for identification of the ecto-ATPase catalytic subunit.

E-type ATPase from chicken oviduct (806 U/mg of protein) exhibited two prominent bands with apparent molecular masses of 80 and 160 kDa under reducing conditions (Strobel, 1993). In the absence of 2-mercaptoethanol the 160-kDa band predominated, suggesting it to be a homodimer of two 80-kDa monomers linked by disulfide bonds. The E-type-ATPase from chicken liver, purified to a specific activity of 1242 U/mg of protein, likewise exhibited two prominent bands of 85 and 170 kDa (Strobel, 1993).

In the highly purified preparation of ecto-ATPase from rabbit skeletal muscle (close to 7000 U/mg of protein) the molecular weight of the catalytic subunit was determined to be 67 kDa and the digitonin-solubilized enzyme appeared to be a dimer of two identical 67-kDa subunits (Treuheit *et al.*, 1992). The same molecular weight was found for highly purified chicken gizzard ecto-ATPase (90 and 1700 U/mg of protein, measured in the absence and the presence of concanavalin A, respectively) (Stout and Kirley, 1994b).

In several chicken tissues immunoreactive bands were identified at positions between 85 and 95 kDa (Cunningham *et al.*, 1993; Stout and Kirley, 1994a) (see Section IV,A, in which the antibodies used in the two investigations are described). An explanation regarding the discrepancy between the molecular mass of approximately 90 kDa, obtained for the gizzard enzyme in these two studies on the one hand, and the molecular mass of 67 kDa observed by Stout and Kirley (1994b) on the other, was proposed by the latter authors (Stout and Kirley, 1994a). They investigated the effects of a variety of SDS buffers and heating conditions on the immunoreactivity of chicken tissue preparations and found that the apparent molecular mass of the immunoreactive band changed depending on the treatment of the sample prior to electrophoresis. Evidently, when urea was absent during this treatment, the ecto-ATPase had an apparent molecular mass of about 67 kDa, and was virtually nonreactive on Western blots toward the anti-peptide antibody used in this study. The immunoreactive bands were observed when the enzyme preparations were boiled in the presence of 8 *M* urea, and the fact that the anti-peptide antibody under these conditions yielded similar Western blot results to the inhibitory antibody described by Cunningham *et al.* (1993) was interpreted as strongly suggesting that the two antibodies recognize the same ecto-ATPase protein (Stout and Kirley, 1994a). Stout and Kirley (1994a) also suggest that T-cadherin is a more abundant protein with the same electrophoretic mobility under the conditions used by Cunningham *et al.* (1993).

The molecular mass for the functional enzyme was estimated by irradiation–inactivation curves to be 70 kDa for the bovine lung enzyme (Picher *et al.*, 1993), 132 kDa for the pig pancreas, and 189 kDa for the bovine aorta enzyme (Côté *et al.*, 1991).

V. Kinetics

Detailed information regarding substrate structure requirements was provided by the investigations of Cusack *et al.* (1983); Pearson *et al.* (1985); Pearson (1986); Welford *et al.* (1986, 1987); Dombrowski et al., (1993); and Beukers *et al.* (1993).

Lack of vanadate sensitivity and P_i product inhibition implies the absence of a phosphorylated intermediate in the catalytic process, and an unsuccessful search for a phosphorylated intermediate has been conducted by several laboratories (Martin and Senior, 1980; Hidalgo *et al.*, 1983; Paz *et al.*, 1988; Valente *et al.*, 1990). Accordingly, partial reactions were not catalyzed by the enzyme, that is, neither P_i-ATP, ADP–ATP, nor P_i–HOH exchange, and the reaction appears to be "irreversible" at all stages (Norton *et al.*, 1986; Sabbadini and Dahms, 1989).

Additional kinetic results will be referred to either of two groups: those relevant to optimal assay conditions are discussed in Section V,A, whereas those that may serve to delimitate subfamilies of ecto-ATPases are discussed in Section V,B.

A. Assay

1. Assay of Ecto-ATPase Activity

To establish that an enzyme is indeed an ectoenzyme, six criteria may be used. The list is partly based on the work of DePicrre and Karnovsky (1974a,b) and given in the review by Pearson (1985): (1) the enzyme acts on extracellular substrate; (2) cellular integrity is maintained; (3) the products are released extracellularly; (4) the enzyme activity is not released to the extracellular medium; (5) the enzyme structure or activity can be modified by nonpenetrating reagents; and (6) no additional enzyme activity is found when cells are broken. This last criterion cannot be met by an ecto-ATPase unless specific inhibitors of intracellular ATPases are included in the assay, and even then intracellular activity may represent ectoenzyme that is in the process of transport to its functional site at the plasma membrane.

Ecto-ATPase activity has been measured in isolated cells, cell cultures, and tissues, and the activity was found to meet these criteria. As the substrate is added to, and the product formation is measured in, the extracellular medium, the most important (first four) criteria are met if a sample containing no cells is drawn at the end of the incubation, and it is shown that hydrolysis does not continue in this sample under incubation conditions. This is sufficient as ATP and ADP do not cross the cell membrane to be hydrolyzed in the cells (Henderson and Paterson, 1973).

For an isolated enzyme it is possible to derive optimal assay conditions from the analysis of kinetic results obtained in experiments in which substrate and ligand concentrations are varied. This is not possible when the enzyme is situated as an integral part of a functioning living cell. Among the multitude of cell membrane channels, some may react and change the physiological state of the cell, if, for example, extracellular cations or anions are varied. Furthermore, as virtually all cells have purinoceptors and react to the addition of extracellular nucleotides, steady state cannot be assumed, not even if the time course of extracellular nucleoside phosphate hydrolysis appears to be linear.

Against this background many researchers chose to assay the ecto-ATPase activity by adding substrate to the cells, maintained under the conditions that were used for cell culture (e.g., DePierre and Karnovsky, 1974a,b; Pearson *et al.,* 1980, and subsequent work by the latter authors; see Section II,A,5). If the buffer contains phosphate it does not inhibit the enzyme activity (Pearson *et al.,* 1980), but phosphate must be replaced if the color assay of P_i release is used. Albumin or fetal calf serum should not be present in the assay as they bind nucleoside phosphates and the available substrate concentration is difficult to calculate in the presence of these proteins.

For the reasons mentioned above it is often impossible to define "optimal" substrate concentrations for the measurement of V_{max}, but relevant information should be obtained, if the hydrolysis rate is measured at substrate concentrations in the physiological range. In most cases this will imply the use of radiolabeled nucleoside phosphates.

The activity of the following enzymes (some with established ectoorientation) may influence the apparent ATP or NTP and ADP or NDP hydrolysis rates, depending on the assay method, that is, whether it measures the rate of P_i or $^{32}P_i$ release, the rate of ADP/NDP production, or the rate of AMP/NMP production:

1. Nucleoside-diphosphate kinase (EC 2.7.4.6):
   ATP + NDP↔ADP + NTP

2. Adenylate kinase (EC 2.7.4.3): ATP + AMP↔ADP + ADP
3. Nucleoside-phosphate kinase (EC 2.7.4.4):
   ATP + NMP↔ADP + NDP
4. 5′-Nucleotidase (EC 3.1.3.5): NMP↔nucleoside + P_i
5. Unspecific alkaline and acid phosphatases (EC 3.1.3.1 and 3.1.3.2, respectively)
6. ATP pyrophosphatase (EC 3.6.1.8):
   ATP + H_2O↔AMP + pyrophosphate

Enzyme 1 was present in cultured pig endothelial and smooth muscle cells (Pearson *et al.*, 1980) whereas enzyme 2 was found in rat brain synaptosomes (Nagy *et al.*, 1989). It is possible to inhibit the latter enzyme, the adenylate kinase, with Ap_5A (adenylyl-(3′-5′)-adenosine) (LeBel *et al.*, 1980). The 5′-nucleotidase (enzyme 4) was inhibited by concanavalin A (for refernces, see Pearson, 1985) and adenosine-α,β-methylenediphosphate (Pearson, 1986). The unspecific phosphatases (enzyme 5) are abundant in isolated granulocytes (DePierre and Karnovsky, 1974b; Pearson *et al.*, 1980). They may be inhibited by levamisole or tetramisol. For exclusion of the pyrophosphatase (enzyme 6) see Battastini *et al.* (1991).

The histochemical method should also be mentioned as a valuable assay of ecto-ATPase activity on intact cells (Wachstein and Meisel, 1957). In the preceding pages there are numerous references to authors who have worked with this method.

2. Assay of E-Type ATPase Activity

If an enzyme is part of a "functional complex" the problems associated with kinetic investigations are almost the same as those discussed above for an enzyme that is part of a functioning cell: substrates and ligands may affect the hydrolytic rate not only by interaction with the catalytic protein, but also by influencing the interaction between the components of the complex, for example, their association/dissociation state or their possible phosphorylation of each other. In Section II the specific activity of the individual tissues was quoted or calculated to the extent that it was possible. In contrast, no effort is made here to quote, calculate, or compare $K_{1/2}$ values. The $K_{1/2}$ indicates the substrate concentration at which a half-maximal hydrolysis rate is obtained, under the specific conditions of the experiment. Under each set of conditions another value is measured, depending on the concentrations of (known and unknown) cofactors and ligands. But many authors showed the influence of various ligands on the substrate activation curves, and these results provide some rules of thumb or patterns that are relevant to a kinetic deduction of optimal assay conditions.

One rule or pattern is illustrated most clearly by S. C. Lin and Way (1984) and Grondal and Zimmermann (1986): the higher the concentration of divalent cation the lower the apparent affinity for ATP, and the higher the ATP concentration the lower the apparent affinity for divalent cations. In addition, ATP in excess of divalent cation is inhibitory (Wallach and Ullrey, 1962; Hidalgo *et al.*, 1983; S. C. Lin and Way, 1984; Nagy *et al.*, 1986; Grondal and Zimmermann, 1986; Van Erum *et al.*, 1988; Homann *et al.*, 1993). Inhibition by divalent cations in excess of ATP was also reported (LeBel *et al.*, 1980; Knowles *et al.*, 1983; Yagi *et al.*, 1989; Papamarcaki and Tsolas, 1990; Battastini *et al.*, 1991), but this latter effect was not visible under all assay conditions (Carraway *et al.*, 1980; Hidalgo *et al,* 1983; Nagy *et al.*, 1986; S. C. Lin and Way, 1984; Grondal and Zimmermann, 1986; Van Erum *et al.*, 1988). Finally, millimolar concentrations of Mg^{2+} decreased the hydrolysis rate at saturating Ca^{2+} (Knowles *et al.*, 1983; Lin, 1985b; Sun *et al.*, 1990; Plesner *et al.*, 1991; Vasconcelos *et al.*, 1993).

According to these rules an assay containing millimolar and equimolar concentrations of ATP and either Ca^{2+} or Mg^{2+} will provide the maximal hydrolysis rate, and these are precisely the assay conditions most widely used.

Moreover the rules predict that possibly more sensitive assay conditions could be found if concentrations of substrate and divalent cations were both reduced, and results in accordance with this prediction were reported by LeBel *et al.*, (1980), Laliberté *et al.* (1982), and Gandhi and Ross (1988).

With the limited purpose of deriving a more sensitive and specific assay for E-type ATPase activity we investigated the influence of substrate and ligands on the activity of enzyme isolated from rat mesenteric small arteries (Plesner *et al.*, 1991). It was found that in the presence of only endogenous divalent cations, increasing concentrations of monovalent cations caused an increase in the ATP hydrolysis rate, with a maximum between 10 and 20 m*M* salts at all substrate concentrations (1–30 μM). Further increasing the concentration of monovalent cation salts decreased the activity. The activation caused by low (millimolar) concentrations of monovalent cations was equilibrium ordered and identical to the pattern obtained with micromolar concentrations of divalent cations in the absence of monovalent cation salts, indicating that the cation activator must bind to the enzyme prior to the binding of substrate and that it is trapped on the enzyme until the reaction cycle is complete. The inhibition obtained with monovalent cation concentrations above 20 m*M* was competitive. Identical results, qualitatively and quantitatively, were found with Na^+, K^+, choline$^+$, Rb^+, and Cs^+, indicating that the E-type ATPase activity lacks specificity not only for substrate and divalent cations, but also for monovalent cations. At 1 m*M* Mg^{2+} the monovalent cation salts inhibit the activity at micromolar substrate concentrations, whereas they do not affect the activity in the presence of millimolar concentrations of both Mg^{2+} and ATP. Insensitivity to Na^+ and K^+ under the latter conditions has been reported by several authors (e.g., Hamlyn and Senior, 1983; Teo *et al.*, 1988; Zhao and Dhalla, 1988; Gandhi and Ross, 1988).

By varying substrate (ATP), divalent cations (Ca^{2+} and Mg^{2+}), and monovalent cation (Na^+) against each other, optimal ligand concentrations were determined, that is, optimal assay conditions were derived [^{32}P]-ATP (50 μM), CaCl (50μM), MgCl (2μM), and NaCl (20mM) (Plesner *et al.*, 1991). This assay can measure 10^{-6} U, which is probably three orders of magnitude less than the amount measured in the colorimetric assay used by most authors. As inhibitors of P-type ATPase (vanadate, 50 μM), F-type ATPase (oligomycin, 6 μM) and V-type ATPase (NEM, 0.5 mM) were included in the assay it is also specific for the E-type activity. In comparing the hydrolysis rate measured in this assay to the hydrolysis rate obtained under assay conditions more commonly used (5 mM ATP and 5 mM Ca^{2+}), the same value was obtained, and its advantage therefore is only its sensitivity.

In some cases it is necessary to exclude the interference of the enzymes mentioned in Section V, A, 1.

B. Delimitation of Subfamilies

Beaudoin and co-workers distinguished three types of E-type ATPases: type I was found in an enzyme preparation from pig pancreas, whereas types II and III were present in the microsome fraction from bovine aorta and bovine lung, respectively (LeBel *et al.*, 1980; Côté *et al.*, 1991, 1992b; Picher *et al.*, 1993). The delimitation was based on heat denaturation curves, ^{60}Co γ irradiation–inactivation curves, and electrophoretic migration patterns under nondenaturing conditions. But in each of the enzyme preparations investigated more than one E-type ATPase would be present, and therefore the measured parameters may not be sufficient for the definition of distinct types.

Vascular endothelial and smooth muscle cells have high ecto-ATPase activities (see Section II,A,6). As the properties of the enzyme for the two cell types seem indistinguishable (Yagi *et al.*, 1991; Côté *et al.*, 1992a), the term vascular ecto-ATPase may be used to designate the subfamily of ecto-ATPases in blood vessels. The vascular ecto-ATPase is found in all tissues, and apparently the T-tubule subfamily of ecto-ATPases is almost as widespread (Beeler *et al.*, 1983, 1985; Cunningham *et al.*, 1993; Stout and Kirley, 1994a). These two subfamilies differ with respect to substrate specificity (ADP is not hydrolyzed by the latter), divalent cation preference (the former is more active with Ca^{2+}, the latter with Mg^{2+},), and probably in several other ways. Therefore, with these two subfamilies present in a tissue, it is difficult to distinguish a possible tissue-specific subfamily and describe its properties.

1. ADP (NDP) as Substrate

In an attempt to define specific characteristics of ecto-ATPases/E-type ATPases (Section I,A) it was stated that they hydrolyze nucleoside tri- and/or diphosphates. This wording was used because it is possible that a) some of these enzymes hydrolyse NTP only, b) some hydrolyze NDP only, and c) some hydrolyze both NTP and NDP. (The latter group of enzymes is often referred to as "apyrases.")

a) As mentioned, the T-tubule ecto-ATPase subfamily probably hydrolyzes only NTP (Sabbadini and Dahms, 1989; Treuheit *et al.*, 1992).

b) One group of researchers (see Section II,A,4) presented studies on inhibition and substrate specificity, indicating that ATP and ADP were hydrolyzed by separate enzymes (Pearson *et al.*, 1980, 1985; Cusack *et al.*, 1983; Pearson and Gordon 1985; Gordon *et al.*, 1986, 1989; Pearson, 1986), for example, ADP catabolism by endothelial cells was inhibited by a wide range of ATP analogs, whereas ATP catabolism was much less affected. L-ATP was catabolized to L-ADP moderately well, whereas L-ADP hydrolysis was not detectable, and different tolerance with respect to L-sugar substitution was also observed in inhibition studies of ATP and ADP hydrolysis by the cells (Pearson, 1986).

It is a matter of controversy whether ATP and ADP are hydrolyzed by separate enzymes in vascular cells, see below. But ADP was hydrolyzed at a higher rate than ATP by the E-type activity of placenta (Kelley *et al.*, 1990; Papamarcaki and Tsolas, 1990), possibly indicating that placenta may contain an E-type ADPase.

c) Most investigators working with tissues other than skeletal muscle found that their data were best explained assuming that nucleoside tri- and diphosphates were hydrolyzed by the same catalytic protein (Harper *et al.*, 1978; Culić *et al.*, 1990; Papamarcaki and Tsolas 1990; Battastini *et al.*, 1991; Yagi *et al.*, 1992). Several studies concentrated on the question, and the authors reached the same conclusion, based on kinetic results (Laliberté and Beaudoin, 1983), radiation inactivation, and heat inactivation of ATPase and ADPase activities (Côté *et al.*, 1991; Picher *et al.*, 1993), half-life measurements of ATPase and ADPase activities (Rodriguez-Pasqual *et al.*, 1992), and the ratio between ATPase and ADPase activity during B cell maturation (Barankiewicz *et al.*, 1988). It should be mentioned also that the highly purified chicken liver and oviduct enzymes (see Sections I,C; II,E,1; III,B) catalyzed the hydrolysis of both ATP and ADP (Strobel, 1993).

Thus, the evidence is strong that ATP and ADP hydrolysis may be catalyzed by the same catalytic protein, and according to several reports (Miura *et al.*, 1987; Yagi *et al.*, 1989, 1991, 1992; Côté *et al.*, 1991, 1992a,b), such a catalytic protein is also present in vascular tissue.

To distinguish the E-type ATPase that hydrolyze both NTP and NDP (the apyrase) from the enzymes that hydrolyze only NTP or NDP, Dhalla and Zhao (1988) defined three characteristics: (1) the ability to hydrolyze NTP and NDP at comparable rates, (2) inhibition (more than 50%) by 10–20 m*M* of azide, and (3) lack of stimulation by concanavalin A. There are however problems with each of these characteristics.

1. The ADP hydrolysis rate was compared to the ATP hydrolysis rate and given as percentage of the latter (V_{ADP}/V_{ATP}) in many studies, but the values obtained in the individual tissues showed large variations, for example, in pancreas ranging from 60 to 90% (LeBel *et al.*, 1980) to 10–30% (Martin and Senior, 1980; Hamlyn and Senior, 1983) and in nervous tissue from 70 to 80% (Grondahl and Zimmermann, 1986; Torres *et al.*, 1990; Sarkis and Salto, 1991) to 20–50% (Nagy *et al.*, 1986; Gandhi and Ross, 1988; Battastini *et al.*, 1991; Hohmann *et al.*, 1993).

2. Knowles *et al.* (1983) found a positive correlation between V_{ADP}/V_{ATP} and inhibition by azide (10 m*M*) studying plasma membranes from several normal and tumor tissues. In many cases, it was seen that the ADP hydrolysis was inhibited slightly more by azide than was the ATP-hydrolysis, and in two studies the difference was substantial (Knowles *et al.*, 1983; Strobel, 1993). But in most studies, the effect of azide was better correlated to the concentration applied than to the capacity to hydrolyze ADP. At V_{ADP}/V_{ATP} = 70–80% in nervous tissue no inhibition was observed at 5 m*M* azide (Sarkis and Salto, 1991) whereas 10 m*M* inhibited 25% (Grondal and Zimmerman, 1986). At V_{ADP}/V_{ATP} 40% in synaptosomes (Battastini *et al.*, 1991) and 20% in heart (Zhao and Dhalla, 1988) azide inhibition was low or insignificant at 1–5 m*M* but increased to 25–30% at 10–20 m*M* of the inhibitor. In the pancreas, 10 m*M* of azide inhibited 50% (LeBel *et al.*, 1980) whereas 1 m*M* was without effect (Hamlyn and Senior, 1983). Some researchers used azide (5 m*M*) in the assay (Nagy *et al.*, 1986; Beukers *et al.*, 1993), and with azide 10 m*M* in the assay. Sun *et al.*, (1990) found V_{ADP}/V_{ATP} to be 50% in vascular smooth muscle.

3. The rat pancreas ecto-ATPase hydrolyzed ADP (V_{ADP}/V_{ATP} = 25%), yet the activity was stimulated by concanavalin A (Hamlyn and Senior, 1983), and a solubilized E-type ATPase from rat mesenteric artery which hydrolyzed ATP and ADP at equal rates was stimulated by a factor of 2 in the presence of the lectin (Plesner, unpublished observation).

Thus, E-type ATPases possibly may be divided into those that hydrolyze NTP only, those that hydrolyze NDP only, and those that hydrolyze both;

however, useful criteria to distinguish the groups have not been defined and many questions regarding ADP-hydrolysis by these enzymes remain to be answered.

In one case it was noticed that the ability to hydrolyze ADP disappeared upon solubilization (Hohmann *et al.*, 1993), but in other studies solubilization increased V_{ADP}/V_{ATP} (Dawson *et al.*, 1986) or solubilization did not affect the ratio (LeBel *et al.*, 1980; Strobel, 1993).

2. Inhibitors

Most studies contain observations on the effect of certain inhibitors (see Pearson, 1985; Dhalla and Zhao, 1988; Dahms and Sabbadini, 1989). So far no compound was found that specifically inhibits the ecto-ATPase or E-type ATPase activity, and the following therefore summarizes only those inhibitor studies that are relevant to the questions raised above.

It was claimed (Dhalla and Zhao, 1988) that azide inhibition is specific for apyrases (this question was discussed in the previous section).

Many authors, searching for the means to label the catalytic unit, encountered the problem that either the enzyme hydrolyzed the probe (see Sections I,C; II,C; and IV, B) or the probe did not inhibit the enzyme activity, for example, FITCH does not (Gandhi and Ross, 1988; Kirley, 1988; Sabbadini and Dahms, 1989; Filippini *et al.*, 1990) and FSBA does not inhibit the T-tubule enzyme from chicken (Sabbadini and Dahms, 1989) whereas it does inhibit E-type ATPases from other tissues (Martin and Senior, 1980; Filippini *et al.*, 1990a; Côté *et al.*, 1992b; Dombrowski *et al.*, 1993; Bajpaj and Brahmi, 1993).

Mercurial sensitivity was shown to be a characteristic of the ecto-ATPase in several human carcinoma lines (Shi and Knowles, 1994). Induction of a mercurial-insensitive enzyme was observed as a consequence of the synergistic actions of EGF and cyclic AMP (Knowles *et al.*, 1985; 1988a) (see Section II,G). The mercurial compound used in these studies was pCMPS which inhibited the mercurial-sensitive ecto-ATPase at concentrations below 0.2 m*M*.

One paper reported that following treatment with the anion transport inhibitor (DIDS, 1 m*M*) the ATPase activity of the bile acid transporter was completely abolished (Sippel *et al.*, 1993). Knowles (1988a) found 40% inhibition with 0.1 m*M* DIDS but Strobel (1993) found no effect with 1 m*M* DIDS on either ATPase or ADPase activity of the immunopurified chicken liver and oviduct enzymes. Likewise SITS 0.1 m*M* (Hamlyn and Senior, 1983) and pyridoxal phosphate anion channel inhibitors (Martin and Senior, 1980) were without effect.

Diethylpyrocarbonate (DEPC) was an effective inhibitor, and DEPC inactivation studies indicate that a histidine residue is essential for the catalytic activity and possibly a tyrosine residue as well (Kirley, 1988; Saborido *et al.*, 1991a).

Several investigators reported that chlorpromazine and trifluoperazine inhibit ecto-ATPase (Medzihradsky *et al.*, 1980) and E-type ATPase activity (Nagy *et al.*, 1986). The inhibition is not specific, however, as equivalent concentrations inhibit the Na^+, K^+-ATPase (L. Plesner, unpublished observation). As suggested by Nagy *et al.* (1986), the inhibition may be due to a nonspecific membrane-desintegrating effect of these highly hydrophobic phenothiazines.

Phorbol esters and diacylglycerol inhibit the T-tubule ecto-ATPase, and promising results (Okamoto *et al.*, 1985; Moulton *et al.*, 1986; Sabbadini and Dahms, 1989) suggesting that the (T-tubule) ecto-ATPase may be an alternative receptor for these lipid second messengers are discussed in Section VI.

Strobel (1993) found that ADP hydrolysis was effectively inhibited by vanadate, but this was not seen in other studies (Sarkis and Salto, 1991; Côté *et al.*, 1992b).

3. Activators

The stimulating effect of concanavalin A on ecto-ATPase and E-type ATPase activities has been mentioned several times above, see also Section V,B,5.

E-type ATPase from rat liver (Emmelot and Bos, 1966) was activated by saponin, and the T-tubule enzyme was also stimulated by another glycoside detergent, digitonin (Saborido *et al.*, 1991a; Cunningham *et al.*, 1993). The effects resemble the effects of lectins, that is, saponin and digitonin activated by normalizing time courses, substrate curves, and temperature dependence of chicken T-tubule and brain E-type ATPase activity (Cunningham *et al.*, 1993). Moreover, activation by concanavalin A was decreased or abolished in the presence of saponin and digitonin (Saborido *et al.*, 1991a).

The activation by concanavalin A, digitonin, and saponin (as well as solubilization with lysolecithin) decreases the substrate affinity (Hidalgo *et al.*, 1983; Beeler *et al.*, 1983; Hamlyn and Senior, 1983; Moulton *et al.*, 1986), that is, the presence of these agents elicits a high-activity/low-affinity form of the enzyme. (Regarding species differences, see Section V,B,5).

It was claimed (Dhalla and Zhao, 1988) that apyrases may not be stimulated by concanavalin A. This question is discussed in Section V,B,1.

Several fatty acyl-CoAs were found to act as potent activators of the chicken T-tubule enzyme (Moulton *et al.*, 1986), see Section VII.

4. Ca^{2+} and Mg^{2+} Activation

The skeletal muscle E-type ATPase has been called the T-tubule Mg^{2+}-ATPase, possibly to distinguish it from the Ca^{2+} pump of the sarcoplasmic reticulum and the Ca^{2+} pump of the plasma membrane. In the early studies on this enzyme, there were conflicting reports on divalent cation preference (Malouf and Meissner, 1979; Hidalgo *et al.*, 1983; Beeler *et al.*, 1985). But summarizing more recent work, Sabbadini and Dahms (1989) were able to conclude that Ca^{2+} is merely acting as a Mg^{2+} congener. Mg^{2+} is the divalent cation routinely used in the assay of this enzyme (Hidalgo *et al.*, 1986; Valente *et al.*, 1990; Saborido *et al.*, 1991a; Treuheit *et al.*, 1992; Cunningham *et al.*, 1993), further supporting that the T-tubule ecto-ATPase subfamily is more active with Mg^{2+} than with Ca^{2+}.

The opposite was found for the vascular E-type ATPase subfamily, characterized in bovine aorta (Miura *et al.*, 1987; Yagi *et al.*, 1989), human umbilical vessels (Yagi *et al.*, 1992), pig vascular smooth muscle cells (Sun *et al.*, 1990), and rat mesenteric small arteries (Plesner *et al.*, 1991).

The presence of both of these subfamilies may explain the fact that convincing data demonstrating a distinct divalent cation preference are not available from any other tissue. Either the maximal hydrolysis rate obtained with the two divalent cations was the same as reported for heart (Tuana and Dhalla, 1988; Zhao and Dhalla, 1991), lymphocytes (Filippini *et al.*, 1990), lung (Picher *et al.*, 1993), brain synaptosomes (Nagy *et al.*, 1986; Grondal and Zimmermann, 1986; Battastini *et al.*, 1991), and pancreas (Martin and Senior, 1980; LeBel *et al.*, 1980), or the results conflicted as was the case in reports on the kidney (Møortl *et al.*, 1984; Van Erum *et al.*, 1988; Culic *et al.*, 1990; Sabolic *et al.*, 1992).

Some authors found that the enzyme was more sensitive toward divalent ions when hydrolyzing ADP than it was when ATP was the substrate (Yagi *et al.*, 1989; LeBel *et al.*, 1980), but the opposite relationship was also reported (Côté *et al.*, 1992b).

An observation by Hohmann *et al.*, (1993) indicated that important questions regarding divalent ion preference have not yet been answered. Before solubilization of the ecto-ATPase of bovine brain synaptosomes, a three-fold higher rate of hydrolysis was measured with Mg^{2+} than with Ca^{2+}, but following solubilization of the enzyme, the ratio (V_{Mg}/V_{Ca}) decreased to close to 1. This could be explained if the procedure failed to solubilize the T-tubule subfamily E-type ATPase activity of the brain, but contradicting this explanation it was found that the ability to hydrolyze ADP was also lost upon solubilization (Hohmann *et al.*, 1993).

5. Substrate Inhibition, Species Differences

ATP in excess of divalent cations is inhibitory (V,A,1), and this phenomenon is probably tissue unspecific. In contrast, some other patterns of substrate inactivation seem to be species dependent.

a. Rat Tissues. Several authors working with rat tissues observed nonlinear time courses that were modulated by nucleotides, lectins, detergents, and cross-linkers. The time course of rat pancreas plasma membrane E-type ATPase activity was nonlinear, but fell off over 15 min to a rate one-third the initial rate (Martin and Senior, 1980). Nonlinear time courses were also observed when the ecto-ATPase activity of intact pancreas cells was measured (Hamlyn and Senior, 1983). With both preparations, the inactivation was increased at an increasing temperature and abolished in the presence of concanavalin A.

In a preparation of rat skeletal muscle low-density vesicles the rate of ATP hydrolysis decayed exponentially, and the rapid decline depended on the presence of ATP or its nonhydrolyzable analog AdoPP(NH)P (Beeler *et al.*, 1983). In the absence of substrate or analog, the enzyme activity was stable, and neither P_i, ADP, nor protein concentration influenced the ATP-dependent inactivation. Once inactivated, removal of ATP from the medium did not immediately restore the original activity. ATP or AdoPP(NH)P-dependent inactivation could be blocked by concanavalin A, wheat germ agglutinin, or rabbit antiserum against the membrane, if added before ATP. Addition of the lectins after ATP addition prevented further inactivation, but did not restore the original activity. Low concentrations of ionic and nonionic detergents increased the rate of ATP-dependent inactivation. Higher concentrations of detergents, which solubilize the membrane completely, inactivated the enzyme activity. Cross-linking the membrane components with glutaraldehyde prevented ATP-dependent inactivation and decreased the sensitivity to detergents (Beeler *et al.*, 1983). The authors propose that ATP regulation requires the mobility of proteins within the membrane. Cross-linking the membrane proteins with lectins, antiserum, or glutaraldehyde prevents inactivation, whereas increasing the mobility with detergents accelerates ATP-dependent inactivation. Of the lectins tested only two were able to prevent ATP-induced inactivation: wheat germ agglutinin, which binds *N*-acetylglucosamine, and concanavalin A, which binds *D*-mannose residues. Lectin protection disappeared on addition of the respective sugars, indicating that the lectins must bind to the membrane in order to prevent inactivation (Beeler *et al.*, 1983). In 1985, Beeler *et al.* demonstrated the presence in most rat tissues of E-type ATPase activity exhibiting ATP-stimulated inactivation. Zhao and Dhalla (1988) worked with rat heart and reported observations almost identical to those described by Beeler *et al.* (1983), and "Mg^{2+}-induced ATP hydrolysis" that exhibited concanavalin A-protected inactivation by ATP was described in rat myometrium (Missiaen *et al.*, 1988a) and other types of rat smooth muscle (Missiaen *et al.*, 1988b).

In measuring nucleoside triphosphate hydrolysis, catalyzed by segments of isolated intact rat mesenteric small arteries, the substrate NTP was shown to induce a modification of the hydrolytic activity that occurred more rapidly the higher the concentration of NTP. The modified system hydrolyzed ATP with a decreased substrate affinity (Juul *et al.*, 1991b). By comparing the characteristics of E-type ATPase isolated from the vessels (Plesner *et al.*, 1991) to the characteristics of the activity of the intact vessel, they showed strong resemblance and identical maximal velocities, but the substrate-induced modification was seen only in the intact preparation (Juul *et al.*, 1991b).

In plasma membrane vesicles prepared from pregnant rat myometrium, Magocsi and Penniston (1991) found E-type ATPase activity, but the activity decreased when the enzyme was incubated at 37°C. Half of the activity disappeared with a half-life of about 2 min. The inactivation was accelerated in the presence of ATP or ADP. The authors found indications that two E-type ATPases were present: a stable enzyme that was inhibited by azide and hydrolyzed XTP and XDP at equal rates, and a labile enzyme that was inhibited by P_i and hydrolyzed only XTP. The latter enzyme was inhibited by pCMPS. The two enzymes differed in affinity for Ca^{2+}, but substrate affinity was the same. Concanavalin A protected the labile enzyme against inactivation, but not when the ATPase activity was measured in whole myometrial cells. It was suggested that the lability of the labile enzyme was due to product inhibition, "probably by inorganic phosphate."

The existence of two E-type ATPases, with the proposed properties, does not explain the results of Beeler et al. (1983, 1985), Zhao and Dhalla (1988), or Juul *et al.* (1991b), as nonhydrolyzable analogs of ATP could induce the inactivation of the enzyme (Beeler *et al.*, 1983) that was not due to the accumulation of products (Beeler *et al.*, 1983; Zhao and Dhalla, 1988; Juul *et al.*, 1991b). In pig pancreas the ADP-hydrolyzing activity was less stable than the ATP-hydrolyzing activity (Laliberté and Beaudoin, 1983), in contrast to the properties of the two enzymes suggested by Magocsi and Penniston (1991).

The observation that ATP modification of the rat ecto-ATPase results in a low substrate affinity may explain the high K_m (10–100 mM) obtained by Meghji *et al.*, (1992), working with rat ventricular myocytes. The lack of saturability observed with the rat cells was in striking contrast to the substrate affinities measured in previous studies on pig tissues (see Section II,A,5).

Nonlinear time courses were reported for the ecto-ATPase activity of rat epididymal spermatozoa (Majumder and Biswas, 1979), and while working with isolated membranes from mammary ascites adenocarcinomas maintained in rats Carraway *et al.* (1980) found that substrate inactivation could be released by concanavalin A.

b. Chicken Tissue Only linear time courses were observed with the chicken T-tubule membrane vesicles, but a highly complex dependency on the substrate was interpreted as combined negative cooperatively and substrate inhibition (Moulton *et al.*, 1986). Concanavalin A abolished both the negative cooperatively and substrate inhibition, if added prior to the substrate. If MgATP was added prior to concanavalin A, no activation was observed, and nucleotide suppression of concanavalin A stimulation required the presence of Mg^{2+}, although hydrolysis of the nucleotide was not necessary, as the nonhydrolyzable analog AMP-PNP also worked. The immunopurified chicken oviduct and liver enzyme did not exhibit substrate inactivation (Strobel, 1993).

c. Rabbit Tissue The time courses obtained with T-tubule ecto-ATPase from rabbit were linear and neither substrate inactivation nor concanavalin A activation was observed (Hidalgo *et al.*, 1983; Horgan and Kuypers, 1988; Kirley, 1988, 1991; Valente *et al.*, 1990; Saborido *et al.*, 1991a). Lack of stimulation by concanavalin A is an inherent property of the rabbit enzyme and not due to LPC or digitonin (see Section V,B,3) used for solubilization, because the chicken gizzard enzyme was still activated by concanavalin A following solubilization with these detergents (Stout and Kirley, 1994b).

To conclude the subject of species differences, it is possible that ecto-ATPases/E-type ATPases from rat tissues are particularly difficult to solubilize, see Section III,A.

VI. Physiological Regulation and Relation to Disease

T-tubule ecto-ATPase was proposed to be an alternative receptor for phorbol esters and diacylglycerol, on the basis of results of detailed investigations (Okamoto *et al.*, 1985; Moulton *et al.*, 1986; Sabbadini and Dahms, 1989): Triton X-100 inhibited the T-tubule ecto-ATPase by 90% at sub critical micelle concentrations (sub-CMCs) and, because concanavalin A protected against the inhibition, it was suggested that Triton X-100 was operating as a substrate mimetic agent at a putative site, where substrate or analogs modulate the hydrolytic activity, that is, Triton X-100 would be operating through a mechanism unrelated to solubilization or native annular lipid replacement. The ability of an amphiphile such as Triton X-100 to mimic the effect of a nucleotide was reported by Huang *et al.* (1985), who presented evidence for Triton X-100 mimicking the effect of ATP at the low-affinity regulatory ATP site of the Na^+, K^+-ATPase.

It was considered a possibility (Moulton *et al.,* 1986) that such a site might be designed to recognize regulatory agents with combined adenine nucleotide and amphipatic properties, in particular fatty acyl-CoAs, and in fact several fatty acyl-CoAs were found to act as potent activators in the physiologically significant sub-CMC range. In contrast, several free fatty acids exerted concanavalin A-modulatable inhibition. These findings led to the evaluation of lipid second messenger effects on the enzyme activity, which was found to respond to traditional protein kinase C (PKC) activators and inhibitors, but in a completely opposite sense to PKC. Phorbol esters and diacylglycerol, prominent activators of PKC, inhibited the ecto-ATPase in submicromolar concentrations, whereas the PKC inhibitor sphingosine increased the activity. Phorbol and diacylglycerol inhibition of the ATPase appeared to be unrelated to either a direct or indirect involvement of PKC for several reasons, for example, ATPase inhibition by phorbol esters and diacylglycerol as well as phorbol binding to the ATPase were blocked by concanavalin A, which is not known to have effects on PKC activity. Attempts to phosphorylate the ATPase with exogenous PKC did not alter ATPase activity. In contrast to concanavalin A, ATP did not protect against phorbol inactivation. The ligand therefore hardly reacts with the catalytic site, but in all probability with the low-affinity concanavalin A-modulated regulatory site. It was suggested that the physiological ligand for this site is the diacylglycerol produced endogenously by a T-tubule membrane-linked phopholipase, and that the T-tubule-enzyme ecto-ATPase is an alternative receptor for phorbol esters and diacylglycerol.

The amount of ecto-ATPase on the secretory cells in human and chicken oviducts could be related to the hormonal status (Section II,E,1) (Anderson and Rosenberg, 1976).

Treadmill training was shown to increase T-tubule ecto-ATPase activity in rats, whereas treatment with anabolic/androgenic steroids increased the activity of the enzyme in a muscle type-specific way (Saborido *et al.,* 1991b).

pp120 was identified as a substrate for several receptor tyrosine kinases (Margolis *et al.,* 1988), and insulin stimulated the ecto-ATPase activity of NIH-3T3 mouse fibroblasts that were cotransfected with a cDNA encoding the insulin receptor and a cDNA encoding pp120/HA4/C-CAM/bile acid transporter (S. M. Najjar, personal communication) (see Section I,B). Apparently there are no other reports on experiments involving ecto-ATPase/E-type ATPase and insulin, but Jarett and Smith (1974) found that rat adipocyte plasma membrane Mg^{2+}-ATPase was increased on treatment with insulin.

Ecto-ATPases in vascular tissues probably play a major role in preventing platelet aggregation and subsequent thrombus formation (Section II,A). "Vigorous" ADPase activity was observed in the pulmonary circulation of several species studied, including humans (Chelliah and Bakhle, 1983). Comparing the ADP breakdown in the pulmonary circulation of lungs from rats with streptozotocin-induced diabetes to ADP breakdown in lungs from normal rats, it was found that less ADP was hydrolyzed in the diabetic lungs at high ADP concentrations (Bakhle and Chelliah, 1983). In diabetes mellitus platelet aggregation is enhanced, and the authors suggested that decreased ADP-hydrolyzing capacity of diabetic endothelium may contribute to the etiology of diabetes-associated vascular disease.

ADPase activity was demonstrated by histochemistry in the glomerular basement membrane of rat kidney, and it was suggested that the enzyme subserves an important antithrombotic role (Bakker *et al.*, 1987). Selective reduction of ADP degradation in intact kidneys (by the addition of competitive substrate, UDP, or the nondegradable analog ADP-β-S) strongly promoted intraglomerular proaggregatory condition (Poelstra *et al.*, 1991).

Trams and Lauter (1974) suggested that the ecto-ATPases may be involved in nonsynaptic information transfer (Section I,B). Rosenblatt *et al.* (1976) observed that the activity of the enzyme was decreased in seizure-prone mice. Trams and Lauter (1978b) raised and cultured glia cells from neonatal, seizure-prone mice and a significant deletion of ecto-ATPase activity was observed when compared to the enzyme activity of normal control cells. In a study by Nagy *et al.* (1990), the ecto-ATPase activity of intact synaptosomes isolated from nonepileptic and epileptic human brain samples was compared. In nonepileptic human brains the synaptosomal ecto-ATPase activities were comparable in all tested brain areas, and comparable activities were also found in different parts of normal brains from rat, mouse, gerbil, and monkey. In contrast, the ecto-ATPase activity of hippocampus samples from epileptic brains was substantially increased (in some cases by 300%) whereas the ecto-ATPase activity of the epileptic temporal cortex was decreased (by approximately 30%) as compared to the activity of the normal brain samples. Because ATP is an excitatory neurotransmitter in the central nervous system (Edwards *et al.*, 1992; Silinsky *et al.*, 1992; Evans *et al.*, 1992), and its metabolic product adenosine is known to participate in the feedback suppression of continued discharge of the presynaptic cell, these findings may concern important elements of seizure development or maintenance in human temporal lobe epilepsy.

The possible role of ecto-ATPases in growth control and cancer cell function is an important issue that was discussed in Sections I,B and II,G.

VII. Concluding Remarks

To avoid the mere listing of experimental results and ease reading, a review should be organized according to some integrating idea. The reason why this article does not meet this requirement may be found in the history and situation of ecto-ATPase research: colleagues entered the field from a variety of otherwise unrelated areas, and it is striking how many groups contributed to this line of research without knowing that they were dealing with an ecto-ATPase. It is still the situation that otherwise acquainted researchers may be unaware that they work on the same enzyme. Therefore, to ease communication, an effort was made to present the multitude of ideas and areas behind experimental results relevant to the field, thus giving up "the one integrating idea" concept.

Lack of communication was a sad characteristic of ecto-ATPase research, and the confusion about nomenclature was an important reason why researchers kept missing relevant publications.

Therefore, in the course of writing this article the question regarding systemic name and EC number was presented to the International Nomenclature Committee. The committee answered, recommending that the EC number should be 3.6.1.15, and the systemic name diphosphate phosphohydrolase. It further suggested that the "comments" to the 3.6.1.15 entry in "Enzyme Nomenclature" (IUBMB, 1992) should include the following sentences: "ectoenzymes in this class are present on the plasma membrane of most mammalian cells. They are dependent on Ca^{2+} or Mg^{2+} for activity." Additional changes in the comments may be desired. It is hoped that this issue will be discussed and consensus reached as to what the comments should contain to identify the ecto-ATPases.

Acknowledgments

The author is grateful to colleagues who provided preprints or information on work in progress, and to professor J. V. Møller and Dr. I. W. Plesner for critically reading the manuscript. I thank Pia Schytz and Dorte Abildskov for typing the references. Research in the author's laboratory was supported by "Fonden til Lægevidenskabens Fremme," "Carlsberg Fonden," and "Aarhus Universitets Forskningsfond".

References

Abood, L. G., and Gerard, R. W. (1954). Enzyme distribution in isolated particulate of rat peripheral nerve. *J. Cell. Comp. Physiol.* **43,** 379–392.

Abraham, E. H., Prat, A. G., Gerweck, L., Seneveratne, T., Arceci, R. J., Kramer, R., Guidotti, G., and Cantiello, H. F. (1993). The multidrug resistance (*mdr1*) gene product functions as an ATP channel. *Proc. Natl. Acad. Sci. U.S.A.* **90,** 312–316.

Acs, G., Ostrowski, W., and Straub, F. B. (1954). Über die Adenylpyrophosphatase-aktivität an der oberflache der Aszites-Krebszellen. *Acta Physiol. Acad. Sci. Hung.* **6,** 260–263.

Ågren, G., Pontén, J., Ronquist, G., and Westermark, B. (1971). Demonstration of an ATPase at the cell surface of intact normal and neoplastic human cells in culture. *J. Cell. Physiol.* **78,** 171–176.

Anderson, B., and Rosenberg, M. D. (1976). Variation of oviductal lumen ATPase with the ovulatory cycle of the hen. *Biol. Reprod.* **14,** 253–255.

Anderson, B., Kim, N. B., Rhea, R. P., and Rosenberg, M. D. (1974). Nucleoside triphosphate hydrolase activitics of avian vitelline membrane and oviductal lumen. *Dev. Biol.* **37,** 306–316.

Apasov, S., Redegeld, F., and Sitkovsky, M. (1993). Cell-mediated cytotoxicity: Contact and secreted factors. *Curr. Opin. Immuno.* **5,** 404–410.

Aurivillius, M., Hansen, O. C., Lazrek, M. B. S., Bock, E., and Öbrink, B. (1990). The cell adhesion molecule cell-CAM 105 is an ecto-ATPase and a member of the immunoglobulin superfamily. *FEBS Lett.* **264,** 267–269.

Bajpai, A., and Brahmi, Z. (1993). Regulation of resting and IL-2-activated human cytotoxic lymphocytes by exogenous nucleotides: Role of IL-2 and ecto-ATPases. *Cell. Immunol.* **148,** 130–143.

Bakhle, Y. S., and Chelliah, R. (1983). Effect of streptozotocin-induced Diabetes on the metabolism of ADP, AMP and adenosine in the pulmonary circulation of rat isolated lung. *Diabetologia* **24,** 455–459.

Bakker, W. W., Willink, E. J., Donga, J., Hulstaert, C. E., and Hardonk, M. J. (1987). Antithrombotic activity of glomerular adenosine diphosphatase in the glomerular basement membrane of the rat kidney. *J. Lab. Clin. Med.* **109,** 171–179.

Banerjee, R. K. (1978). Viral adenosine triphosphatase. *Experientia* **34,** 1430–1432.

Banerjee, R. K. (1979). Characterization of an adenosine triphosphatase from myeloblasts infected with the avian myeloblastosis virus. *Eur. J. Biochem.* **97,** 59–64.

Banerjee, R. K. (1981). Ecto-ATPase. *Mol. Cell. Biochem.* **37,** 91–99.

Banerjee, R. K., and Racker, E. (1977). Solubilization, purification, and characterization of a nucleoside triohosphatase from avian myeloblastosis virus. *J. Biol. Chem.* **252,** 6700–6706.

Barankiewicz, J., Dosch, H. -M., and Cohen, A. (1988). Extracellular Nucleotide catabolism in human B and T lymphocytes. *J. Biol. Chem.* **263,** 7094–7098.

Bartles, J. R., Feracci, H. M., Stieger, B., and Hubbard, A. L. (1987). Biogenesis of the rat hepatocyte plasma membrane in vivo: Comparison of the pathways taken by apical and basolateral proteins using subcellular fractionation. *J. Cell Biol.* **105,** 1241–1251.

Battastini, A. M. O., da Rocha, J. B. T., Barcellos, C. K., Dias, R. D., and Sarkis, J. J. F. (1991). Characterization of an ATP diphosphohydrolase (EC 3.6.1.5) in synaptosomes from cerebral cortex of adult rats. *Neurochem. Res.* **16,** 1303–1310.

Bean, B. P. (1992). Pharmacology and electrophysiology of ATP-activated ion channels. *Trends Pharmacol. Sci.* **13,** 87–91.

Beaudoin, A. R., Vachereau, A., Grondin, G., St-Jean, P., Rosenberg, M. D., and Strobel, R. (1986). Microvesicular secretion, a mode of cell secretion associated with the presence of an ATP-diphosphohydrolase. *FEBS Lett.* **203,** 1–2.

Beeler, T. J., Gable, K. S., and Keffer, J. M. (1983). Characterization of the membrane bound Mg^{2+}-ATPase of rat skeletal muscle. *Biochim. Biophys. Acta* **734,** 221–234.

Beeler, T. J., Wang, T., Gable, K., and Shirley, L. (1985). Comparison of the rat microsomal Mg-ATPase of various tissues. *Arch. Biochem. Biophys.* **243,** 644–654.

Beukers, M. W., Pirovano, I. M., van Weert, A., Kerkhof, C. J. M., Ijzerman, A. P., and Soudijn, W. (1993). Characterization of ecto-ATPase on human blood cells. *Biochem. Pharmacol.* **46,** 1959–1966.

Binet, L., and Burstein, M. (1950). Poumon et action vasculaire de l'adenosine-triphosphate. *Presse Med.* **58,** 1201–1203.

Birch-Machin, M. A., and Dawson, A. P. (1988). Ca^{2+} transport by rat liver plasma membranes: The transporter and the previously reported Ca^{2+}-ATPase are different enzymes. *Biochim. Biophys. Acta* **944,** 308–314.

Bo, X., Simon, J., Burnstock, G., and Barnard, E. A. (1992). Solubilization and molecular size determination of the P_{2X} purinoceptor from rat vas deferens. *J. Biol. Chem.* **267,** 17581–17587.

Bonting, S. L. (1970). Sodium-potassium activated adenosinetriphosphatase and cation transport. *In* "Membranes and Ion Transport" (E. E. Bittar, ed.), Vol. 1, pp.257–263. Wiley (Interscience), London.

Born, G. V. R. (1962). Aggregation of blood platelets by adenosine diphosphate and its reversal. *Nature* (*London*) **194,** 927–929.

Born, G. V. R., and Cross, M. J. (1963). The aggregation of blood platelets. *J. Physiol.* (*London*) **168,** 178–195.

Brashear, R. E., and Ross, J. C. (1969). Disappearance of adenosine disphosphate in vivo. *J. Lab. Clin. Med.* **73,** 54–59.

Burnstock, G. (1988). Sympathetic purinergic transmission in small blood vessels. *Trends Pharmacol. Sci.* **9,** 116–117.

Burnstock, G., and Kennedy, C. (1986). A dual function for adenosine 5′-triphosphate in the regulation of vascular tone. Excitatory cotransmitter with noradrenaline from perivascular nerves and locally released inhibitory intravascular agent. *Circ. Res.* **58,** 319–330.

Busse, D., Pohl, B., Bartel, H., and Buschmann, F. (1980). The Mg^{2+}-dependent adenosine triphosphatase activity in the brush border of rabbit kidney cortex. *Arch. Biochem. Biophys.* **201,** 147–159.

Carraway, C. A. C., Corrado, F. J., IV, Fogle D. D., and Carraway, K. L. (1980). Ectoenzymes of mammary gland and its tumours. *Biochem. J.* 191, 45–51.

Chambers, D. A., Salzman, E. W., and Neri, L. L. (1967). Characterization of "Ecto-ATPase" of human blood platelets. *Arch. Biochem Biophys.* **119,** 173–178.

Che, M., Nishida, T., Gatmaitan, Z., and Arias, I. M. (1992). A nucleoside transporter is functionally linked to ectonucleotidases in rat liver canalicular membrane. *J. Biol. Chem.* **267,** 9684.

Chelliah, R., and Bakhle, Y. S. (1983). The fate of adenine nucleotides in the pulmonary circulation of isolated lung. *Q. J. Exp. Physiol. Cogn. Med. Sci.* **68,** 289–300.

Cheung, P. H., Thompson, N. L., Early, K., Culić, O., Hixson, D., and Lin, S. H. (1993a). Cell-CAM105 isoforms with different adhesion functions are coexpressed in adult rat tissues and during liver development. *J. Biol. Chem.* **286,** 6139–6146.

Cheung, P. H., Culić, O., Qiu, Y., Earley, K., Thompson, N., Hixson, D. C., and Lin, S. -H. (1993b). The cytoplasmic domain of C-CAM is required for C-CAM-mediated adhesion function: studies of a c-CAM transcript containing an unspliced intron. *Biochem. J.* **295,** 427–435.

Cheung, P. H., Luo, W., Qiu, Y., Zhang, X., Earley, K., Millirons, P., and Lin, S.-H. (1993c). Structure and function of C-CAM1. *J. Biol. Chem.* **268,** 24303–24310.

Coade, S. B., and Pearson, J. D. (1989). Metabolism of adenine nucleotides in human blood. *Circ. Res.* **65,** 531–537.

Côté, Y. P., Picher, M., St-Jean, P., Beliveau, R., Potier, M., and Beaudoin, A. R. (1991). Identification and localization of ATP-diphosphohydrolase (Apyrase) in bovine aorta: Relevance to vascular tone and platelet aggregation. *Biochim. Biophys. Acta* **1078,** 187–191.

Côté, Y. P., Filep, J. G., Battistini, B., Gauvreau, J., Sirois, P., and Beaudoin, A. R. (1992a). Characterization of ATP-diphosphohydrolase activities in the intima and media of the bovine aorta: evidence for a regulatory role in platelet activation in vitro. *Biochim. Biophys. Acta* **1139,** 133–142.

Côté, Y. P., Ouellet, S., and Beaudoin, A. R. (1992b). Kinetic properties of Type-II ATP diphosphodrolase from the tunica media of the bovine aorta. *Biochim. Biophys. Acta* **1160,** 246–250.

Cote, Y. P., Pavate, C. T., and Beaudoin, A. R. (1992c). the control of nucleotides in blood vessels: Role of the ATP diphosphohydrolase (Apyrase). *Curr. Top. Pharmacol.* **1,** 83–93.

Culić, O., Sabolić, I., and Žanić-Grubisić, T. (1990). The stepwise hydrolysis of adenine nucleotides by ectoenzymes of rat renal brush-border membranes. *Biochim. Biophys. Acta* **1030,** 143–151.

Culić, O., Huang, Q. H., Flanagan, D., Hixson, D., and Lin, S. H. (1992). Molecular cloning and expression of a new rat liver cell-CAM105 isoform. *Biochem. J.* **285,** 47–53.

Cummins, J., and Hydén, H. (1962). Adenosine triphosphate levels and adenosine triphosphatases in neurons. Glia and neuronal membranes of the vestibular nucleus. *Biochim. Biophys. Acta* **60,** 271–283.

Cunningham, H. B., Yazak, P. J., Domingo, R. C., Oades, K. V., Bohlen, H., Sabbadini, R. A., and Dahms, A. S. (1993). The skeletal muscle transverse tubular Mg-ATPase: Identity with Mg-ATPases of smooth muscle and brain. *Arch. Biochem. Biophys.* **303,** 32–43.

Cusack, N. J., Pearson. J. D., and Gordon, J. L. (1983). Stereoselectivity of ectonucleotidases on vascular endothelial cells. *Biochem. J.* **214,** 975–981.

Dawson, J. M., Cook, N. D., Coade, S. B., Baum, H., and Peters, T. J. (1986). Demonstration of plasma-membrane adenosine diphosphatase activity in rat lung. *Biochim. Biophys. Acta* **856,** 566–570.

DePierre, J. W., and Karnovsky, M. L. (1974a). Ecto-enzymes of the guinea pig polymorphonuclear leukocyte. I. *J. Biol. Chem.* **249,** 7111–7120.

DePierre, J. W., and Karnovsky, M. L. (1974b). Ecto-Enzymes of the guinea pig polymorphonuclear leukocyte. II. *J. Biol. Chem.* **249,** 7121–7129.

De Souza, L. R., and Reed, J. K. (1991). The involvement of ecto-ATPase activity in the phosphorylation of intracellular proteins by the addition of extracellular [^{32}P]ATP in PC12 cells. *Neurochem. Int.* **19,** 581–592.

Dhalla, N. S., and Zhao, D. (1988). Cell membrane Ca^{2+}/Mg^{2+} ATPase. *Prog. Biophys. Mol. Biol.* **52,** 1–37.

Dhalla, N. S., and Zhao, D. (1989). Possible role of sarcolemmal Ca^{2+}/Mg2+ ATPase in heart function. *Magnesium Res.* **2,** 161–172.

Dombrowski, K. E., Trevillyan, J. M., Cone, C., Lu, Y., and Phillips, C. A. (1993). Identification and partial characterization of an ectoATPase expressed by human natural killer cells. *Biochemistry* **32,** 6515–6522.

Dosne, A. M., Legrand, C., Bauvois, B., Bodevin, E., and Caen, J. P. (1978). Comparative degradation of adenylnucleotides by cultured endothelial cells and fibroblasts. *Biochem. Biophys. Res. Commun.* **85,** 183–189.

Dubyak, G. R. (1991). Signal transduction by P_2-purinergic receptors for extracellular ATP. *Am. J. Respir. Cell Mol. Biol.* **4,** 295–300.

Dubyak, G. R., and El-Moatassim, C. (1993). Signal transduction via P_2-purinergic receptors for extracellular ATP and other nucleotides. *Am. J. Physiol.* **265,** C577–C606.

Dubyak, G. R., and Fedan, J. S. (eds.) (1990). Biological actions of extracellular ATP. *Ann. N.Y. Acad. Sci.* **603,** 1–542.

Dzhandzhugazyan, K., and Bock, E. (1993). Demonstration of (Ca^{2+}-Mg^{2+})-ATPase activity of the neural cell adhesion molecule. *FEBS Lett.* **336,** 279–283.

Edlund, M., Gaardsvoll, H., Bock, E., and Öbrink B. (1993. Different isoforms and stock-specific variants of the cell adhesion molecule C-CAM (Cell-CAM 105) in rat liver. *Eur. J. Biochem.* **213,** 1109–1116.

Edwards, F. A., Gibb, A. J., and Colquhoun, D. (1992). ATP receptor-mediated synatpic currents in the central nervous system. *Nature* (*London*) **359,** 144–147.

Ehrlich, Y. H., Davis, T. B., Bock, E., Kornecki, E., and Lenox, R. H. (1986). Ecto-protein kinase activity on the external surface of neural cells. *Nature* (*London*) **320,** 67–70.

El-Moatassim, C., Dornand, J., and Mani, J. C. (1992). Extracellular ATP and cell signalling. *Biochim. Biophys. Acta* **1134,** 31–45.

Emmelot, P., and Bos, C. J. (1966). Studies on plasma membranes. III. Mg-ATPase, Na^+, K^+-Mg^{2+}-ATPase and 5′-nucleotidase activity of plasma membranes isolated from rat liver. *Biochim. Biophys. Acta* **120,** 369–382.

Engelhardt, W. A. (1957). Enzymes as structural elements of physiological mechanisms. *Proc. Int. Symp. Enzyme Chem.,* (Tokyo and Kyoto) **2,** 163–166.

Enyedi, A., Minami, J., Caride, A. J., and Penniston, J. T. (1988). Characteristics of the Ca^{2+}-pump and Ca^{2+}-ATPase in the plasma membrane of rat myometrium. *Biochem. J.* **252,** 215–220.

Epstein, M. A., and Holt, S. J. (1963a). The localization by electron microscopy of HeLa cell surface enzymes splitting adenosine triphosphate. *J. Cell Biol.* **19,** 325–336.

Epstein, M. A., and Holt, S. J. (1963b). Electron microscope observations on the surface adenosine triphosphatase-like enzymes of HeLa cells infected with herpes virus. *J. Cell Biol.* **19,** 337–347.

Ernster, L., and Jones, L. C. (1962). A study of the nucleotide tri- and diphosphatase activities of rat liver microsomes. *J. Cell Biol.* **15,** 563.

Essner, E., Novikoff, A. B., and Masek, B. (1958). Adenosine triphosphatase and 5-nucleotidase activities in the plasma membrane of liver cells as revealed by electron microscopy. *J. Biophys. Biochem. Cytol.* **4,** 711–715.

Etheredge, E., Haaland, J. E., and Rosenberg, M. D. (1971). The functional properties of ATPases bound to and solubilized from the membrane complex of the hen's egg. *Biochim. Biophys. Acta* **233,** 145–154.

Evans, R. J., Derkach, V., and Surprenant, A. (1992). ATP mediates fast synaptic transmission in mammalian neurons. *Nature* (*London*) **357,** 503–505.

Filippini, A., Taffs, R. E., Agui, T., and Sitkovsky, M. V. (1990). Ecto-ATPase activity in cytolytic t-lymphocytes. *J. Biol. Chem.* **265,** 334–340.

Fleetwood, G., Coade, S. B., Gordon, J. L., and Pearson, J. D. (1989). Kinetics of adenine nucleotide catabolism in the coronary circulation of the rat. *Am. J. Physiol.* **256,** H1565-H1572.

Foresta, C., Rossato, M., and Di Virgilio, F. (1992). Extracellular ATP is a trigger for the acrosome reaction in human spermatozoa. *J. Biol. Chem.* **267,** 19443–19447.

Gaarder, A., Jonsen, J., Laland, S., Hellem, A., and Owren, P. A. (1961). Adenosine diphosphate in red cells as a factor in the adhesiveness of human blood platelets. *Nature* (*London*) **192,** 531–532.

Gandhi, C. R., and Ross, D. H. (1988). Characterization of a high-affinity Mg^{2+}-independent Ca^{2+}-ATPase from rat brain synaptosomal membranes. *J. Neurochem.* **50,** 248–256.

Gao, J. -P., and Knowles, A. F. (1993). The epidermal growth factor/cAMP-inducible EctoCA^{2+}-ATPase of human hepatoma Li-7A cells is similar to rat liver ectoATPase/hepatocyte cell adhesion molecule (Cell-CAM 105). *Arch. Biochem. Biophys.* **303,** 90–97.

Ghijsen, W., Gmaj, P., and Murer, H. (1984). Ca^{2+}-stimulated, Mg^{2+}-independent ATP hydrolysis and the high affinity Ca^{2+}-pumping ATPase. Two different activities in rat kidney basolateral membranes. *Biochim. Biophys. Acta* **778,** 481–488.

Gloor, S., Antonicek, H., Sweadner, K. J., Pagliusi, S., Frank, R., Moos, M., and Schachner, M. (1990). The adhesion molecule on glia (AMOG) is a homologue of the beta subunit of the Na,K-ATPase. *J. Cell Biol.* **110,** 165–174.

Gmaj, P., Murer, H., and Carafoli, E. (1982). Localization and properties of a high-affinity (Ca^{2+} + Mg^{2+})-ATPase in isolated kidney cortex plasma membranes. *FEBS Lett.* **144,** 226–230.

Gordon, E. L., Pearson, J. D., and Slakey, L. L. (1986). The hydrolysis of extracellular adenine nucleotides by cultured endothelial cells from pig aorta. *J. Biol. Chem.* **261,** 15496–15504.

Gordon, E. L., Pearson, J. D., Dickinson, E. S., Moreau, D., and Slakey, L. L. (1989). The hydrolysis of extracellular adenine nucleotides by arterial smooth muscle cells. *J. Biol. Chem.* **264,** 18986–18992.

Gordon, J. L. (1986). Extracellular ATP: Effects, sources and fate. *Biochem. J.* **233,** 309–319.

Grondal, E. J. M., and Zimmermann, H. (1986). Ectonucleotidase activites associated with cholinergic synaptosomes isolated from *Torpedo* electric organ. *J. Neurochem.* **47,** 871–881.

Gropp, A., Hupe, K., and Hellweg, H. R. (1958). Zum Nachweis der Adenosine Triphosphatase (ATPase) an *in vitro* gezüchten Carcinomzellen. *Naturwissenschaften* **45,** 394.

Gutmann, H. R., Chow, Y. M., Vessella, R. L. Schuetzle, B., and Kaplan, M. E. (1993). The kinetic properties of the ecto-atpase of human peripheral blood lymphocytes and of cronic lymphatic leukemia cells. *Blood* **62,** 1041–1046.

Haaland, J. E., and Rosenberg, M. D. (1969). Activation of membrane associated ATPase in hen's egg after ovulation. *Nature (London)* **223,** 1275–1276.

Haaland, J. E., Etheredge, E., and Rosenberg, M. D. (1971). Isolation of an ATPase from the membrane complex of the hen's egg. *Biochim. Biophys. Acta* **233,** 137–144.

Hamlyn, J. M., and Senior, A. E. (1983). Evidence that Mg^{2+}- or Ca^{2+}-activated adenosine triphosphatase in rat pancreas is a plasma-membrane ecto-enzyme. *Biochem. J.* **214,** 59–68.

Harlan, J., DeChatelet, L. R., Iverson, D. B., and McCall, C. (1977). Magnesium-dependent adenosine triposphatase as a marker enzyme of the plasma membrane of human polymorphonuclear leukocytes. *Infect. Immun.* **15,** 436–443.

Harper, F., Lamy, F., and Calvert, R. (1978). Some properties of a Ca^{2+}- and (or) Mg^{2+}-requiring nucleoside di- and triphosphatase(s) associated with the membranes of rat pancreatic zymogen granules. *Can. J. Biochem.* **56,** 565–576.

Hellewell, P. G., and Pearson, J. D. (1987). Adenine nucleotides and pulmonary endothelium. *In* "Pulmonary endothelium in Health and Disease" (U. Ryan, ed.), pp. 327–348. Dekker, New York.

Hempling, H. G., Stewart, C. C., and Gasić, G. (1969). The effect of exogenous ATP on the electrolyte content of TA_3 ascites tumor cells. *J. Cell. Physiol.* **73,** 133–140.

Henderson, J. F., and Paterson, A. R. P. (1973). "Nucleotide Metabolism." Academic Press, New York.

Herbert, E. (1956). A study of the liberation of orthophosphate from adenosine triphosphate by the stromata of human erythrocytes. *J. Cell. Comp. Physiol.* **47,** 11–36.

Herrmann, H. -J., Moritz, V., and Kühne, C. H. (1992). Structural wall tissue alterations of the microvasculature in the course of spontaneous hypertension of rats. *Int. J. Microcirc. Clin. Exp.* **11,** 1–20.

Hidalgo, C., Gonzalez, M. E., and Lagos, R. (1983). Characterization of the Ca^{2+}- or Mg^{2+}-ATPase of transverse tubule membranes isolated from rabbit skeletal muscle. *J. Biol. Chem.* **258,** 13937–13945.

Hidalgo, C., Parra, C., Riquelme, G., and Jaimovich, E. (1986). Transverse tubules from frog skeletal muscle. Purification and properties of vesicles sealed with the inside-out orientation. *Biochim. biophys. Acta* **855,** 79–88.

Hinoda, Y., Neumaier, M., Hefta, S. A., Drzeniek, Z., Wagener, C., Shively, L., Hefta, L. J. F., Shively, J., and Paxton R. J. (1988). Molecular cloning of a cDNA coding biliary glycoprotein I: Primary structure of a glycoprotein immunologically crossreactive with carcinoembryonic antigen. *Proc. Natl. Acad. Sci. U.S.A.* **85,** 6959–6963; correction: **86,** 1668 (1989).

Hohmann, J., Kowalewski, H., Vogel, M., and Zimmermann, H. (1993). Isolation of a Ca^{2+} or Mg^{2+}-activated ATPase (ecto-ATPase) from bovine brain synaptic membranes. *Biochim. Biophys. Acta* **1152,** 146–154.

Horgan, D. J., and Kuypers, R. (1988). Biochemical properties of purified transverse tubules isolated from skeletal muscle triads. *Arch. Biochem. Biophys.* **260,** 1–9.

Huang, N., Ahmed, A. H., Wang, D., and Heppel, L. A. (1992). Extracellular ATP stimulates increases in Na^+/K^+ pump activity, intracellular pH and uridine uptake in cultures of mammalian cells *Biochem. Biophys. Res. Commun.* **182,** 836–843.

Huang, W. H., Kakar, S. S., and Askari, A. (1985). Mechanisms of detergent effects on membrane-bound (Na^+ + K^+)-ATPase*. *J. Biol. Chem.* **260,** 7356–7361.

Ilsbroux, I., Vanduffel, L., Teuchy, H., and De Cuyper, M. (1985). An azide-insensitive low-affinity ATPase stimulated by Ca^{2+} or Mg^{2+} in basal-lateral and brush border membranes of kidney cortex. *Eur. J. Biochem.* **151,** 123–129.

Jarett, L., and Smith, R. M. (1974). The stimulation of adipocyte plasma membrane magnesium ion-stimulated adenosine triposphatase by insulin and concanavalin A. *J. Biol. Chem.* **249,** 5195–5199.

Jones, B. M. (1966). A unifying hypothesis of cell adhesion. *Nature* (*London*) **212,** 362–365.

Jones, B. M., and Kemp, R. B. (1970a). Aggregation and Electrophoretic mobility studies on dissociated cells. *Exp. Cell Res.* **63,** 301–308.

Jones, B. M., and Kemp, R. B. (1970b). Inhibition of cell aggregation by antibodies directed against actomyosin. *Nature* (*London*) **226,** 261–262.

Jones, P. C. T. (1966). A contractile protein model for cell adhesion. *Nature* (*London*) **212,** 365–369.

Juul, B., Aalkjær, C., and Plesner, L. (1991a). Nucleotides and rat mesenteric small arteries. *In* "Resistance Arteries: Structure and Function" (M. J. Mulvany, C. Aalkjær, A. M. Heagerty, N. C. B. Nyborg, and S. Strandgaard, eds.), pp. 91–95 Elsevier, Amsterdam and New York.

Juul, B., Luscher, E. M., Aalkjær, C., and Plesner, L. (1991b). Nucleotide hydrolytic activity of isolated intact rat mesenteric small arteries. *Biochim. Biophys. Acta* **1067,** 201–207.

Juul, B., Aalkjær, C., and Plesner, L. (1992). Effects of ATP and UTP on $Ca^{2+}{}_i$, membrane potential and force in isolated rat small arteries. *J. Vasc. Res.* **29,** 385–395.

Juul, B., Plesner, L., and Aalkjær, C. (1993). Effect of ATP and related nucleotides on the tone of isolated rat mesenteric resistance arteries. *J. Pharmacol. Exp. Ther.* **264,** 1234–1240.

Kang, J. -J., Cunningham, H. B., Jachec, C., Priest, A., Dahms, A. S., and Sabbadini, R. A. (1991). Direct effects of phorbol esters and diacylglycerols on the T-tubule Mg^{2+}-ATPase. *Arch. Biochem. Biophys.* **290,** 214–223.

Karasaki, S. (1992). Subcellular localization of surface adenosine triphosphatase activity in prenoplastic liver parenchyma. *Cancer Res.* **32,** 1703–1712.

Karasaki, S., and Okigaki, T. (1976). Surface membrane nucleoside triphosphatase activity and tumorigenicity of cultured liver epithelial cells. *Cancer Res.* **36,** 4491–4499.

Karasaki, S., Simard, A., and de Lamirande, G. (1977). Surface morphology and nucleoside phosphatase activity of rat liver epithelial cells during oncogenic transformation in vitro. *Cancer Res.* **37,** 3516–3525.

Karasaki, S., Suh, M. H., Salas, M., and Raymond, J. (1980). Cell surface adenosine 5′-triphosphatase as an *in vitro* marker of the lineage and cytodifferentiation of oncogenic epithelial cells from rat liver parenchyma. *Cancer Res.* **40,** 1318–1328.

Kelley, L. K., and Smith, C. H. (1987). Use of GTP to distinguish calcium transporting ATPase activity from other calcium dependent nucleotide phosphatases in human placental basal plasma membrane. *Biochem. Biophys. Res. Commun.* **148,** 126–132.

Kelley, L. K., Borke, J. L., Verma, A. K., Kumar, R., Penniston, J. T., and Smith, C. H. (1990). The calcium-transporting ATPase and the calcium- or magnesium-dependent nucleotide phosphatase activities of human placental trophoblast basal plasma membrane are separate enzyme activities. *J. Biol. Chem.* **265,** 5453–5459.

Kennedy, C. (1990). P_1- and P_2-purinoceptor subtypes: An update. Arch. Int. Pharmacodyn. Ther. **303,** 30–50.

Kinne-Saffran, E., and Kinne, R. (1974). Localization of a calcium-stimulated ATPase in the basal-lateral plasma membranes of the proximal tubule of rat kidney cortex. *J. Membr. Biol.* **17,** 263–274.

Kirley, T. L. (1988). Purification and characterization of the Mg^{2+}-ATPase from rabbit skeletal muscle transverse tubule. *J. Biol. Chem.* **263,** 12682–12689.

Kirley, T. L. (1991). The Mg^{2+}-ATPase of rabbit skeletal-muscle transverse tubule is a highly glycosylated multiple-subunit enzyme. *Biochem. J.* **278,** 375–380.

Knowles, A. F. (1988a). Differential expression of Ecto-Mg^{2+}-ATPase and EctoCa^{2+}-ATPase activities in human hepatoma cells. *Arch. Biochem. Biophys.* **263,** 264–271.

Knowles, A. F. (1988b). Inhibition of growth and induction of enzyme activities in a clonal human hepatoma cell line (Li-7A): Comparison of the effects of epidermal growth factor and an anti-epidermal growth factor receptor antibody. *J. Cell. Physiol.* **134,** 109–116.

Knowles, A. F. (1990). Synergistic Modulation of EctoCa^{2+}-ATPase activity of hepatoma (Li-7A) cells by epidermal growth factor and cyclic AMP. *Arch. biochem. Biophys.* **283,** 114–119.

Knowles, A. F. (1995). A dissection of the rat liver ectoATPase/C-CAM/pp120/Bile acid transporter gene. *Biochem. Biophys. Res. Commun.* (in press).

Knowles, A. F., and Leng, L. (1984). Purification of a low affinity Mg^{2+}(Ca^{2+})-ATPase from the plasma membranes of a human oat cell carcinoma. *J. Biol. Chem.* **259,** 10919–10924.

Knowles, A. F., and Murray, S. L. (1990). Effect of inhibition of protein synthesis, RNA Synthesis, and tyrosine kinase on the induction of ecto-ATPases of human hepatoma (Li-7A) Cells. *Ann. N.Y. Acad. Sci.* **603,** 519–522.

Knowles, A. F., Isler, R. E., and Reece, J. F. (1983). The common occurrence of ATP diphosphohydrolase in mammalian plasma membranes. *Biochim. biophys. Acta.* **731,** 88–96.

Knowles, A. F., Salas-Prato, M., and Villela, J. (1985). Epidermal growth factor inhibits growth while increasing the expression of an ecto-Ca^{2+}-ATPase of a human hepatoma cell line. *Biochem. Biophys. Res. Commun.* **126,** 8–14.

Knowles, M. R., Clarke, L. L., and Boucher, R. C. (1991). Activation by extracellular nucleotides of chloride secretion in the airway epithelia of patients with cystic fibrosis. *N. Engl. J. Med.* **325,** 533–538.

Kragballe, K., and Ellegaard, J. (1978). ATP-ase activity of purified human normal T- and B-lymphocytes. *Scand. J. Haematol.* **20,** 271–279.

Kriho, V., Pappas, G. D., and Becker, R. P. (1990). Electron microscopy and biochemical determinations of chromaffin cell Ecto-ATPase activity. *Proc. Int. Congr. Electron Microsc., 12th,* Seattle, *1990,* p. 936.

Laliberté, J. F., and Beaudoin, A. R. (1983). Sequential hydrolysis of the γ- and β-phosphate groups of ATP by the ATP diphosphohydrolase from pit pancreas. *Biochim. Biophys. Acta* **742.** 9–15.

Laliberté, J. F., St-Jean, P., and Beaudoin, A. R. (1982). Kinetic effects of Ca^{2+} and Mg^{2+} on ATP hydrolysis by the purified ATP diphosphohydrolase. *J. Biol. Chem.* **257,** 3869–3874.

Lambert, M., and Christophe, J. (1978). Characterization of (Mg,Ca)-ATPase activity in rat pancreatic plasma membranes. *Eur. J. Biochem.* **91,** 485–492.

LeBel, D., Poirier, G. G., Phaneuf, S., St-Jean, P., Laliberté J. F., and Beaudoin, A. R. (1980). Characterization and purification of a calcium-sensitive ATP diphosphohydrolase from pig pancreas. *J. Biol. Chem.* **255,** 1227–1233.

Lim, Y. P., Callanan, H., Lin, S. H., Thompson, N. L., and Hixson, D. C. (1993). Preparative mini-slab gel continuous elution electrophoresis: Application for the separation of two isoforms of rat hepatocyte cell adhesion molecule, cell-CAM 105, and its associated proteins. *Anal. Biochem.* **214,** 156–164.

Lin, S. C., and Way, E. L. (1984). Characterization of calcium-activated and magnesium-activated ATPases of Brain Nerve Endings. *J. neurochem.* **42,** 1697–1706.

Lin, S. -H. (1985a). Novel ATP-dependent calcium transport component from rat liver plasma membranes. *J. Biol. Chem.* **260,** 7850–7856.

Lin, S. -H. (1985b). The rat liver plasma membrane high affinity (Ca^{2+}-Mg^{2+})-ATPase is not a calcium pump. *J. Biol. Chem.* **260,** 10976–10980.

Lin, S. -H. (1989). Localization of the ecto-ATPase (ecto-nucleotidase) in the rat hepatocyte plasma membrane. *J. biol. Chem.* **264,** 14403–14407.

Lin, S. -H., and Fain, J. N. (1984). Purification of ($Ca^{2+}Mg^{2+}$)-ATPase from rat liver plasma membranes. *J. Biol. Chem.* **259,** 3016–3020.

Lin, S. -H., and Guidotti, G. (1989). Cloning and expression of a cDNA coding for a rat liver plasma membrane ecto-ATPase. *J. Biol. Chem.* **264,** 14408–14414.

Lin, S. -H., and Russell W. E. (1988). Two Ca^{2+}-dependent ATPases in rat liver plasma membrane. *J. Biol. Chem.* **263,** 12253–12258.

Lin, S. -H., Culić, O., Flanagan, D., and Hixson, D. C. (1991). Immunochemical characterization of two isoforms of rat liver ecto-ATPase that show an immunological and structural identity with a glycoprotein cell-adhesion molecule with M_r 105000. *Biochem. J.* **278,** 155–161.

Lipkin, V. M., Khramtsov, N. V., Andreeva, S. G., Moshnyakov, M. V., Petukhova, G. V., Rakitina, T. V., Feshchenko, E. A., Ishchenko, K. A., Mirzoeva, S. F., Chernova, M. N., and Dranytsyna, S. M. (1989). Calmodulin-independent bovine brain adenylate cylase. *FEBS Lett.* **254,** 69–73.

Lipkin, V. M., Surina, E. A., Petukhova, G. V., Petrov, V. M., Mirzoeva, S. F., and Rakitina, T. V. (1992). Relationship of neural cell adhesion molecules (N-CAMs) with adenylate cyclase. *FEBS Lett.* **304,** 1, 9–11.

Lüthje, J., Schomburg, A., and Ogilvie, A. (1988). Demonstration of a novel ecto-enzyme on human erythrocytes, capable of degrading ADP and of inhibiting ADP-induced platelet aggregation. *Eur. J. Biochem.* **175,** 285–289.

Lutty, G. A., and McLeod, S. (1992). A new technique for visualization of the human retinal vasculature. *Arch. Ophthalmol. Chicago,* **110,** 267–276.

Luu-The, V., Goffeau, A., and Thinés-Sempoux, D. (1987). Rat liver plasma membrane Ca^{2+}- or Mg^{2+}-activated ATPase. Evidence for proton movement in reconstituted vesicles. *Biochim. Biophys. Acta* **904,** 251–258.

Magocsi, M., and Penniston, J. T. (1991). Ca^{2+} or Mg^{2+} nucleotide phosphohydrolases in myometrium: Two ecto-enzymes. *Biochim. Biophys. Acta* **1070,** 163–172.

Majumder, G. C. (1981). Enzymic characteristics of ecto-adenosine triphosphatase in rat epididyman intact spermatozoa. *Biochim. J.* **195,** 103–110.

Majumder, G. C., and Biswas, R. (1979). Evidence for the occurrence of an ecto-(adenosine triphosphatase) in rat epididymal spermatozoa. *Biochem. J.* **183,** 737–743.

Malouf, N. N., and Meissner, G. (1979). Localization of a Mg^{2+}- or Ca^{2+}-activated ("basic") ATPase in skeletal muscle. *Exp. Cell Res.* **122,** 233–250.

Malouf, N. N., and Meissner, G. (1980). Cytochemical localization of a "basic" ATPase to canine myocardial surface membrane. *J. Histochem. Cytochem.* **28,** 1286–1294.

Mant, M. J., and Parker, K. R. (1981). Two platelet aggregation inhibitors in tsetse (glossina) saliva with studies of roles of thrombin and citrate in invitro platelet aggregation. *Br. J. Haematol.* **48,** 601–608.

Margolis, R. N., Taylor, S. I., Seminara, D., and Hubbard, A. L. (1988). Identification of pp120, and endogenous substrate for the hepatocyte insulin receptor tyrosine kinase, as an integral membrane glycoprotein on the bile canalicular domain. *Proc. Natl. Acad. Sci. U.S.A.* **85,** 7256–7259.

Margolis, R. N., Schell, M. J., Taylor, S. I., and Hubbard, A. L. (1990). Hepatocyte plasma membrane ecto-ATPase is a substrate for tyrosine kinase activity of the insulin receptor. *Biochem. Biophys. Res. Commun.* **166,** 562–566.

Marsh, J. B., and Haugaard, N. (1957). Adenosine triphosphatase activity of intact muscle cells. *Biochim. Biophys. Acta* **23,** 204–205.

Martin, S. S., and Senior, A. E. (1980). Membrane adenosine triphosphatase activities in rat pancreas. *Biochim. Biophys. Acta* **602,** 401–418.

Martin, W., Cusack, J., Carleton, J. L., and Gordon, J. L. (1985). Specificity of the P_2-purinoceptor that mediates endothelium-dependent relaxation of the pig aorta. *Eur. J. Pharmacol.* **108,** 295–299.

Mason, R. G., and Saba, S. R. (1969). Platelet ATPase activities. I. Ectogation. *Am. J. Pathol.* **55,** 215–223.

Matsukawa, R. (1990). Separation of Ca^{2+}-ATPase from Mg^{2+}-ATPase in plasma membrane-rich fraction of bovine parotid gland. *Arch. Biochem. Biophys.* **280,** 362–368.

Medzihradsky, F., Cullen, E. I., Lin, H.-L., and Bole, G. G. (1980). Drug-sensitive ecto-ATPase in human leukocytes. *Biochem. Pharmacol.* **29,** 2285–2290.

Meghji, P., Pearson, J. D., and Slakey, L. L. (1992). Regulation of extracellular adenosine production by ectonucleotidases of adult rat ventricular myocytes. *Am. J. Physiol.* **263,** H40–H47.

Merten, M. D., Breittmayer, J.-P., Figarella, C., and Frelin, C. (1993). ATP and UTP increase secretion of bronchial inhibitor by human tracheal gland cells in culture. *Am. J. Physiol.* **265,** L479–L484.

Minami, J., and Penniston, J. T. (1987). Ca^{2+} Uptake by corpus-luteum plasma membranes. *Biochem. J.* **242,** 889–894.

Missiaen, L., Wuytack, F., and Casteels, R. (1988a). Effect of ovarian steroids on membrane ATPase activities in microsomes (microsomal fractions) from rat myometrium. *Biochem. J.* **250,** 571–577.

Missiaen, L., Wuytack, F., and Casteels, R. (1988b). Characterization of the Mg^{2+}-activated ATPase activity in smooth-muscle membranes. *Biochem. J.* **250,** 579–588.

Miura, Y., Hirota, K., Arai, Y., and Yagi, K. (1987). Purification and partial characterization of adenosine diphosphatase activity in bovine aorta microsomes. *Thromb. Res.* **46,** 685–695.

Montague, D. J., Staunton, D., and Peters, T. J. (1984). Tissue distribution of adenosine diphosphatase activity in the rat. *Enzyme* **31,** 21–26.

Mörtl, M., Busse, D., Bartel, H., and Pohl, B. (1984). Partial purification and characterization of rabbit-kidney brushborder (Ca^{2+} or Mg^{2+})-dependent adenosine triphosphatase. *Biochim. Biophys. Acta* **776,** 237–246.

Moseley, R. H., Jarose, S., and Permoad, P. (1991). Adenosine transport in rat liver plasma membrane vesicles. *Am. J. Physiol.* **261,** G716–G722.

Moulton, M. P., Sabbadini, R. A., Norton, K. C., and Dahms, A. S. (1986). Studies on the transverse tubule membrane Mg-ATPase. *J. Biol. Chem.* **261,** 12244–12251.

Mughal, S., Cuschieri, A., and Al-Bader, A. A. (1989). Intracellular distribution of Ca^{2+}-Mg^{2+} adenosine triphosphatase (ATPase) in various tissues. *J. Anat.* **162,** 111–124.

Murray, S. L., and Knowles, A. F. (1990). Butyrate induces an EctoMg^{2+}-ATPase activity in Li-7A human hepatoma cells. *J. Cell. Phys.* **144,** 26–35.

Nagy, A. (1986). Enzymatic characteristics and possible role of synaptosiomal ecto-adenosine triphosphatase from mammalian brain. *In* "Cellular Biology of Ectoenzymes" (G. W. Kreutzberg *et al.,* eds.), pp 49–59. Springer-Verlag, Berlin.

Nagy, A., and Delgado-Escueta, A. V. (1984). Rapid preparation of synaptosomes from mammalian brain using nontoxic isoosmotic gradient material (Percoll). *J. Neurochem.* **43,** 1114–1123.

Nagy, A. K., Shuster, T. A., and Rosenberg, M. D. (1983). Adenosine triphosphatase activity at the external surface of chicken brain synaptosomes. *J. Neurochem.* **40,** 226–234.

Nagy, A. K., Shuster, T. A., and Delgado-Escueta, A. V. (1986). Ecto-ATPase of mammalian synaptosomes: Identification and enzymic characterization. *J. Neurochem.* **47,** 976–986.

Nagy, A. K., Shuster, T. A., and Delgado-Escueta, A. V. (1989). Rat brain synaptosomal ATP: AMP-phosphotransferase activity. *J. Neurochem.* **53,** 1166–1172.

Nagy, A. K., Houser, C. R., and Delgado-Escueta, A. V. (1990). Synaptosomal ATPase activities in temporal cortex and hippocampal formation of humans with focal epilepsy. *Brain Res.* **529,** 192–201.

Najjar, S. M., Accili, D., Philippe, N., Jernberg, J., Margolis, R., and Taylor, S. I. (1993). pp120 Ecto-ATPase, an endogenous substrate of the insulin receptor tyrosine kinase, is expressed as two variably spliced isoforms. *J. Biol. Chem.* **268,** 1201–1206.

Norton, K., Moulton, M., Rose, R., Sabbadini, R., and Dahms, A. S. (1986). Reaction of the transverse tubule membrane Mg-ATPase with nucleotidemimetic inhibitors. *Biophys. J.* **49,** 561a.

Novikoff, A. B., and Essner, E. (1960). The liver cell. *Am. J. Med.* **29,** 102–131.

Novikoff, A. B., Hecht, L., Podber, E., and Ryan, J. (1952). Phosphatases of rat liver. I. The dephosphorylation of adenosine triphosphate. *J. Biol. Chem.* **194,** 153–170.

Novikoff, A. B., Drucker, J., Shin, W.-Y., and Goldfischer, S. (1961). Further studies on the apparent adenosine triphosphatase activity of cell membranes in formol-calcium-fixed tissues. *J. Histochem. Cytochem.* **9,** 434–451.

Novikoff, A. B., Essner, E., Goldfischer, S., and Heus, M. (1962a). Nucleoside phosphatase activities of cytomembranes. *In* "The Interpretation of Ultrastructure" (R. J. C. Harris, ed.), pp. 149–192. Academic Press, New York.

Novikoff, A. B., de-Thé, G., Beard, D., and Beard, J. W. (1962b). Electron microscopic study of the ATPase activity of the bai strain a (myeloblastosis) avian tumor virus. *J. Cell Biol.* **15,** 451–462.

Öbrink, B. (1991). C-CAM (Cell-CAM 105)—a member of the growing immunoglobulin superfamily of cell adhesion proteins. *BioEssays* **13,** 227–234.

Ocklind, C., and Öbrink, B. (1982). Intercellular adhesion of rat hepatocytes. *J. Biol. Chem.* **257,** 6788–6795.

Odin, P., Tingström, A., and Öbrink, B. (1986). Chemical characterization of cell-CAM 105, a cell-adhesion molecule isolated from rat liver membranes. *Biochem. J.* **236,** 559–568.

Odin, P., Asplund, M., Busch, C., and Öbrink, B. (1988). Immunohistochemical localization of Cell CAM 105 in rat tissues: Appearance in epithelia, platelets, and granulocytes. *J. Histochem. Cytochem.* **36,** 729–739.

Ohnishi, T., and Kimura, S. (1976). Contact-mediated changes in ATPase activity at the surface of primary cultured hepatoma cells. *Histochemistry* **49,** 107–112.

Ohnishi, T., and Yamaguchi, K. (1978). Effects of db-cAMP and theophylline on cell surface adenosine triphosphatase activity in cultured hepatoma cells. *Exp. Cell Res.* **116,** 261–268.

Okamoto, V. R., Moulton, M. P., Runte, E. M., Kent, C. D., Lebherz, H. G., Dahms, A. S., and Sabbadini, R. A. (1985). Characterization of transverse tubule membrane proteins: Tentative identication of the Mg-ATPase. *Arch. Biochem. Biophys.* **237,** 43–54.

Olmo, N., Turnay, J., Risse, G., Deutzmann, R., von der Mark, K., and Lizarbe, M. A. (1992). Modulation of 5′-nucleotidase activity in plasma membranes and intact cells by the extracellular matrix proteins laminin and fibronectin. *Biochem. J.* **282,** 181–188.

Olsen, M., Krog, L., Edvardsen, K., Skovgaard, L. T., and Bock, E. (1993). Intact transmembrane isoforms of the neural cell adhesion molecule are released from the plasma membrane. *Biochem. J.* **295,** 833–840.

Olsson, R. A., and Pearson, J. D. (1990). Cardiovascular purinoceptors. *Physiol. Rev.* **70,** 761–845.

Papamarcaki, T., and Tsolas, O. (1990). Identification of ATP diphosphohydrolase activity in human term placenta using a novel assay for AMP. *Mol. Cell. Biochem.* **97,** 1–8.

Paz, C. A. O., González, D. A., and Alonso, G. L. (1988). Demonstration of the simultaneous activation of Ca^{2+}-independent and Ca^{2+}-dependent ATPases from rat skeletal muscle microsomes. *Biochim. Biophys. Acta* **939,** 409–415.

Pearson, J. D. (1985). Ectonucleotidases: Measurement of activities and use of inhibitors. *Methods Pharmacol.* **6,** 83–107.

Pearson, J. D., (1986). Ectonucleotidases of vascular endothelial cells: Characterization and possible physiological roles. *In* "Cellular Biology of Ectoenzymes" (G. W. Kreutzberg *et al.*, eds.), pp 17–26. Springer-Verlag, Berlin.

Pearson, J. D., and Gordon, J. L. (1979). Vascular endothelial and smooth muscle cells in culture selectivily release adenine nucleotides. *Nature (London)* **281,** 384–386.

Pearson, J. D., and Gordon, J. L. (1985). Nucleotide metabolism by endothelium. *Annu. Rev. Physiol.* **47,** 617–627.

Pearson, J. D., Carleton, J. S., and Gordon, J. L. (1980). Metabolism of adenine nucleotides by ectoenzymes of vascular endothelial and smooth-muscle cells in culture. *Biochem. J.* **190,** 421–429.

Pearson, J. D., Coade, S. B., and Cusack, N. J. (1985). Characterization of ectonucleotidases on vascular smooth-muscle cells. *Biochem. J.* **230,** 503–507.

Pershadsingh, H. A., and McDonald, J. M. (1980). A high affinity calcium-stimulated magnesium-dependent adenosine triphosphatase in rat adipocyte plasma membranes. *J. Biol. Chem.* **255,** 4087–4093.

Phillips, S. A., Perrotti, N., and Taylor, S. I. (1987). Rat liver membranes contain a 120 kDa glycoprotein which serves as a substrate for the tyrosine kinases of the receptors for insulin and epidermal growth factor. *FEBS Lett.* **212,** 141–144.

Picher, M., Côté, Y. P., Beliveau, R., Potier, M., and Beaudoin, A. R. (1993). Demonstration of a novel type of ATP-diphosphohydrolase (EC 3.6.1.5) in the bovine lung. *J. Biol. Chem.* **268,** 4699–4703.

Plesner, L., Juul, B., Skriver, E., and Aalkjær, C. (1991). Characterisation of Ca^{2+} or Mg^{2+}-dependent nucleoside tirphosphatase from rat mesenteric small arteries. *Biochim. Biophys. Acta* **1067,** 191–200.

Poelstra, K., Baller, J. F. W., Hardonk, M. J., and Bakker, W. W. (1991). Demonstration of antithrombotic activity of glomerular adenosine diphosphatase. *Blood* **78,** 141–148.

Pommier, G., Ripert, G., Azoulay, E., and Depieds, R. (1975). Effect of concanavalin a on membrane-bound enzymes from mouse lymphocytes. *Biochim. Biophys. Acta* **389,** 483, 494.

Prat, A. G., Abraham, E. H., Amara, J., Gregory, R., Ausiello, D. A., and Cantiello,

H. F. (1992). Expression of CFTR results in the appearance of an ATP channel. *J. Gen. Physiol.* **100,** 33a.

Premont, R. T. (1991). A bovine brain cDNA purported to encode colmodulin-insensitive adenylyl cyclase has extensive identity with neural cell adhesion molecules (N-CAMs). *FEBS Lett.* **295,** 230–231.

Ranscht, B., and Dours-Zimmermann, M. T. (1991). T-cadherin, a novel cadherin cell adhesion molecule in the nervous system lacks the conserved cytoplasmic region. *Neuron* **7,** 391–402.

Redegeld, F., Filippini, A., Trenn, G., and Sitkovsky, M. V. (1993). Possible role of extracellular ATP in cell-cell interactions leading to CTL-mediated cytotoxicity. *In* "Cytotoxic Cells" (M. Sitkovsky and P. Henkart, eds.), pp. 307–313. Birkhäuser, Boston.

Rhea, R. P., and Rosenberg, M. D. (1971). ATPase activities of inner and outer regions of the vitelline membrane during development. *Dev. Biol.* **26,** 616–626.

Rhea, R. P., Anderson, B., Kim, N. B., and Rosenberg, M. D. (1974). Biochemical, electron microscopic, and cytochemical studies of ATPase localization in avian, murine, and human oviducts. *Fertil. Steril.* **25,** 788–808.

Ribeiro, J. M. C., and Garcia, E. S. (1980). The salivary and crop apyrase activity of *Rhodnius prolixus*. *J. Insect Physiol.* **26,** 303.

Ribeiro, J. M. C., Rossignol, P. A., and Spielman, A. (1984). Role of mosquito saliva in blood vessel location. *J. Exp. Biol.* **108,** 1–7.

Ribeiro, J. M. C., Makoul, G. T., Levine, J., Robinson, D. R., and Spielman, A. (1985). Antihemostatic, antiinflammatory, and immunosuppressive properties of the saliva of tick, *Ixodes dammini*. *J. Exp. Med.* **161,** 322–344.

Richardson, P. J., Brown, S. J., Bailyes, E. M., and Luzio, J. P. (1987). Ectoenzymes control adenosine modulation of immunoisolated cholinergic synapses. *Nature (London)* **327,** 232–234.

Riordan, J. R., Slavik, M., and Kartner, N. (1977). Nature of the lectin-induced activation of plasma membrane Mg^{2+} ATPase. *J. Biol. Chem.* **252,** 5449–5455.

Robinson, C. W., Jr., Kress, S. C., Wagner, R. H., and Brinkhous, K. M. (1965). Platelet agglutination and deagglutination with a sulfhydryl inhibitor, methyl mercuric nitrate: Relationships to platelet ATP-ase. *Exp. Mol. Pathol.* **4,** 457–464.

Rodriguez-Pascual, F., Torres, M., and Miras-Portugal, M. T. (1992). Studies on the turnover of ecto-nucleotidases and ectodinucleoside polyphosphate hydrolase in cultured chromaffin cells. *Neurosci. Res. Commun.* **11,** 101–107.

Rodriguez-Pascual, F., Torres, M., and Miras-Portugal, M. T. (1993). 8-Azido-adenine nucleotides as substrates of ecto-nucleotidases in chromaffin cells: Inhibitory effect of photoactivation. *Arch. Biochem. Biophys.* **306,** 420–426.

Ronquist, G., and Ågren, G. K. (1975). A Mg^{2+}- and Ca^{2+}-stimulated adenosine triphosphatase at the outer surface of Ehrlich ascites tumor cells. *Cancer Res.* **35,** 1402–1406.

Ronquist, G., and Brody, I. (1985). The prostasome: Its secretion and function in man. *Biochim. Biophys. Acta* **822,** 203–218.

Rorive, G., and Kleinzeller, A. (1972). The effect of ATP and Ca^{2+} on the cell volume in isolated kidney tubules. *Biochim. Biophys. Acta* **274,** 226–239.

Rosenberg, M. D., Gusovsky, T., Cutler, B., Berliner, A. F., and Anderson, B. (1977). Purification and characterization of an extracellular ATPase from oviductal secretions. *Biochim. Biophys. Acta* **482,** 197–212.

Rosenberg, M. D., Cutler, B., Gusovsky, T., Okuniewicz, J., Segal, I., and Sondag, J. (1980). Binding to cell surfaces by antibodies against ExoATPase. *Enzyme* **25,** 276–280.

Rosemblatt, M. S., and Scales, D. J. (1989). Morphological, immunological and biochemical characterization of purified transverse tubule membranes isolated from rabbit skeletal muscle. *Mol. Cell. Biochem.* **87,** 57–69.

Rosenblatt, D. E., Lauter, C. J., and Trams, E. G. (1976). Deficiency of a Ca^{2+}-ATPase in brains of seizure prone mice. *J. Neurochem.* **27,** 1299–1304.

Rothstein, A., and Meier, R. (1948). The relationship of the cell surface to metabolism I Phosphatases in the cell surface of living yeast cells. *J. Comp. Cell. Physiol.* **32,** 77–95.

Rothstein, A., Meier, R. C., and Scharff, T. G. (1953). Relationship of cell surface to metabolism. IX. Digestion of phosphorylated compounds by enzymes located on surface of intestinal cell. *Am. J. Physiol.* **173,** 41–46.

Saba, S. R., Rodman, N. F., and Mason, R. G. (1969). Platelet ATPase activities. II. ATPase activities of isolated platelet membrane fraction. *Am. J. Pathol.* **55,** 225–233.

Sabbadini, R. A., and Dahms, A. S. (1989). Biochemical properties of isolated transverse tubular membranes. *J. Bioenerg. Biomemb* **21,** 163–213.

Sabbadini, R. A., and Okamoto, V. R. (1983). The distribution of ATPase activities in purified transverse tubular membranes. *Arch. Biochem. Biophys.* **223,** 107–119.

Sabolić, I., Culić, O., Lin, S. H., and Brown, D. (1992). Localization of ecto-ATPase in rat kidney and isolated renal cortical membrane vesicles. *Am. J. Physiol.* **262,** F217–F228.

Saborido, A., Moro, G., and Megias, A. (1991a). Transverse tubule Mg^{2+} ATPase of skeletal muscle. *J. Biol. Chem.* **266,** 23490–23498.

Saborido, A., Vila, J., Molano, F., and Megias, A. (1991b). Effect of anabolic steroids on mitochondria and sarcotubular system of skeletal muscle. *J. Appl. Physiol.* **70**(3), 1038–1043.

Salzman, E. W., Chambers, D. A., and Neri, L. L. (1966). Possible mechanism of aggregation of blood platelets by andenosine diphosphate. *Nature (London)* **210,** 167–169.

San, R. H. C., Laspia, M. F., Soiefer, A. I., Maslansky, C. J., Rice, J. M., and Williams, G. M. (1979). A survey of growth in soft agar and cell surface properties as markers for transformation in adult rat liver epithelial-like cell cultures. *Cancer Res.* **39,** 1026–1034.

Sarkis, J. J. F., and Salto, C. (1991). Characterization of a synaptosomal ATP diphosphydrolase from the electric organ of torpedo marmorata. *Brain Res. Bull.* **26,** 871–876.

Sarkis, J. J. F., Guimaraes, J. A., and Ribeiro, J. M. C. (1986). Salivary apyrase of *Rhodnius prolixus*. *Biochem. J.* **233,** 885–891.

Seifert, R., and Schulz, G. (1989). Involvement of pyrimidinoceptors in the regulation of cell functions by uridine and uracil nucleotides. *Trends Pharmacol. Sci.* **10,** 365–369.

Shi, X.-J., and Knowles, A. F. (1994). Prevalence of the mercurial sensitive Ecto-ATPase in human small cell lung carcinoma: Characterization and partial purification. *Arch. Biochem. Biophys.* **315,** 177–184.

Silinsky, E. M., Gerzanich, V., and Vanner, S. M. (1992). ATP mediates excitatory synaptic transmission in mammalian neurones. *Br. J. Pharmacol.* **106,** 762–763.

Sippel, C. J., Ananthanarayanan, M., and Suchy, F. J. (1990). Isolation and characterization of the canalicular membrane bile acid transport protein of rat liver. *Am. J. Physiol.* **258,** G728–G737.

Sippel, C. J., Suchy, F. J., Ananthanarayanan, M., and Perlmutter, D. H. (1993). The rat liver Ecto-ATPase is also a canalicular bile acid transport protein. *J. Biol. Chem.* **268,** 2083–2091.

Sippel, C. J., McCollum, M. J., and Perlmutter, D. H. (1994). Bile acid transport by the rat liver canalicular bile acid transport/Ecto-ATPase protein is dependent on ATP but not on its own Ecto-ATPase activity. *J. Biol. Chem.* **269,** 2820–2826.

Skou, J. C. (1957). The influence of som cations on an adenosine triphophatase from peripheral nerves. *Biochim. Biophys. Acta* **23,** 394–401.

Slakey, L. L., Earls, J. D., Guzek, D., and Gordon, E. L. (1986). Regulation of the hydrolysis of adenine nucleotides at the surface of cultured vascular cells. *In* "Cellular Biology of Ectoenzymes" (G. W. Kreutzberg *et al.*, eds.), pp 27–34. Springer-Verlag, Berlin.

Smith, J. J. B., Cornish, R. A., and Wilkes, J. (1980). Properties of a calcium-dependent apyrase in the saliva of the blood-feeding bug, *Rhodnius prolixus*. *Experientia* **36,** 898–900.

Smith, U., and Ryan, J. W. (1970). An electron microscopic study of the vascular endothelium as a site for bradykinin and adenosine 5′-triphosphate inactivation in rat lung. *Adv. Exp. Med. Biol.* **8,** 249–261.

Smolen, J. E., and Weissmann, G. (1978). Mg^{2+}-ATPase as a membrane ecto-enzyme of human granulocytes. *Biochim. Biophys. Acta* **512,** 525–538.

Stefanović, V., Ciesielski-Treska, J., and Mandel, P. (1977). Neuroblasts-glia interaction in tissue culture as evidenced by the study of ectoenzymes. Ecto-ATPase activity of mouse neuroblastoma cells. *Brain Res.* **122,** 313–323.

Stefanović, V., Ledig, M., and Mandel, P. (1976). Divalent cation activated ecto-nucleoside triphosphatase activity of nervous system cells in tissue culture. *J. Neurochem.* **27,** 799–805.

Stewart, C. C., Gasić, G., and Hempling, H. G. (1969). Effect of exogenous ATP on the volume of TA_3 ascites tumor cells. *J. Cell. Physiol.* **73,** 125–132.

Stout, J. G., and Kirley, T. L. (1994a). Tissue distribution of ecto-Mg-ATPase in adult and embryonic chicken. *Biochem. Mol. Biol. Int.* **32,** 745–753.

Stout, J. G., and Kirley, T. L. (1994b). Purification and characterization of the ecto-Mg-ATPase of chicken gizzard smooth muscle. *J. Biochem. Biophys. Methods* (in press).

Strobel, R. (1993). Purification and immunochemical characterization of ecto-ATP-diphosphohydrolase from chicken oviduct and liver. Ph. D. Thesis, University of Minnesota, Minneapolis.

Strobel, R., and Rosenberg, M. (1992). Immunoaffinity chromatographic purification of chicken ectoATP diphosphohydrolase. *Ann. N. Y. Acad. Sci.* 487–489.

Su, C. (1983). Purinergic neurotransmission and neuromodulation. *Annu. Rev. Pharmacol. Toxoicol.* **23,** 397–411.

Sun, H.-T., Yoshida, Y., and Imai, S. (1990). A Ca^{2+}-activated, Mg^{2+}-dependent ATPase with high affinities for both Ca^{2+} and Mg^{2+} in vascular smooth muscle microsomes: Comparison with plasma membrane Ca^{2+}-pump ATPase. *J. Biochem.* (*Tokyo*) **108,** 730–736.

Tenny, S. R., and Rafter, G. W. (1968). Leukocyte adenosine triphosphatases and the effect of endotoxin of their activity. *Arch. Biochem. Biophys.* **126,** 53–58.

Teo, T. S., Thiyagarajah, P., and Lee, M. K. (1988). Characterization of a high affinity Ca^{2+}-stimulated, Mg^{2+}-dependent ATPase in the rat parotid plasma membrane. *Biochim. Biophys. Acta* **945,** 202–210.

Tingström, A., Blikstad, I., Aurivillius, M., and Öbrink, B. (1990). C-CAM (Cell-CAM 105) is an adhesive cell surface glycoprotein with homophilic binding properties. *J. Cell Sci.* **96,** 17–25.

Torres, M., Pintor, J., and Miras-Portugal, M. T. (1990). Presence of ectonucleotidases in cultured chromaffin cells: Hydrolysis of extracellular adenine nucleotides. *Arch. Biochem. Biophys.* **279,** 37–44.

Trams, E. G. (1974). Evidence for ATP action on the cell surface. *Nature* (*London*) **252,** 480–482.

Trams, E. G., and Lauter, C. J. (1974). On the sidedness of plasma membrane enzymes. *Biochim. Biophys. Acta* **345,** 180–197.

Trams, E. G., and Lauter, C. J. (1978a). A comparative study of brain Ca^{2+}-ATPases. *Comp. Biochem. Physiol. B* **59B,** 191–194.

Trams, E. G., and Lauter, C. J. (1978b). Ecto-ATPase deficiency in glia of seizure-prone mice. *Nature* (*London*) **271,** 270–271.

Traverso-Cori, A., Chaimovich, H., and Cori, O. (1965). Kinetic studies and properties of potato apyrase. *Arch. Biochem. Biophys.* **109,** 173–184.

Treuheit, M. J., Vaghy, P. L., and Kirley, T. L. (1992). Mg^{2+}-ATPase from rabbit skeletal-muscle transverse tubules is 67-kilodalton glycoprotein. *J. Biol. Chem.* **267,** 11777-11782.

Tuana, B. S., and Dhalla, N. S. (1988). Purification and characterization of Ca^{2+}/Mg^{2+} ecto-ATPase from rat heart sarcolemma. *Mol. Cell. Biochem.* **81,** 75–88.

Turrini, F., Sabolic, I., Zimolo, Z., Moewes, B., and Burckhardt, G. (1989). Relation of ATPases in rat renal brush-border membranes to ATP-driven H^+ secretion. *J. Membr. Biol.* **107,** 1–12.

Valente, A. P. C., Barrabin, H., Jorge, R. V., Paes, M. C., and Scofano, H. M. (1990). Isolation and characterization of the Mg^{2+}-ATPase from rabbit skeletal muscle sarcoplasmic reticulum membrane preparations. *Biochim. Biophys. Acta* **1039,** 297–304.

Van Erum, M., Martens, L., Vanduffel, L., and Teuchy, H. (1988). The localization of (Ca^{2+} or Mg^{2+})-ATPases in plasma membranes of renal proximal tubular cells. *Biochim. Biophys. Acta.* **937,** 145–152.

Vara, F., and Serrano, R. (1981). Purification and characterization of a membrane-bound ATP diphosphohydrolase from *Cicer arietinum* (chick-pea) roots. *Biochem. J.* **197,** 637–643.

Vasconcelos, E. G., Nascimento, P. S., Meirelles, M. N. L., Verjovski-Almeida, S., and Ferreira, S. T. (1993). Characterization and locatization of an ATP-disphophohydrolase on the external surface of the tegument of *Schistosoma mansoni*. *Mol. Biochem. Parasitol.* **58,** 205–214.

Vassort, G., Scamps, F., Puceat, M., and Clement, O. (1992). Multiple effects of extracellular ATP in cardiac tissues. *News Physiol. Sci.* **7,** 212–215.

Vestal, D. J., and Ranscht, B. (1992). Glycosyl phosphatidylinositol-anchored T-cadherin mediates calcium-dependent, homophilic cell adhesion. *J. Cell Biol.* **119,** 451–461.

von Kugelgen, I., and Starke, K. (1991). Noradrenaline-ATP cotransmission in the sympathetic nervous system. *Trends Pharmacol. Sci.* **12,** 319–324.

Wachstein, M., and Meisel, E. (1957). Histochemistry of hepatic phosphatases at a physiologic pH. *Am. J. Clin. Path* **27,** 13.

Wallach, D. F. H., and Ullrey, D. (1962). The hydrolysis of ATP and related nucleotides by Ehrlich ascites carcinoma cells. *Cancer Res.* **22,** 228–234.

Wang, T. Y., Hussey, C. V., Sasse, E. A., and Hause, L. L. (1977). Platelet aggregation and the ouabain-insensitive ATPase. Ecto-ATPase, reflection of membrane integrity. *Am. J. Clin. Pathol.* **67,** 528–532.

Weiss, B., and Sachs, L. (1977). Differences in surface membrane ecto-ATPase and ecto-AMPase in normal and malignant cells. *J. Cell. Physiol.* **93,** 183–188.

Welford, L. A., Cusack, N. J., and Hourani, S. M. O. (1986). ATP analogues and the guinea-pig taenia coli: A comparison of the structure-activity relationships of ectonucleotidases with those of the P_2-purinoceptor. *Eur. J. Pharmacol.* **129,** 217–224.

Welford, L. A., Cusack, N. J., and Hourani, S. M. O. (1987). The structure-activity relationships of ectonucleotidases and of excitatory P_2-purinoceptors: Evidence that dephosphorylation of ATP analogues reduces pharmacological potency. *Eur. J. Pharmacol.* **141,** 123–130.

Westfall, D. P., Sedaa, K. O., Shinozuka, K., Bjur, R. A., and Buxton, I. L. (1990). ATP as a cotransmitter. *Ann. N.Y. Acad. Sci.* **603,** 300–310.

White, J. G., and Krivit, W. (1965). Fine structural localization of adenosine triphosphatase in human platelets and other blood cells. *Blood,* **26,** 554–568.

White, T. D. (1988). Role of adenine compounds in autonomic neurotransmission. *Pharmacol. Ther.* **38,** 129–168.

Yagi, K., Arai, Y., Kato, N., Hirota, K., and Miura, Y. (1989). Purification of ATP diphosphohydrolase from bovine aorta microsomes. *Eur. J. Biochem.* **180,** 509–513.

Yagi, K., Shinbo, M., Hashizume, M., Shimba, L. S., Kurimura, S., and Miura, Y. (1991). ATP diphosphohydrolase is responsible for Ecto-ATPase and Ecto-ADPase activities in bovine aorta endothelial and smooth muscle cells. *Biochem. Biophys. Res. Commun.* **180,** 1200–1206.

Yagi, K., Kato, N., Shinbo, M., Shimba, L. S., and Miura, Y. (1992). Purification and characterization of adenosine diphosphatase from human umbilical vessels. *Chem. Pharm. Bull.* **40,** 2143–2146.

Yamaguchi, K., and Ohnishi, T. (1977). Surface ATPase activity at cell-cell contacts in hepatic parenchymal cells and in cAMP-treated hepatoma cell in monolayer culture. *Histochemistry* **54,** 191–199.

Yazaki, P. J., Cunningham, H. B., Kang, J. J., Jachec, C., Hunt, S. A., Domingo, R. C., Sabbadini, R. A., and Dahms, A. S. (1992). Characterization of a muscle membrane ATPase glycoprotein. *In* "Excitation-Contraction Coupling in Skeletal, Cardiac, and Smooth Muscle" (G. B. Frank *et al.,* eds.), pp. 425–426. Plenum, Press, New York.

Yoshida, Y., Sun, H.-T., and Immai, S. (1990). Two high affinity (Ca^{2+} + Mg^{2+})-ATPases of vascular smooth muscle plasma membrane preparation. Their relation to the Ca^{2+}-pumping ATPase. *Am. J. Hypertens.* **3,** 2495–2525.

Zhao, D., and Dhalla, N. S. (1988). Characterization of rat heart plasma membrane Ca^{2+}/Mg^{2+} ATPase. *Arch. Biochem. Biophys.* **263,** 281–292.

Zhao, D., and Dhalla, N. S. (1991). Purification and composition of Ca^{2+}/Mg^{2+} ATPase from rat heart plasma membrane. *Mol. Cell. Biochem.* **107,** 135–149.

Zhao, D., Elimban, V., and Dhalla, N. S. (1991). Characterization of the purified rat heart plasma membrane Ca^{2+}/Mg^{2+} ATPase. *Mol. Cell. Biochem.* **107,** 151–160.

Molecular Genetic Approaches to the Study of Human Craniofacial Dysmorphologies

Gudrun E. Moore
The Action Research Laboratory for the Molecular Biology of Fetal Development, Institute of Obstetrics and Gynaecology, Queen Charlotte's and Chelsea Hospital, Royal Postgraduate Medical School, London W6 OXG, United Kingdom

Craniofacial dysmorphologies are common, ranging from simple facial disfigurement to complex malformations involving the whole head. With the advent of gene mapping and cloning techniques, the genetic element of both simple and complex human craniofacial dysmorphologies can be investigated. For many of the dysmorphic syndromes, it is possible to find families that display a particular phenotype in either an autosomal dominant, recessive, or X-linked manner. This article focuses on a subgroup of craniofacial dysmorphologies, covering these three main inheritance patterns, that are being studied using molecular biology techniques: DiGeorge syndrome, Treacher Collins syndrome, Greig cephalopolysyndactyly syndrome, acrocallosal syndrome, amelogenesis imperfecta, and X-linked cleft palate with ankyloglossia. Once the mutated or deleted gene or genes for each syndrome have been cloned, patterns of normal and abnormal craniofacial development should be elucidated. This should enhance both diagnosis and treatment of these common and disfiguring disorders.

KEY WORDS: Positional cloning, DiGeorge syndrome, Treacher Collins syndrome, Greig syndrome, Acrocallosal syndrome, Amelogenesis imperfecta, X-linked cleft palate.

I. Introduction

Craniofacial dysmorphologies can be categorized as a subset of congenital malformations. A dysmorphology is a description of a developmental state that has been caused by a disorder of morphogenesis that is thought to be intrinsic. This should be differentiated from a disruption or deformation that is almost always extrinsic and a result of a mechanical constraint *in utero*. Deformations of the head and neck are common but usually spontaneously correct themselves in the first few days after birth. Two percent of live-born infants have extrinsically caused deformations that are a result of factors such as abnormal fetal presentation, small maternal size, fetal crowding, and amniotic rupture. This group of deformities is not reviewed in this article, as their causations are almost always extrinsic and environmental rather than genetic. The majority of craniofacial dysmorphologies are congenital malformations diagnosed at birth. There are congenital malformations that show an exclusive craniofacial defect such as cleft palate, but many others have multiple associated abnormalities including cleft lip and palate, such as Patau syndrome (trisomy 13). It is difficult to state categorically how many different separate carniofacial dysmorphologies exist because their etiology is not fully understood and many might become one when their etiology is completely elucidated. There are hundreds that have been separately documented (Gorlin *et al.*, 1990). They can be grouped by their morphology alone from simple to complex, i.e., isolated cleft lip, and cleft lip plus secondary cleft palate. Alternatively they can be grouped by their pattern of Mendelian inheritance, i.e., all autosomal dominant (cleft lip alone, cleft lip plus cleft palate, cleft palate alone, etc.). This will leave many syndromes unclassified and these can then be referred to only as sporadic, multifactorial, or of unknown etiology. The best method of classification combines both morphology and genetic predisposition, as an understanding of both elements is nescessary to explain the ultimate cause of the disorder.

This review attempts to present examples of craniofacial dysmorphology that have a clear genetic component. For those of us who work on these selected models, the long-term aims are to clone the genes involved and thus gain an insight into the mechanisms of the other more common and apparently nongenetic forms of the disorders.

II. Head Development and Causes of Congenital Malformations in the Human Fetus

A. Head Development in the Human Fetus

The critical time period for head development is between weeks 4 and 8 postconception. The head of the human fetus first begins to develop at the beginning of the fourth week. The ends of the embryo (which consists of a cylindrical trilaminar disk) fold to produce the *head fold* cranially and the *tail fold* caudally. The neural folds develop into brain and the forebrain extends out from the oropharyngeal membrane overhanging the heart. The oropharyngeal membrane separates the foregut from the primitive mouth. At 24 days the first mandibular and second hyoid branchial arches are distinct. The first branchial arch gives rise to the mandible and maxilla of the jaw. By 26 days, three pairs of branchial arches and the otic pits (inner ear primordia) are visible. By 28 days, four branchial arches and the lens placodes are present (Fig. 1). During the fifth week the brain develops rapidly, and as a consequence growth of the head is greater than that of other regions. The second hyoid arch forms a depression, the cervical sinus. In the sixth week, swellings arise in the groove between the first two arches. The groove becomes the external acoustic meatus and the swellings fuse to form the auricle of the external ear. By the eighth week the head is twice as large as the body, the neck is established, and the eyelids are distinct, with the eyes able to open.

1. Development of the Branchial Arches

Many craniofacial congenital malformations arise from incorrect development of the branchial arches. As neural crest cells migrate in the head and neck region early in the fourth week, the branchial arches start to become visible. By the end of this week all four arches are visible. The fifth and sixth arches are small and not clearly discernible. Each arch has a mesenchymal core containing migrating cranial neural crest cells, external ectoderm, and internal endoderm. Each arch also has an artery, a cartilage bar, a nerve, and a muscle component. The external part

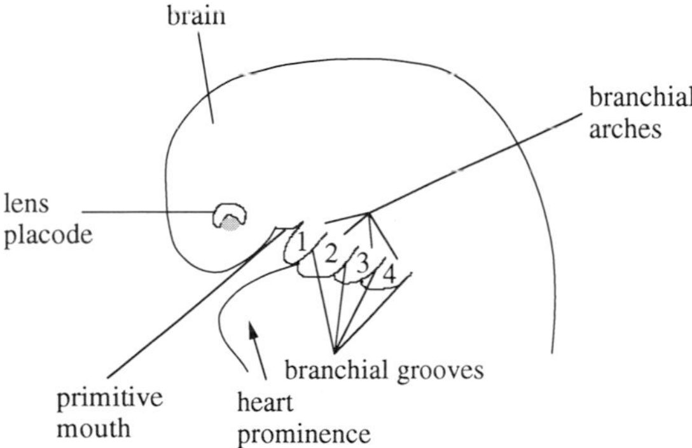

FIG. 1. Head development at 28 days.

(ectoderm) of the arch forms branchial grooves and the internal part (endoderm) pharyngeal pouches. Where they meet a branchial membrane is formed. The branchial arches contribute to the formation of the tongue, face, jaws, palate, neck, nasal cavities, mouth, larynx, and pharynx. All the grooves and membranes disappear during development, apart from the first, which forms the external acoustic meatus and tympanic membrane, respectively. The pharngeal pouches give rise to the tympanic cavity and mastoid antrum, the auditory tube, the palatine tonsils, the thymus, and parathyroid glands.

2. Development of the Face

Facial development occurs between the fifth to eighth week postconception. There are five facial primordia that appear in the fourth week around the primitive mouth (Fig. 2): the frontonasal prominence, developing from mesenchymal proliferation ventral to the forebrain, forming the cranial boundary of the mouth; the two maxillary prominences of the first branchial arch, forming the lateral aspects of the mouth; the two mandibular prominences of the first branchial arch, forming the caudal boundaries of the mouth. Congenital abnormalities arising from failure of fusion or developmental of these prominences will lead to malformation of the face.

During the fourth week, the lower jaw or mandible results from a merging of the two mandibular prominences. By the end of the fourth week, thickenings of surface ectoderm occur on each side of the lower part of

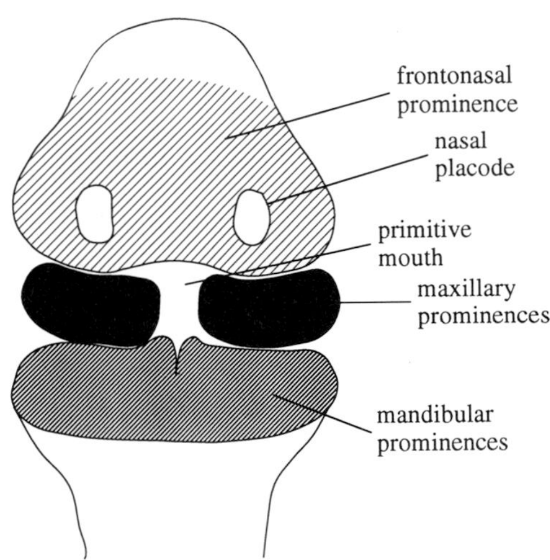

FIG. 2. Facial primordia at 28 days.

the frontonasal prominences, giving rise to the nasal placodes, which lie in depressions called nasal pits. In between, the mesenchyme proliferates, producing the medial and lateral nasal prominences. The maxillary prominences increase in size and grow toward each other and toward the medial nasal prominences. Separation of these tissues is achieved through the nasolacrimal groove forming the sides of the nose. At the end of the fifth week the eyes point forward on the face and the external ear develops. During week 6 the medial nasal prominences merge, forming the intermaxillary segment, which give rise to the philtrum of the lip, gum, and primary palate. The lateral part of the lip and secondary palate arise from the maxillary prominences merging laterally with the mandibular prominences. Mesenchyme from the second branchial arch invades the lips and cheeks to form the facial muscles and the facial nerve, which is also derived from the same arch. The muscles of mastication are formed from the first pair of arches and are innervated by the trigeminal nerve from the same arches. The forehead and the dorsum and apex of the nose are derived from the frontonasal prominences; the sides of the nose from the lateral nasal prominences; the nasal septum and philtrum from the medial nasal prominences; and the lower lip, lower cheek, and chin from the mandibular prominences.

3. Development of the Tongue

The tongue starts to develop at the end of the fourth week from an elevation on the floor of the pharynx. The median tongue bud develops two swellings on either side, called the distal tongue buds. These arise from mesenchyme derived from the ventromedial regions of the first pair of branchial arches. The distal tongue buds grow and merge over the median tongue bud, forming the anterior, oral part of the tongue. The posterior part of the tongue develops from two caudal elevations: the copula, from the second branchial arch, and the hypobranchial eminence from the third and fourth branchial arches. The hypobranchial eminence eventually overgrows the copula, which disappears. The tongue musculature derives from myoblasts that migrate from the occipital somites. The innervation of the tongue arises from several branchial arches: the lingual branch of the mandibular part of the trigeminal nerve from the first branchial arch supplies the oral part of the mucosal sensory supply; the chorda tympani branch of the facial nerve from the second branchial arch supplies the taste buds in the front two-thirds of the tongue; the glossopharyngeal nerve of the third branchial arch supplies the vallate papillae in the anterior two-thirds of the tongue and the pharyngeal posterior part of the tongue; the laryngeal branch of the vagus nerve from the fourth arch supplies a small area of the tongue to the front of the epiglottis.

4. Development of the Palate

Palatogenesis starts at the end of the fifth week, with complete palatal fusion occurring by the twelfth week (Fig. 3). The palate arises from two primordia: primary and secondary. The primary palate arises from the inner segment of the maxilla, formed from the medial nasal prominences. The primary palate contains the incisors and forms a small part of the hard palate situated anterior to the incisive foramen. The secondary palate includes both hard and soft palate, extending posteriorly from the primary palate. The internal parts of the maxillary prominences form two horizontal, mesodermal shelflike structures, the lateral palatine processes, initially projecting downward on each side of the tongue. During the seventh week, as the head expands, the tongue moves downward, leaving the lateral processes to grow and move into position over the tongue. These then fuse together in the middle and then with the primary palate and nasal septum by the ninth week. Bone develops in the primary palate, first extending into the lateral processes to form hard palate. The back of the palate remains without underlying bone, forming the soft palate and uvula.

5. Development of the Eyes and Ears

The development of the eyes and ears is complex, and although extremely important in the complete picture of head development, the craniofacial malformations detailed in Section IV do not focus on these structures.

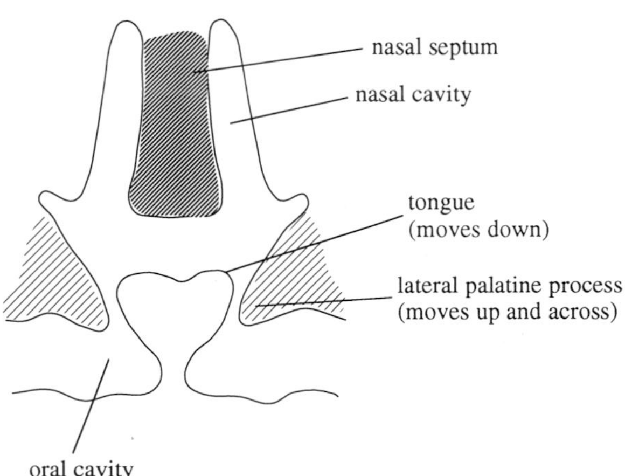

FIG. 3. Cross-section through the fetal head at 42 days, showing the lateral palatine shelves in relation to the tongue and nasal septum.

For the sake of space and the balance of this article they have been omitted (for review, see Moore, 1988).

6. Development of the Tooth

Teeth develop from endoderm and ectoderm, enamel being produced from oral ectoderm and mesenchyme giving rise to all other dentally related tissues. Defective formation of enamel and dentin, and shape and number differences are the common malformations found in teeth. Two sets of teeth develop, the primary or deciduous dentition and the secondary or permanent dentition. Underlying mesenchyme derived from the neural crest induces the overlying ectoderm, first in the anterior mandibular region, then in the anterior maxillary region, and finally progressing posteriorly in both jaws. Tooth development, which is continuous for many years after birth, can be split into three stages called bud, cap, and bell. During the sixth week the oral epithelium thickens, folding inward into the underlying mesenchyme, forming dental laminae. Proliferation within these laminae produces swellings called tooth buds, which grow into the mesenchyme. There are 10 deciduous tooth buds in each jaw. During the cap stage, the internal surface of the tooth bud becomes infiltrated by condensed mesenchyme called the dental papillae, giving rise to the dentin and dental pulp. The ectoderm of this cap stage tooth later produces enamel called the enamel organ. Mesenchyme around the dental papillae and the enamel organ condenses, forming a follicle that goes on to form the cementum and the periodontal ligament. With the differentiation of the enamel organ the tooth becomes bell shaped. Cells in the dental papilla differentiate into odontoblasts producing predentin. This later calcifies to give dentin. Cells adjacent to the dentin differentiate into ameloblasts that produce enamel over the dentin. Formation of the enamel and dentin starts at the cusp of the tooth, and progresses downward to the root. The epithelial root sheath is formed from inner and outer layers of enamel at the neck region of the tooth. This sheath grows into the mesenchyme, giving rise to root formation. The jaws ossify as the teeth develop, with the teeth held in bony sockets. The permanent teeth develop in a manner similar to the deciduous teeth, 6 to 7 years later.

7. Development of the Brain

Brain development is complex and once again, to save space and to maintain the balance of this article, only the briefest summary is given. The cranial end of the neural tube gives rise to the forebrain, midbrain, and hindbrain. The forebrain consists of the cerebral hemispheres and diencephalon. The midbrain goes on the form the adult midbrain, with the hindbrain giving rise to the pons, cerebellum, and medulla oblongata. The neural tube lumen forms the ventricles of the brain. (For more details on the development of the head and neck, see Moore, 1988.)

B. Causes of Congenital Malformations

The causes of congenital malformations are mostly unknown (60%), but a proportion can be attributed to single gene defects (7.5%), chromosomal aberrations (6%), and multifactorial traits (20%), with maternal factors such as infection and diabetes making up the remainder. The two largest groups are the multifactorial and unknown. Even those that are deemed multifactorial are difficult to study, as they are composed of a proportion of genetic factors combined with environmental factors.

From a genetic viewpoint the easiest malformations to approach are the single-gene defects. These can be split into those mutations associated with the autosomes that are either recessive or dominant in phenotype and those on the X chromosome (recessive or dominant). There are many examples of single-gene defects that cause craniofacial dysmorphology. Examples of these are dealt with in depth in Section IV. The second group that can give insight into the role of specific genes in the development of the head and neck consists of those craniofacial dysmorphologies that are found as a result of a chromosomal aberration, whether numerical or structural.

1. Numerical Chromosomal Abnormalities

a. Autosomes *Aneuploidy* is defined as a chromosome number that is not an exact multiple of the haploid number. The majority of aneuploids have one extra chromosome of a pair, a trisomy, or one less, a monosomy. In most cases an extra chromosome is not compatible with life. In miscarriage trisomy for nearly all the chromosomes are found except for chromosome 1 (Boue *et al.,* 1975). Trisomy of chromosome 1 must therefore be incompatible with implantation. The three most common trisomies that are compatible with life are trisomy 21 (Down syndrome), trisomy 18 (Edwards syndrome), and trisomy 13 (Patau syndrome). All three of these syndromes have multiple associated malformations and all include craniofacial abnormalities (Table I).

This implies that there will be many genes on these chromosomes that are important for normal development of the head and neck. By selecting candidate genes from these particular chromosomes that have already been cloned, expression analysis can be carried out in affected tissues

TABLE I

Malformations of the Head and Neck Found in Down, Edwards, and Patau Syndromes[a,b]

Malformation	Syndrome		
	Down	Edwards	Patau
Craniofacial			
Flat nasal bridge	61		
Prominent nasal bridge			100
Flat occiput	76		75
Prominent occiput		91	
Scalp defects			75
Sloping forehead			100
Micrognathia		96	84
Eyes			
Upslanting palpebral fissure	79		
Narrow palpebral fissure		80	
Epicanthic folds	48		56
Brushfield spots	53		
Strabismus	22		
Nystagmus	11		
Ocular hypotelorism			83
Micropthalmia		82	76
Iris coloboma			33
Ears			
Dysplastic	53		
Absent lobules	70		
Malformed ears		88	80
Mouth			
Open mouth	61		
Small mouth		86	
Fissured lips	56		
Protruding tongue	42		
Macroglossia	43		
Furrowing of tongue	61		
Narrow palate	67		
Cleft palate	0.5		69
High arched palate		87	72
Cleft lip	0.5		58
Irregular teeth	71		
Neck			
Broad, short	53		79
Loose skin, nape		86	59

[a] Adapted from Gorlin *et al.* (1990, Chapter 3).

[b] Numbers represent percentages observed.

from trisomic abortions and compared with normal age-matched tissues (Loughna *et al.*, 1993).

b. Sex Chromosomes Monosomy X (45,X) is found in 18% of chromosomally abnormal abortuses but only in 1 in 20,000 live births. Turner, (1938), described the syndrome at birth as a condition of sexual infantilism, short stature, webbing of the neck, and cubitus valgus. Other craniofacial features include a low hairline and a characteristic facial appearance with epicanthic folds, ptosis of the eyelids, prominent ears, and micrognathia (Jensen, 1985). The palate is highly arched in 36% of cases and cleft palate can occur at a higher than normal frequency (Horowitz and Morishima, 1974).

Trisomy for the sex chromosomes is also quite common, with an incidence of 1 in 1000 births each (XXY, XYY, and XXX). In general these are associated with few craniofacial dysmorphologies and as a consequence are not dealt with here in depth.

Polyploidy (extra haploid sets) is also found. Triploidy is a common cause of fetal wastage prior to and during the second trimester (20%) (Bendon *et al.*, 1988). Most newborns die in the first few days of life but a few survive to early infancy. True triploidy is estimated to occur in 1 in 2500 births with 60% being 69,XXY and the rest mostly 69,XXX (Gosden *et al.*, 1976). There are associated craniofacial abnormalities, such as a large posterior fontanel that is always found and asymmetry of the occipitoparietal calvaria in 50% of cases. Micropthalmia, iris and choroid colobomas, and epicanthic folds are also seen in 25% of triploidy cases (Bendon *et al.*, 1988).

Not surprisingly, an abnormal chromosome number, whether it be due to a single whole chromosome or a complete additional haploid set, causes a huge range of dysmorphic features including those affecting the head and neck. Analysis of these complex malformations using molecular techniques is difficult as so much genetic material has been duplicated or is missing.

2. Structural Chromosomal Abnormalities

Structural chromosomal rearrangements result from chromosome breakage followed by abnormal reconstitution. Some structural changes, although abnormal, can be stable and undergo cell division. If they occur in germ cells they are then passed to the next generation. Stable types of structural abnormalities include deletions, duplications, inversions, translocations, insertions, and isochromosomes.

a. Deletions When a chromosome breaks a piece can be lost or deleted. When the centromere is deleted an acentric unstable chromosome may result. Deletions have been reported for all the autosomes and the sex chromosomes. Almost all have many associated malformations including craniofacial abnormalities. When the deletion is small and can be cytogenetically localized, the phenotype seen can be correlated with genes that are missing or interrupted. This localization can then give positional information about a gene or genes that may play a specific role in normal development. As documented in more detail in Section IV, the association of a specific deletion on chromosome 22 with DiGeorge syndrome has led to some exciting developments in cloning a family of genes important in craniofacial and neural crest development.

b. Duplications When a piece of a chromosome replicates itself the result will be a duplicated region. Duplications do not give rise to such severe phenotypes as deletions because genetic material is not lost. Like deletions, duplications have been reported for all the chromosomes and have associated malformations including craniofacial abnormalities. For example, duplication of 9p has classic craniofacial stigmata. The facies has a high, broad forehead, large fontanel, and open metopic suture in childhood, mild microbrachycephaly, flat occiput, enophthalmus, mild hypertelorism, and divergent strabismus. The eyes are small, and deep set, with downward slanting palpebral fissures. The nose has a globular tip, broad nasal root, and short philtrum. The large mouth has a lower everted lip. The pinnae are large, low set, with abnormal anthelix. The neck is short and webbed, with a low hairline (Young *et al.*, 1982). Like deletions, duplications can pinpoint chromosomal regions that contain specific genes that when disrupted or duplicated lead to malformation.

c. Inversions If a chromosome breaks in two places and the fragmented material inverts and reforms, an inversion has occurred. Inversions do not appear to lead to abnormal phenotypes unless the site of the inversion disrupts a gene or genes. However, as inversions are inherited the problem manifests itself in the next generation. Crossing over occurring between homologous chromosomes within the inverted region can lead to an imbalance in the genetic material transferred.

d. Translocations There are two main types of translocation. The first is the balanced reciprocal translocation in which pieces of chromosome are swapped between two nonhomologous chromosomes. These do not usually result in abnormal phenotypes but can, like inversions, be a problem in the next generation in the production of unbalanced gametes. The second type of translocation is the Robertsonian translocation that arises when two acrocentric chromosomes fuse at the centromere, losing their short arms. Like balanced translocations and inversions, the abnormal phenotype caused by Robertsonian translocations arises when the gametes are formed. Six types of gametes are possible, of which only one is normal. Two will be disomic for one of the two translocated chromosomes and two will be nullisomic. The remaining gamete will carry the same translocation as its parent. For example, if the Robertsonian translocation is between chromosome 14 and 21, there is a one-sixth chance of a trisomy 21 offspring or a monosomy 21 offspring. These fetuses will have abnormal phenotypes similar to those of Down syndrome or monosomy 21, respectively.

3. Multifactorial Inheritance

It is estimated that the cause of at least 20% of congenital malformations is multifactorial. This means that although there is a genetic part to their etiology, there is also an environmental component. Both types of insult are needed for the malformation to occur. There may be many genes and many environmental factors that each play a minor causative role.

The multifactorial-threshold model was developed to try to explain how these two types of insult might interact. Most malformations are stage specific, that is, if an insult occurs at a certain stage then subsequent developmental stages cannot be reached and normal development is arrested. At this arresting stage the continuously variable distribution becomes divided into a normal and an abnormal class separated by a threshold (Fig. 4). The simplest way to analyze potential multifactorial traits is

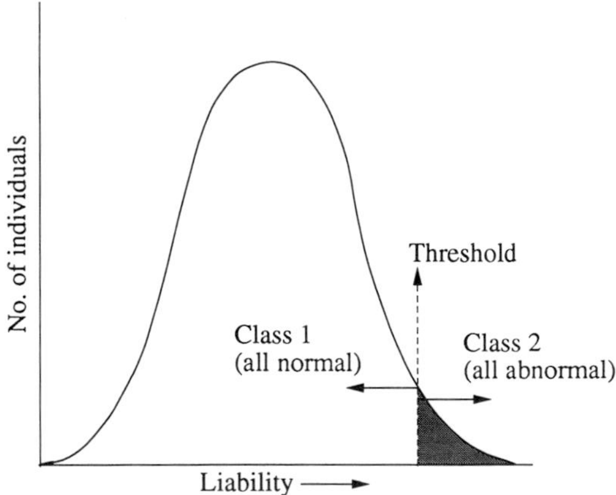

FIG. 4. A continuously variable distribution divided into abnormal and normal classes by a threshold.

to assess the frequency of the malformation in the general population and compare it with that seen in various categories of relatives within the affected families. Using this information the empirical risk to other potential family members can be assessed.

In an attempt to separate the role of genes from that of the environment, the concept of heritability has been developed. Heritability is defined as the proportion of the total phenotype variance of a trait that is caused by additive genetic variance. The liability of an individual to a threshold trait is taken as a combination of three values: g (the contribution of genetic factors with variance G), b (the contribution of environment within the family with variance B), and e (the contribution of random environmental factors with variance E):

$$\text{Heritability } (h^2) = \frac{G}{(G + B + E)} = G/V$$

Heritability is a measure of whether the role of genetic factors in producing a malformation is large or small. The nearer the value is to 1, the greater the genetic role.

One way of estimating heritability is by comparison of the ratio of the concordance rate of a condition in monozygotic (MZ) and dizygotic (DZ) twin pairs with the incidence in the general population. Monozygotic twins have identical genetic information whereas dizygotic twins are like siblings. For cleft lip with or without cleft palate the concordance in MZ twins is 0.3 whereas in DZ twins it is 0.05. There is a much higher chance for the monozygous twins to share the dysmorphology, indicating that a genetic input is playing a part in the etiology.

III. Techniques of Gene Analysis

A. Linkage Studies in Families

The development of recombinant DNA technology has revolutionized the study of inherited diseases. Among many other important possibilities, the characterization of mutant genes should provide the basis for early prenatal diagnosis, information that might lead to better treatments than had been previously available, and the possibility of gene therapy. Over the last 15 years several hundred disease genes have been identified, mostly through some knowledge of the protein product. However, the vast majority of single-gene disorders give no obvious clues that allow a functional approach to cloning. The technique of positional cloning has become an important tool with cystic fibrosis, myotonic dystrophy, and

Huntington's disease among a list of about 20 disease genes cloned in this way (Table II).

The technique of positional cloning, as its name implies, sets out to determine a precise chromosomal localization for the disease gene. The localization is then successively narrowed down until the correct gene can be identified and characterized. Central to this process is linkage mapping. Linkage analysis is used to follow the inheritance of polymorphic DNA marker alleles of known chromosomal localization compared with the inheritance of a disease allele, in affected families. When a DNA marker is located a large distance away from the disease gene, the inheritance of the alleles that track with the disease allele will appear random owing to the effect of recombination. When a DNA marker is located close to the disease gene, opportunity for recombination between the two loci is small, and inheritance of the marker alleles will tend to follow the inheritance of the disease.

Once families in which the disease gene is segregating have been collected, and data from polymorphic markers have been obtained, the analysis is usually performed with the aid of specific computer programs. Two-point and multipoint linkage analyses are most commonly used to predict a localization (e.g., see Fig. 5). Two-point linkage analysis compares

TABLE II

Examples of Genes Identified by Positional Cloning

Year	Disease	Chromosome	Ref.
1986	Chronic granulomatous	Xp	Royer-Pokora *et al.* (1986)
1986	Duchenne muscular dystrophy	Xp	Monaco *et al.* (1986)
1989	Cystic fibrosis	7q	Riordan *et al.* (1989)
1990	Wilms' tumor	11p	Rose *et al.* (1990)
1990	Testes determining factor	Y	Sinclair *et al.* (1990)
1991	Fragile X	Xq	Verkerk *et al.* (1991); Fu *et al.* (1992)
1991	Aniridia	11p	Ton *et al.* (1991)
1992	Myotonic dystrophy	19q	Aslanidis *et al.* (1992)
1992	Norrie syndrome	Xp	Berger *et al.* (1992); Chen *et al.* (1992)
1992	Menkes syndrome	Xq	Vulpe *et al.* (1993); Chelly *et al.* (1993); Mercer *et al.* (1993)
1993	X-linked agammaglobulinemia	Xq	Vetrie *et al.* (1993)

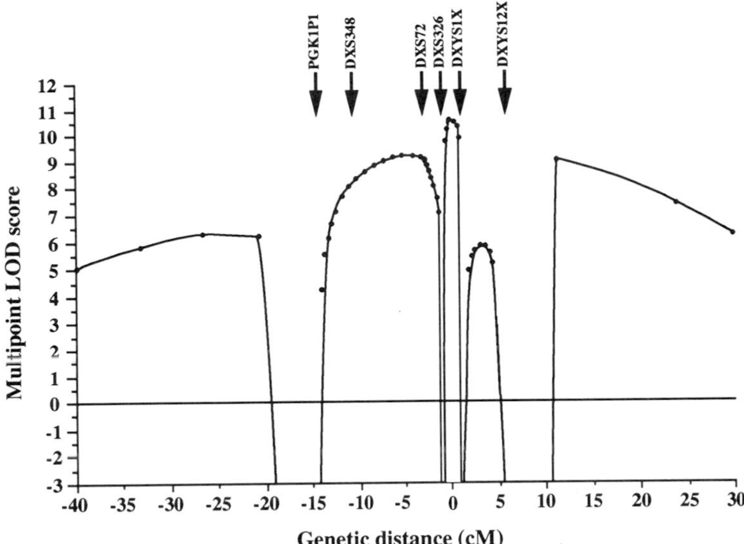

FIG. 5. Multipoint analysis showing a maximum lod score of 10.54 between DXS326 and DXYS1X for the CPX locus. The position of 0 cm is arbitrarily placed midway between DXS326 and DXYS1X.

information at the chosen marker locus with that of the disease locus. The statistical analysis calculates a lod (logarithm of the odds) score at a distance (θ) from the marker locus. Calculations are performed for a range of distances, with the peak score giving an indication of informativity of the probe and the detection of recombinations between the disease locus and the marker locus (e.g., if the peak score is at $\theta = 0$, no recombinations are detected). It is generally accepted that a lod score of > 3.0 (1000 : 1) is significant for linkage and $<$-2.0 (1 : 100) is significant for exclusion. Once a linkage has been detected, a selection of informative markers in the vicinity is used in a multipoint analysis to indicate the most likely position of the disease locus with respect to these markers.

Until recently, the most common form of polymorphic DNA marker used in this type of study was anonymous DNA fragments that detect two or three allele restriction fragement length polymorphisms (RFLP) using Southern blots of patient DNA. These random DNA fragments are assigned chromosomal localizations using a variety of techniques, including *in situ* hybridization to metaphase chromosomes and hybridization to panels of somatic cell hybrids containing limited numbers of human chromosomes or even fragments of specific human chromosomes maintained in hamster cells. For many chromosomal regions, unequivocal probe orders have been established using recombination analysis in standard sets of human pedigrees or hybridization to collections of patient DNA with well-defined deletions or translocations. This has established a set of markers at defined intervals throughout the genome, a valuable resource for linkage mapping. In 1989, Weber and May found that an abundant class of simple repeat sequences, "microsatellites," that are distributed throughout the genome were highly polymorphic. As a consequence, linkage analysis has become a much more rapid and a less labor-intensive task to perform. These di-, tri-, and tetranucleotide repeats are easily assayable by polymerase chain reaction (PCR), allowing much more rapid typing than conventional RFLPs, and are generally more informative. Mapping with microsatellites has become the method of choice with >2000 polymorphic repeats mapped in the human genome, giving an average resolution >5 cM (Weissenbach *et al.,* 1992). Important factors when considering a linkage study are the availability, size, and structure of affected families and, in particular, the number of affected individuals. To achieve an lod score of 3, one needs a minimum of 11 informative meiotic events. Given that few probes will ever be fully informative, at least 20 meiotic events will usually be required to give significant linkage or exclusion information. To achieve this, it is usual to collect numbers of small families that are carefully diagnosed to ensure the all have the same phenotype or, where possible, large families where genetic heterogeneity is less likely to be a problem.

Once linkage has been found, it is worth checking whether any known genes have already been assigned to the chromosomal location identified. It is often possible, from a knowledge of the disease, to speculate on the type of gene that might be involved. Known genes mapping in the area can then be assessed as to whether they are likely candidates for the disease gene and investigated further. There have already been several examples of disease genes that have been identified as a consequence of candidate genes with appropriate expression patterns that "coincidentally" map to regions identified by linkage studies. These include the rhodopsin gene for retinitis pigmentosa (Dryja *et al.,* 1990), fibrillin for Marfan syndrome (Maslen *et al.,* 1991; Lee *et al.,* 1991; Dietz *et al.,* 1991), amyloid precursor protein for early onset Alzheimer's disease (Goate *et al.,* 1991), and *HuP2* for Waardenburg syndrome (Tassabehji *et al.,* 1992; Baldwin *et al.,* 1992).

B. From Genetic Distance to Real Physical Distance

Largely depending on the size of the available pedigrees and a certain amount of luck, linkage analysis may narrow the distance between the two closest flanking markers to <1–2 cM. It has been estimated that in the human genome, 1 cM is approximately equivalent to 1 million base pairs although this conversion will vary at any given locus depending on the local frequency of recombination. It may still be possible to narrow the genetic distance further by isolating new polymorphic markers from within the region. However, the limit of genetic resolution may have been reached with available family members. At this stage physical characterization of the remaining interval must commence. This will serve two purposes, first to provide new polymorphic markers that might allow the region to be narrowed further and second to provide cloned DNA covering the region from which candidate genes can be isolated.

1. Long-Range Restriction Map

Pulsed-field gel electrophoresis (PFGE) is a technique that has been devised for separating large DNA fragments with a resolution of up to 10 Mb (Schwartz and Cantor, 1984). Such fragments are generated from total human DNA following digestion with restriction enzymes that have recognition sequences that are represented only rarely in the genome. Southern blots are produced from gels containing a selection of single and double digests with these rare-cutter enzymes. Probes from the area under study are successively hybridized to these filters and a restriction map is built up both by analyzing the size of the fragments that are detected and by identifying fragments that contain more than one probe. Generation of a map in this way provides several valuable resources. First, the map allows the physical size of the area to be determined; second, the map provides a means of both ordering and orientating new clones generated from the area; third, it identifies clusters of rare-cutter sites that are often found at the 5′ end of genes (Bird, 1986). This latter finding may therefore give some clues to the location of some of the genes in the area.

2. Chromosome Jumping and Walking

To construct a detailed map, it is usually necessary to obtain new probes internal to the flanking markers so that data from both ends will join up. For this purpose "jumping" and "linking" libraries have been constructed (Collins and Weissman, 1984; Marx, 1985). Jumping libraries contain a representative selection of individual clones that contain the DNA adjacent to a rare-cutter restriction site joined to DNA adjacent to the next rare-cutter site along the genome, with the intervening DNA deleted. Characterization of these clones therefore can represent jumps of several hundred kilobases from the starting point. The linking library is then used to generate clones that overlap between each jump. Jumps are not directional, although this can be determined using the PFGE map. More recently yeast artifical chromosome (YAC) libraries have been used for the same purpose (Burke *et al.*, 1987). Yeast artificial chromosome clones consist of long tracts of human DNA (100–2000 kb) cloned into a vector that provides a yeast origin of replication and telomeres. These clones can then be propagated in yeast host cells as artificial chromosomes. Because of their large insert size, YAC clones provide an efficient way of moving along the chromosome from a given starting point. Yeast artificial chromosome end fragments can be easily isolated to provide new probes both to rescreen the library for the purpose of chromosome walking and for use with the developing PFGE map (Riley *et al.*, 1990). As this procedure is repeated, a set of overlapping or contiguous clones called a "contig" is built up (Fig. 6).

One major problem with YAC overlap mapping occurs as a consequence of cocloning of noncontiguous fragments in the same clone. Therefore, when end fragments are isolated to initiate the next "walk" along the genome, the investigator can mistakenly jump to a different part of the genome. Yeast artificial chromosome clones are often chimeric and they are also susceptible to deletions and rearrangements such that in some libraries up to 40% of the clones are not authentic contiguous representations of the genome. A useful strategy to counter this problem is sequence-tagged site (STS) content mapping devised by Olsen *et al.* (1989). When a probe is used to screen a YAC library, usually several overlapping clones containing the probe sequence are identified. The chance of this happening is enhanced if several different libraries are available for screening. End fragment clones are identified for each and sequenced so that PCR primers can be made. These STSs are then used to screen each of

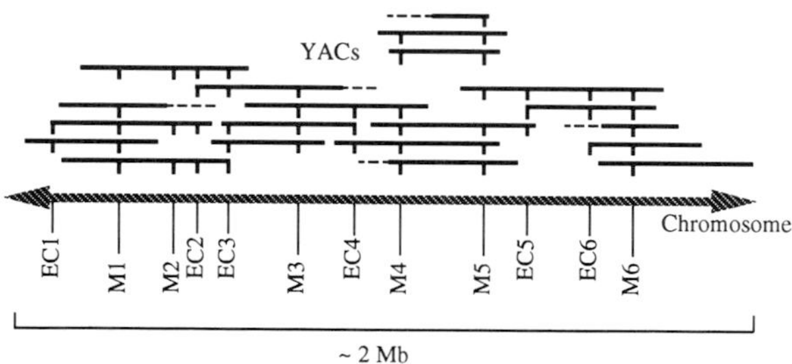

FIG. 6. A schematic diagram of a series of overlapping YACs. Bold lines represent YAC clones. M1–6 represent gene clones or anonymous DNA sequences used to identify YAC clones from library. EC1–6 represent end clones generated from YACs to confirm YAC overlaps and integrity of DNA inserts. Dotted lines represent chimeric regions of YAC clones.

the YACs isolated, allowing the clones to be ordered according to content. As contigs are built up following subsequent screenings of the libraries, bogus clones or regions of clones are identified by failure of the STSs to fit into the consensus pattern. An additional check of the integrity of the YAC clones can be provided by fluorescence *in situ* hybridisation (FISH). The YAC clones are labeled with a fluorescent dye and hybridized to metaphase chromosomes. Localization to a single chromosome region does not rule out internal rearrangements of the clone but makes cocloning artifacts unlikely.

C. Identification of Gene Sequences

The identification of novel, region-specific gene sequences has proved to be a major problem for the positional cloner. The most commonly used techniques have included hybridization of short genomic segments to DNA of other species to identify sequences that are evolutionarily conserved (Monaco *et al.*, 1986); direct screening of cDNA libraries or Northern blots with phage or cosmid clones; cloning hypomethylated CpG islands (Estivill *et al.*, 1987); or searching a DNA sequence for open reading frames (Gray *et al.*, 1982), enhancers (Weber *et al.*, 1984), or promoters (Allen *et al.*, 1988). Although these techniques have been successful, they are generally inefficient. Techniques have been developed to identify exons in "exon trapping" assays (Duyk *et al.*, 1990; Hamaguchi *et al.*, 1992) (Fig. 7). Cloned DNA, usually in the form of cosmids, is shotgun subcloned into the intron of an exon–intron–exon cassette in specially constructed vectors. The constructs are then transfected into human COS cells and allowed to express the insert as RNA, using a promoter sequence in the vector. The RNA is processed by the host cells, splicing out introns and capturing exons by virtue of the exon/intron splice sites in the cassette. Trapped exon sequences are then identified by a simple PCR assay using flanking primers designed to the cassette exons. The technique works best on cosmid DNA, which requires further steps to move from the YAC contigs and means a great deal of tedious repetition in the search of large areas. Also, it has often proved difficult to show that a trapped sequence actually derives from a coding sequence as some trapped products derive from cryptic splice sites. However, this technique has been used successfully in the identification of several genes, including that responsible for Huntington disease (Huntington's Disease Collaborative Research Group, 1993).

Possibly the most promising techniques currently available use the YAC clones directly. Yeast artificial chromosome clones, after suppression of repetitive sequences, can themselves be used to screen cDNA libraries,

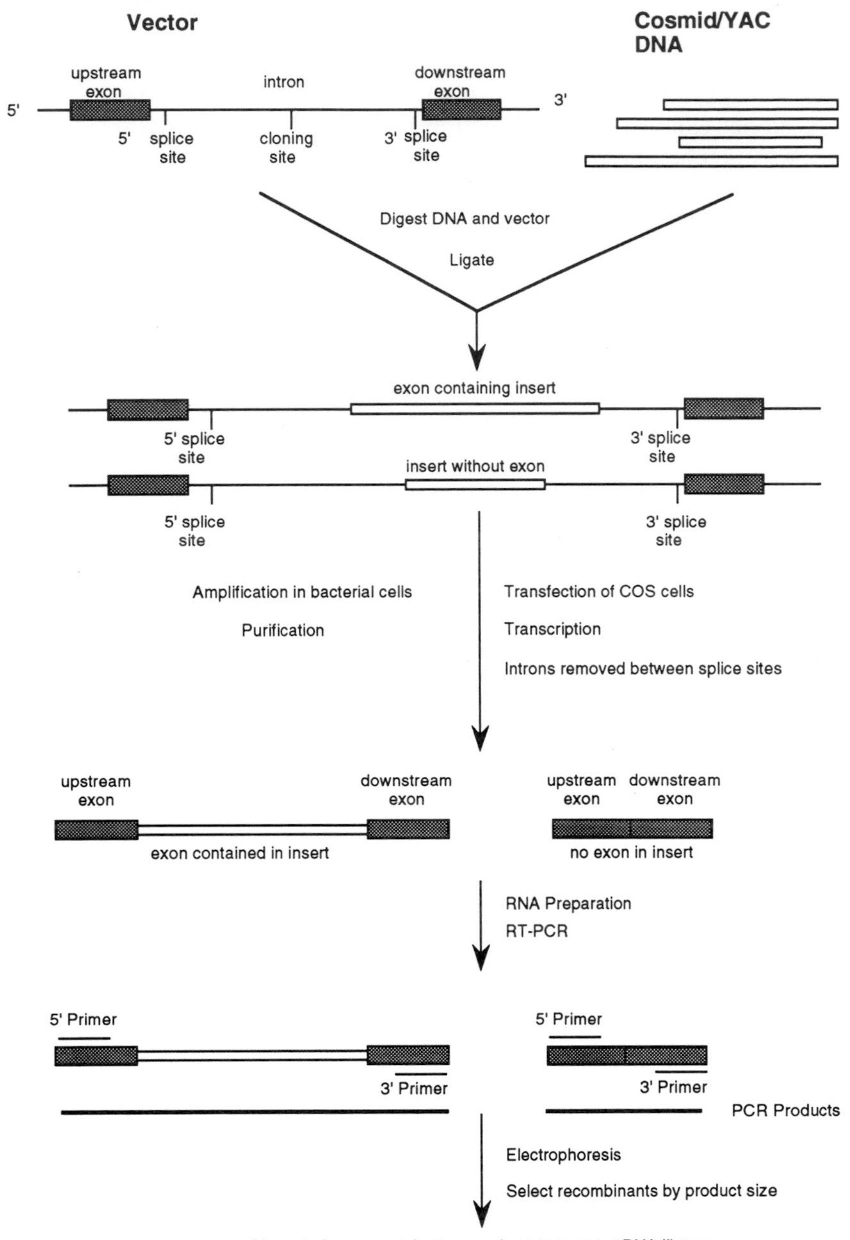

FIG. 7. A schematic diagram describing the method of exon trapping.

although this method tends not to be very sensitive. A better approach is to immobilize the YACs and use amplified cDNA as the probe before using PCR to capture the YAC-specific transcripts (Parimoo *et al.*, 1991; Lovett *et al.*, 1991). These protocols have been modified slightly to increase specificity (Vetrie *et al.*, 1993). To establish clone integrity, two or more overlapping YACs are used to screen a cosmid library prior to cDNA selection. Use of a chromosome-specific cosmid library helps to avoid detecting positives resulting from coligation artifacts. Resulting positives are then gridded out on filters. Inserts from a suitable cDNA library (chosen by the criteria of tissue and/or developmental stage) are amplified by PCR, using primers directed from the vector ends. The cDNA inserts are labeled and hybridized to the YAC-specific cosmid grids. The filters are washed to remove nonspecific sequences and the specifically bound sequences are eluted. The recovered cDNAs are reamplified and subcloned. Localization of the selected cDNAs can then be checked to ensure they are region specific and the clones characterized as candidates for the disease gene.

D. Identification and Detection of Mutations

Candidates for the disease gene are those that map within the confines of the closest flanking recombinants, as defined by individuals in the families under study. The expression pattern of such genes is investigated using Northern analysis and *in situ* hybridization. Good candidates are expected to be expressed in tissues appropriate to the defect and at suitable developmental stages. Candidate genes are then sequenced and compared with both DNA and protein databases so that functionality predictions can be made. A variety of techniques is currently available for the detection of DNA mutations that might give rise to the defect. These include single-strand conformation polymorphism (SSCP) detection (Orita *et al.*, 1989), heteroduplex analysis (White *et al.*, 1992), denaturing gradient gel electrophoresis (DGGE) (Sheffield *et al.*, 1989), chemical cleavage mismatch analysis (Cotton *et al.*, 1988), and direct sequence comparison. The first four techniques offer sensitive means of screening stretches of DNA up to about 500 bp in length, sequencing then being required once a mutation has been located.

SSCP analysis relies on the electrophoretic mobility of single-stranded DNA through a gel medium. The mobility of single-stranded DNA is dependent not only on molecular size but also on the adopted conformation, which is determined by the sequence present. Even a single base change will alter the conformation and affect mobility through the gel. In practice, regions of the cDNA clone under study will be amplified from both normal and affected individuals using PCR. The PCR products are denatured and analyzed on a nondenaturing polyacrylamide gel. Products are visualized using autoradiography with end-labeled PCR primers.

Heteroduplex analysis is carried out in much the same way. In this case, it is necessary to analyze heterozygous individuals to detect mutations. The PCR products are amplified either from carrier individuals, or from homozygous normal and affected individuals whose PCR products have been mixed. The amplified DNA is then heated to denature the double strands and allowed to reanneal prior to loading on a nondenaturing polyacrylamide gel. Where annealing occurs between a mutant and a normal strand, heteroduplexes are formed and a loop of single-stranded DNA occurs at the site of the mismatch. Such structures sufficiently alter the mobility characteristics that differences can easily be detected.

DGGE, as the name implies, relies on differing mobility characteristics when electrophoresing double-stranded DNA through a gel containing a denaturing formamide gradient. When DNA is run on a denaturing gel, it migrates in a double-stranded form until a critical concentration of denaturant is reached that causes the strands to separate, and retards progress through the gel. When comparing DNA fragments, a point mutation will alter the point at which the DNA dissociates, giving marked differences in the mobility characteristics. Chemical cleavage mismatch analysis theoretically detects all possible mismatches using a combination of hydroxylamine, which detects cytosine mismatches, and osmium tetroxide, which detects thymine mismatches. Target DNA, again usually a PCR product, is end labeled, denatured, and allowed to form heteroduplexes. The samples are then split and treated separately with each of the two chemicals. Following incubation with piperidine, the samples are analyzed on denaturing polyacrylamide gels. Because cleavage occurs only when a mismatch is present, homoduplexes have a single band the size of the original fragment, whereas heteroduplexes will have been cleaved by one or other of the chemicals. The size of the resulting fragments will indicate the position of the base differences within the fragment.

To obtain PCR fragments for the above-described screening methods, it is necessary either to have an appropriate cDNA library for the individuals to be tested (which is usually not practical) or to use reverse transcriptase PCR (RT-PCR) to amplify cDNA directly from the patient's RNA. Because this method is extremely sensitive, it may be possible to amplify very low abundance mRNAs that were only transiently expressed during a particular developmental stage (Piatak *et al.*, 1993).

Gaining formal proof of identification of the causative mutation is not easy. However, finding a mutation that will alter the protein-coding region is a good start. The mutation should be present in all affected individuals, not in those unaffected, and present in half-dose in carriers of recessive conditions. Proof that the mutation is causative usually relies on a biological assay, where introduction of the mutated gene into a suitable host gives the expected phenotype.

E. Expression Analysis

Expression of a gene occurs once the DNA has been transcribed into messenger RNA and that mRNA has been translated into protein. As stated above, to understand fully the function of a gene and its mutation, expression of the gene must be studied. Most simply this can be done by purifying the total RNA from a relevant tissue source and analyzing it by Northern blotting. This blot can be probed with the gene of interest to establish transcript size, and to some extent, copy number abundance. There are several methods for extracting RNA from tissues that are based on similar principles (Chomczynski and Sacchi, 1987). The tissue is homogenized in a solution containing guanidinium, phenol, and chloroform. The guanidinium is a strong denaturant that will destroy proteins, including active RNases. The phenol and chloroform remove proteins and lipids and separate them from the aqueous phase, which contains the nucleic acids. Total RNA can then be separated by a careful adjustment of the precipitation conditions using ethanol. RNA is single stranded but often forms secondary structure. This can be prevented by heating to 65°C and by using gel conditions containing a denaturant such as formaldehyde. Once the gel has been run, the RNA can be visualized using ethidium bromide. Total RNA contains 90% ribosomal RNA, the 28S and 18S bands of which can be seen clearly on these gels at approximately 4 and 2 kb in size. Many mRNAs contain a tract of adenines (A) at the end of their sequence. These poly(A) tails can be selected, using columns containing oligo(dT) to trap the mRNA. This allows a higher concentration of mRNA to be purified and can be useful for analyzing low copy number transcripts and transcripts that have sizes similar to the ribosomal bands. Transcript bands on Northern blots can be densitometrically analyzed against control, housekeeping genes (e.g., β-actin), to assess abundance and to visualize changes in expression levels.

Another method for assessing the actual concentration of a transcript is to use ribonuclease protection assays (Zinn *et al.*, 1983). In this case, an antisense probe is synthesized and hybridized *in vitro* to total RNA. The riboprobe will anneal only to the specific mRNA under analysis. The RNA is then treated with an RNase that destroys all the remaining single-stranded RNA, leaving the RNA of interest intact. The original riboprobe is radioactivly labeled, allowing the products to be visualized by autoradiography, and its concentration can be calculated using the amount of radioactivity or band density present.

If very low copy transcripts are present, a newer technique, reverse transcriptase PCR, can be used (Piatak *et al.*, 1993). In this case oligonucleotides to the expressing sequence are made and, once the mRNA has been transcribed into cDNA using reverse transcriptase, amplification can be carried out using PCR. If this is done in a cycle-controlled, linear manner, transcripts can be visualized that are normally too low to be detected by Northerns analysis or by ribonuclease protection assay.

Cell and tissue type localization of mRNAs can be visualized in tissue sections, using *in situ* hybridization (Haffner and Willison, 1987). The tissue section is fixed to avoid RNA degradation, and then hybridized with a radiolabeled riboprobe or specific oligonucleotides to detect the mRNA of interest. After autoradiography, silver grains on the section indicate the presence of hybridized probe and mRNA localization. Only antisense probes will hybridize; sense probes therefore act as a control to assess nonspecific hybridization.

F. Transgenics

Another important way of assessing the phenotype of an unknown gene is to produce a transgenic mouse strain. It is possible to inject the gene of interest into recently fertilized mouse eggs, usually into the male pronucleus. The embryo is then placed in a foster mother or back into its own mother after *in vitro* growth to the blastocyst stage. A small proportion of these pregnancies is successful, and some of these will contain the gene of interest incorporated into the genome of the offspring. The gene will be incorporated randomly, often in varying copy number at one or more insertion sites, and usually with no chromosome specificity. The foreign gene is then found in all cell lines of the offspring, including the germline. This transgenic animal can then be used to study the phenotype of the gene inserted, and to carry out any drug therapy if the gene inserted was one conferring a disease state. Many transgenic lines have been produced particularly for single-gene conditions like cystic fibrosis (Snouwaert *et al.*, 1992; Dorin *et al.*, 1992). The technique has also been to study the "knock-out" of important genes for assessment of their role in development.

IV. Human Models in the Study of Molecular Aspects of Craniofacial Dysmorphologies

Since the arrival of molecular biology, the genetic element of both simple and complex craniofacial dysmorphologies has been approached using human material. For many dysmorphic syndromes, it is now possible to find subgroups of families that display the phenotype in an autosomal dominant, recessive, or X-linked manner. In some cases all three types of inheritance are found separately, giving evidence to the involvement of several separate mutated genes in the dysmorphic pathway. By selecting these "genetic" families, experimental approaches used in the analysis and cloning of genes mutated in purely genetic disorders, for example, cystic fibrosis (Riordan *et al.,* 1989), myotonic dystrophy (Aslanidis *et al.,* 1992), and Duchenne muscular dystrophy (Monaco *et al.,* 1986), are being successfully employed. This article presents a small subgroup of craniofacial dysmorphologies that cover the three main inheritance patterns (autosomal dominant, recessive, and X-linked) and are being studied using molecular biology techniques, as detailed examples.

A. Molecular Approaches to Craniofacial Dysmorphologies Displaying a Mixed Inheritance Pattern

1. DiGeorge Syndrome

a. Clinical Features and Incidence of DiGeorge Syndrome In 1965, DiGeorge reported the syndrome as an association of absence of the thymus with an additional absence of the parathyroid glands resulting in thymic deficiency. Later, in 1966, Good named the syndrome after DiGeorge (Taitz *et al.,* 1966). The condition has also been referred to as III-IV pharyngeal pouch syndrome (Robinson, 1975) and DiGeorge anomaly (Opitz and Lewin, 1987). In most cases the syndrome occurs sporadically; however, autosomal dominant (Raatika *et al.,* 1981) and recessive modes of inheritance (Steele *et al.,* 1972) have been reported from family studies. The apparent recessives are most probably due to affected but undiagnosed parents, or germline mosaicism. As detailed in Table III, the condition has also been associated with several chromosomal abnormalities. It can be induced by teratogens, alcohol (Ammann *et al.,* 1982), and retinoic acid (Lammer *et al.,* 1985) and has been found to be associated with several other dysmorphologies such as arhinencephaly (Conley *et al.,* 1979) and CHARGE (Pagon *et al.,* 1981). The minimal incidence of DiGeorge syndrome is 1 in 20,000 births (Muller *et al.,* 1988).

The DiGeorge sequence has varying clinical severity, possibly depending on the timing of an insult to the fate of the neural crest cells during the fourth to seventh week of gestation. The most minor expression is the absence of the thymus and/or parathyroid glands (Lischner, 1972). The most severe cases display cardiovascular abnormalities and additionl craniofacial anomalies (Conley *et al.,* 1979) (Fig. 8). Craniofacial abnor-

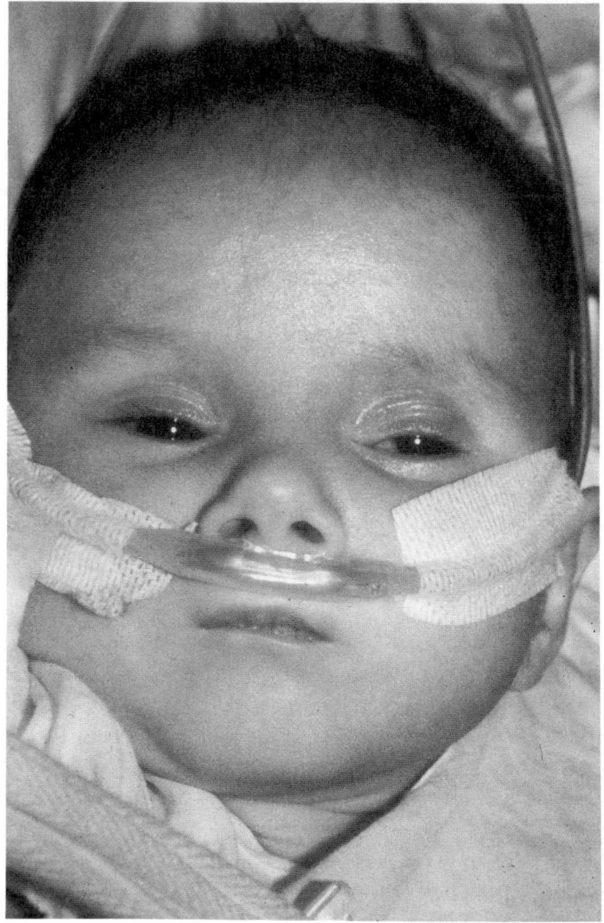

FIG. 8. A typical DiGeorge patient facies (Wilson *et al.*, 1991).

malities are found in approximately 60% of individuals, and include micrognathia, deep and low set posteriorly angulated ears, anteverted nostrils, blunting and clefting of the nose, and hypertelorism. Malformations of the secondary palate, short philtrum, choanal atresia, abnormal middle ear, eye anomalies, and central nervous system malformations have also been described. There is a poor prognosis for affected individuals, including hypocalcemia with seizures, cardiac failure, infections, and developmental delay. Those who survive infancy with a partial phenotype exhibit moderate to severe mental retardation.

b. Molecular Genetic Analysis of DiGeorge Syndrome With a syndrome that appears to cover all possibilities in terms of inheritance and many different chromosomes (Table III), it is difficult to pinpoint one location on the genome that might be even partly responsible for the phenotype. However, the most commonly associated chromosome appears to be chromosome 22. In 1981, De la Chapelle found four members of one family with an unbalanced translocation and monosomy of chromosome 22pter-q11. In 1982, Kelley *et al.* found three unrelated DiGeorge patients with a similar deletion from 22pter-q11, with the remaining q11-qter translocated to different autosomes. Others also found translocations involving chromosome 22 and other alternative autosomes (Greenberg *et al.,* 1984; Augusseau *et al.*, 1986). With other studies, no chromosome abnormalities were seen (Rohn *et al.,* 1984; Keppen *et al.,* 1988). Greenberg *et al.* in 1988 studied 27 DiGeorge individuals and found only 5 with chromosomal anomalies. Three had monsomy 22q11, one had monosomy 10p13, and one had monosomy 18p21.33. This led to the general conclusion that the phenotype was heterogeneous, with monosomy 22q11 as a major player but 10p13 and other autosomes also taking part. Monosomy 17p13 has also been reported and Fukushima *et al.* (1992) suggested 4q as another site.

Using high resolution chromosome banding, Wilson *et al.* (1992a) found monosomy resulting from an interstitial deletion of 22q11.21-q11.23 in 9 of 30 DiGeorge infants. Carey *et al.* (1992), using closely linked DNA probes, demonstrated hemizygosity for 22q11 in 21 of 22 cases. This evidence strongly suggested that this area on chromosome 22 contained genes that, when deleted or disrupted, would lead to the dysmorphic DiGeorge sequence.

TABLE III

Chromosomal Associations with DiGeorge Syndrome

Chromosomal abnormality	Ref.
Duplication (1q)	Van den Berghe *et al.* (1973)
Duplication (8q)	Townes and White (1978)
Deletion (5p)	Taylor and Josikef (1981)
Deletion (10p)	Greenberg *et al.* (1986)
Deletion (17p)	Greenberg *et al.* (1988)
Deletion (22q)	De la Chapelle *et al.* (1984); DeCicco *et al.* (1973); Greenberg *et al.* (1984); Kelley *et al.* (1982); Rosenthal *et al.* (1972)

The next steps were to map physically the critical region on chromosome 22. Fibison *et al.* (1990) mapped several DNA markers to the region including the BCR and BCR-like loci. Scambler *et al.* (1990) used various translocation breakpoints from DiGeorge patients to map over 60 DNA markers. The combination of their work confirmed that the DiGeorge sequence breakpoints occur proximal to the BCR locus and a 22q11 breakpoint found in two derivative-22 patients. It is also clear that many of the deletions found would be submicroscopic. The clinical severity of DiGeorge syndrome in any one family described with a 22q11 microdeletion can also vary considerably. The presence of 22q11 microdeletions has also been found in patients with isolated heart defects (D. I. Wilson *et al.*, 1991, 1992b; Driscoll *et al.*, 1992a).

With regard to the monosomy of 22q11 seen in five families studied, four affected individuals failed to inherit the maternal allele and one failed to inherit the paternal. In addition to this, six maternal and five paternally derived unbalanced translocations have been reported for DiGeorge individuals in the literature. It can be concluded that imprinting or "parent of origin" does not play an active part in the phenotype (Driscoll *et al.*, 1992b).

DiGeorge patients with the more severe phenotype overlap with the phenotype of velocardiofacial syndrome (VCFS). Several authors have suggested that they are one and the same abnormality and that both may be the result of a mutation in the same gene (Stevens *et al.*, 1990; Kelly *et al.*, 1993). Scambler *et al.* (1992) have shown that VCFS, like DiGeorge syndrome, is due to monosomy 22q11. In 1993, Kelly *et al.* studied 12 patients with VCFS and all had monosomy for a part of 22q11. However, Driscoll *et al.* (1992a) found an interstitial deletion of 22q11.21-q11.23 in only 3 of 15 patients with VCFS. The other 12 had apparently normal chromosomes when examined using these cytogenetic banding techniques, but some were deleted when checked with specific DNA probes.

In 1993, Wilson *et al.* suggested the acronym CATCH 22 for the cluster of disorders that were associated with aberrations on the q arm of chromosome 22. CATCH stands for the spectrum of disorders: cardiac, abnormal facies, thymic hypoplasia, cleft palate, and hypocalcemia. The incidence for the whole group is now considered to be 1 in 5000 live births (Scambler, 1993). Palacios *et al.* (1993) looked at the volume density and number of thyrocalcitonin-immunoreactive C cells from the thyroid glands of 11 DiGeorge anomaly patients. They found decreased cell number and size, supporting the hypothesis that the pathogenetic factors involved included a neural crest disturbance.

The next step in the pursuit of genes that cause DiGeorge syndrome or CATCH 22, when haploinsufficient, was to continue to map the areas on 22q11 around individual translocation breakpoints. From the analysis of over 100 deletion patients a small, critical region of 300 kb was defined. This region was carefully selected from various genomic libraries for further analysis. From within the DNA sequence of this shortest region of overlap, Halford *et al.* (1993) have identified a gene name TUPLE1, which is expressed during human and murine embryogenesis. The TUPLE1 gene encodes a protein containing repeat motifs similar to the WD40 domains found in the β-transducin/enhancer of split (TLE) family, which are transcriptional regulators (Williams and Trumbly, 1990). When a part of the human TUPLE1 gene was hybridized to a Northern blot containing several human fetal mRNAs, a transcript of 3.4 kb was detected in heart, brain, lung, liver, and kidney, with a second, smaller 3.2-kb transcript found in liver. From this and other expression data, the authors propose that this gene is involved in the pathogenesis of DiGeorge syndrome and CATCH 22.

It is important now to demonstrate that the gene is truly involved in the development of the tissues that are dysmorphic in CATCH 22. This can be done by making transgenic mice with mutated forms of the gene or by knocking out the gene completely. There are still 2 patients with DiGeorge syndrome (DGS), 6 patients with VCFS, and 25 with conotruncal heart defects that have no detectable deletion or gross rearrangements within 22q11. Point mutations within the TUPLE genes could give rise to this group and have yet to be ruled out. There are still three other chromosomal loci that need investigation: those at 10p13, 17p, and 4q. Submicroscopic deletions at any of these loci or any teratogenic phenocopy may be the cause of the defects in these cases.

B. Molecular Approaches to Craniofacial Dysmorphologies Displaying an Autosomal Dominant Inheritance Pattern

1. Mandibulofacial Dysostosis (Treacher Collins Syndrome, Franceschetti–Zwahlen–Klein Syndrome)

a. Clinical Features and Incidence of Mandibulofacial Dysostosis The syndrome was first described in 1846–1847 by Thomson and Toynbee (1847) but credit for its discovery is usually given to Berry in 1889 and Treacher Collins in 1960 after their detailed reports of the features of the phenotype. In 1940s reviews of the disorder by Franceschetti and Klein (1949) and Franceschetti *et al.* (1949), the syndrome was termed *mandibulofacial dysostosis*. Of 63 cases, 27 had a family history of occurrence of the disorder. Mandibulofacial dysostosis has autosomal inheritance with variable expressivity (Table IV). Of the 60% that represent new mutations the paternal proband is older than average (Jones *et al.*, 1975). It affects

TABLE IV
Craniofacial Structures and Their Observed Malformation in Mandibulofacial Dysostosis[a]

Malformation	Structure affected
Often absent	Malar bones Paranasal sinuses Infraorbital foramen External auditory canal Auditory ossicles, cochlear and vestibular apparatus Stapes and oval window Middle ear and epitympanic space Parotid salivary glands
Nonfusion	Zygomatic arches
Hypoplastic	Zygomatic process frontal bone Lateral pterygoid plates and muscles Supraorbital rims Mandibular condyle Mastoid and mastoid antrum Alar cartilages
Hyperteloric	Orbits

[a] Adapted from Gorlin *et al.* (1990, p. 649).

1 in 50,000 live births (Rovin *et al.,* 1964; Frazen *et al.,* 1967) and approximately 400 cases have been published.

Mandibulofacial dysostosis arises from malformations in structures derived from the first and second pharyngeal arch, groove, and pouch (see Fig. 1). Clinical features include malformation of the facies, skull, eyes, ears, nose, and palate. Sufferers with mandibulofacial dysostosis have a characteristic facies usually with bilateral, symmetric abnormalities (Fig. 9). The narrow face displays hypoplastic supraorbital rims and zygomas, downward sloping palprebral fissures, sunken cheekbones, malformed pinnae, receding chin, and a downturned mouth. A quarter of cases also show a piece of hair that extends upward toward the cheek.

b. Molecular Genetic Analysis of Mandibulofacial Dysostosis Once familes have been identified with a clear autosomal dominant pattern of inheritance, rather than carrying out linkage analysis to all 22 autosomes, the first approach is to select candidate genes or regions for analysis. A candidate gene could be one showing spatial and temporal expression relevant to the development of neural crest tissue, for example, a specific growth factor.

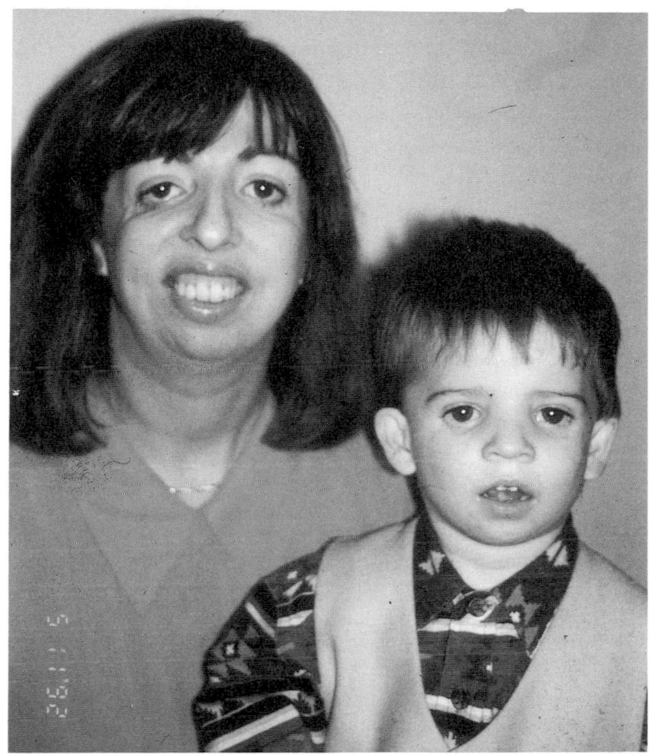

FIG. 9. Typical mandibulofacial dysostosis facies in mother and son.

Candidate regions can be selected on the basis of known chromosomal anomalies within the selected families. It can be hypothesized that these deletions or translocations have caused the syndrome by disrupting or removing critical genes at the translocation or deletion breakpoint. Other families without obvious chromosmal anomalies may well have microdeletions of the same genes or important point mutations that are not visible cytogenetically.

In the case of mandibulofacial dysostosis syndrome, two balanced translocations and one deletion have been described, indicating five chromosomes (5, 13, 6, 16, and 4) as potentially containing genes critical in mandibulofacial development. In 1983, Balestrazzi *et al.* described a girl with a balanced *de novo* translocation t(5;13)(q11;p11). In 1991, Dixon *et al.* (1991a) identified a mother and her two children displaying mandibulofacial dysostosis; they had a balanced translocation at t(6;16)(p21.31;p13.11). Also in 1991, Jabs *et al.* (1991a) observed a patient with a *de novo* deletion in the region 4p15.32-p14.

Studies were subsequently performed to exclude or link the chromosomal areas indicated in the translocations and deletion. Using the family with the translocation at t(6;16)(p21.31;p13.11) as a starting point, Dixon *et al.* (1991a) first defined the translocation breakpoints using *in situ* hybridization. For chromosome 6, region 6p21.31, loci in the HLA complex, probes p45. IDPB003/HLA-DPB2 and pRS5.10/HLA class 1 chain were used and for chromosome 16, region 16p13.11, probes pACHF1.3.2/D16S8 and VK45/D16S131 were used. Once these flanking markers had defined the break points, 12 additional families with mandibulofacial dysostosis were studied, none of whom had any visible chromosomal abnormality. Using informative DNA probes from the two defined breakpoints, pairwise and multipoint linkage analysis was carried out in these families.

The mutated gene or genes that cause mandibulofacial dysostosis were excluded from the translocation breakpoint at 6p21.31 using probe pDCH1, which defines a class II DQ α chain within the MHC complex. In addition, using a probe F13A1 at the factor 13A1 locus 25–35 cM from the HLA complex, mandibulofacial dysostosis could be excluded from all of chromosome 6p except the telomere.

Five DNA markers (D16S8, D16S96, D16S79, D16S131, and D16S75), including the two flanking the translocation breakpoint (D16S8 and D16S131), were used to exclude the site of mandibulofacial dysostosis from chromosome 16p13.11, using pairwise and multipoint linkage analysis. This exclusion from 6p and 16p13.11 was further confirmed by the subsequent birth within the translocation family of a third affected child who did not exhibit the translocation.

The focus of Dixon's research group turned to chromosome 5q11 on the basis of the second known translocation family of Balestrazzi *et al.* (1983) at t(5;13)(q11;p11). The mutations were thought more likely to be on chromosome 5 than 13, owing to the relative paucity of genes on the short arm of chromosome 13. At the same time Jabs *et al.* pursued the deletion family from chromosome 4p15.32-p14. Using DNA prepared from the deletion patient in comparison with the DNA of his parents, several markers (D4S18, D4S19, D4S20, D4S22, and D4S23) were mapped to the deletion. Other chromosome 4p DNA probes (D4S10, D4S15, D4S16, D4S26, D4S356, D4S95, RAF1P1, QDPR, and HOX7) were also tested and were outside the deletion. Three of these probes (D4S18, D4S23, and QDPR) were informative and were used for linkage analysis in eight families with mandibulofacial dysostosis. Negative lod scores were obtained, excluding this region as a candidate for these mutated genes (Jabs *et al.*, 1991a). Meanwhile Dixon *et al.* (1991b) had found a positive linkage to chromosome 5q31-34 in 12 families with mandibulofacial dysostosis. Multipoint analysis placed the mutation responsible for the syndrome between the gene for the glucocorticoid receptor (map position 5q33-35) and the anonymous DNA marker D5S22 (map position 5q34-ter), with a lod score of 9.1. Following a personal communication from M. J. Dixon, Jabs *et al.* (1991b) also tested their eight families with mandibulofacial dysostosis and found a maximum lod score of 8.65 with marker D5S210 (map position 5q31.3-q33.3) at $\theta = 0.02$.

The first critical step toward the elucidation of the mutated gene or genes for mandibulofacial dysostosis has therefore been made. In addition to localization Dixon *et al.* (1991b) have been able to exclude two potential candidate genes from the region. The gene for the glucocorticoid receptor (GRL) (Giuffra *et al.,* 1988; Thériault *et al.,* 1989) and the extracellular matrix glycoprotein osteonectin (SPARC) (Swaroop *et al.*, 1988) have been shown to have recombinants in the mandibulofacial dysostosis families studied by Dixon *et al.* (1991b). In 1993, Dixon *et al.* further refined the locus to 5q32-q33.1, using three microsatellite markers with fluorescence *in situ* data indicating that SPARC distally flanks the critical region.

In 1993, Loftus *et al.* completed a combined genetic and radiation hybrid map of the mandibulofacial dysostosis locus. The previous data had placed the mandibulofacial dysostosis locus between marker D5S519 (proximal) and SPARC (distal). To define more accurately the distance between these markers a genetic map of 15 loci that gave a sex-averaged map of 29 cM was integrated with a radiation hybrid map containing 13 loci. Eight loci were common to both maps and allowed the D5S519–SPARC interval to be estimated at 880 kb.

Two genes, RPS14 (encoding a ribosomal protein S14; Rhoads *et al.,* 1986) and ANX6 (encoding the calcium-binding protein p68; Davies *et al.,* 1989), are in the region of interest and should be analyzed carefully in the families.

If the mutation is not found in one of the more obvious candidate genes in the region, the next step is to continue to fine map the remaining 880-kb area and analyze all the coding sequences for mutations and expression changes in the mandibulofacial dysostosis families.

2. Greig Cephalopolysyndactyly Syndrome

a. Clinical Features and Incidence of Greig Cephalopolysyndactyly Syndrome In 1926, Greig described a mother and daughter with a combination of digital malformations and strangely shaped skulls. More than 40 other cases have been reported that have been given various names, including frontodigital syndrome (Marshall and Smith, 1970), Hootnick–Holmes syndrome (1972), Noack-type acrocephalopolysyndactyly (Gnamey and Farriaux, 1971), frontonasal dysplasia (Kwee and Lindhout, 1983), median cleft face syndrome (Ide and Holt, 1975), and acrofacial dysostosis (Korting and Ruther, 1954). Autosomal dominant inheritance with variable expressivity has been demonstrated in several families.

Clinical features are extremely varied, with half the cases displaying frontal bossing, a larger than average head, and, as a consequence, a broad forehead. Hypertelorism and a broad nasal bridge are also common. Craniosynostosis is also seen (Hootnick and Holmes, 1972; Kwee and Lindhout, 1983). Dysmorphic extremities include broad thumbs and halluces and soft tissue syndactyly of fingers and toes. Postaxial polydactyly of the hands is common, but it is rare in the feet. Preaxial polydactyly is the most common anomaly seen in the feet. Syndactyly can also be seen as an almost constant feature in the feet, with or without associated preaxial polydactyly. Bone age also seems advanced (Tommerup and Nielsen, 1983). Intelligence is rarely abnormal, but agenesis of the corpus callosum and hydrocephaly have been reported (Hootnick and Holmes, 1972).

b. Molecular Genetic Analysis of Greig Cephalopolysyndactyly Syndrome As with Treacher Collins syndrome, the mode of inheritance in Greig cephalopolysyndactyly syndrome (GCPS) is autosomal and the molecular approach to determine the gene or genes involved has been similar. After analysis of several cases with deletions of 7p21, Baccichetti *et al.* (1982) suggested this might be a critical region in GCPS. Both deletions and translocations have been reported for GCPS, with chromosome 7 always involved. Tommerup and Nielsen (1983) found the translocation t(3;7)(p21.1;p13) segregating through a four-generation family, associated with GCPS. Kruger *et al.* (1989) also showed segregation of GCPS with the translocation t(6:7)(q27:p13) in a separate four-generation kindred. Vortkamp *et al.* (1991) also found a third translocation of t(6;7)(q12;p13) associated with GCPS.

Sage *et al.* (1987) began the search for the genes involved in GCPS using molecular techniques to analyze the breakpoints on chromosomes 3 and 7. In 1988, Brueton *et al.* investigated seven GCPS families with no visible chromosomal abnormality and found a positive linkage of 3.17 ($\theta = 0$) to the epidermal growth factor receptor gene (EGFR) localized to chromosome 7p13-p11. Drabkin *et al.* (1989) continued to pursue the translocation families and identified two DNA sequences flanking the translocation points that showed no recombination with GCPS. Pulsed-field analysis also showed the EGFR locus to be linked to the two markers, but no evidence of positive linkage was found between GCPS and EGFR in these translocation families. Linkage was found, however, to the T cell γ receptor locus (TCRG) by this method.

The work of Drabkin *et al.* (1989) was further corroborated by analysis of deletion GCPS patients. In one GCPS individual with a deletion of

7p13-p11.2, EGFR was found to be deleted, but in another with a deletion of 7p14.2-p12.3 the EGFR genes were present (Rosenkranz *et al.,* 1989). They could conclude from this that the EGFR gene was in band 7p12.3-p12.1, whereas the critical GCPS region was more distal at 7p13-p12.3. The first report of a GCPS patient with an interstitial deletion of 7(p13p14)pat was by Pettigrew *et al.* (1989, 1991). Using high-resolution chromosome analysis, they confirmed the above localization. The deleted chromosome was found to be paternal in origin. By reviewing reports of patients with overlapping clinical features of GCPS, many were found with 7p13 deletions. Wagner *et al.* (1990) studied two cases of GCPS associated with cytogenetically visible deletions of 7p using gene probes assigned to the proposed GCPS locus. One displayed loss of the TCRG cluster, and both showed hemizygosity for PGAM2, whereas both Hox 1.4 and interferon β_2 were normal. This placed PGAM2 and GCPS in 7p13-p12.3 with TCRG distal at 7p14.2-p13 and Hox 1.4 and interferon β_2 distal at 7p14.2.

Vortkamp *et al.* (1992) carried out experiments on the three balanced translocation families, and found that two of three translocations interrupted the GLI3 gene, a zinc finger gene of the GLI–Kruppel family localized to 7p13. This gene is expressed in many tissues including testis, myometrium, placenta, colon, and lung, Its protein product shows sequence-specific DNA-binding capability (Ruppert *et al.,* 1988, 1990). Two of the translocations have split the GLI3 gene within the first third of the coding region, thereby truncating the normal protein. In the third patient, the translocation breakpoint was 10 kb downstream of the 3′ end of GLI3. In view of the potential of this gene to act a transcription regulator, it is a good candidate for involvement in limb or craniofacial development.

C. Molecular Approaches to Craniofacial Dysmorphologies Displaying an Autosomal Recessive Inheritance Pattern

1. Acrocallosal Syndrome

a. Clinical Features and Incidence of Acrocallosal Syndrome In 1980, Schinzel and Schmid found two unrelated patients who shared features of postaxial polydactyly, hallux duplication, macrocephaly, absence of the corpus callosum, and severe mental retardation. Nelson and Thompson reported a further two cases in 1982. In 1982, in a summary of seven sporadic cases, Schinzel raised the possibility that the newly termed acrocallosal syndrome (ACLS) might be similar to the Greig cephalopolysyndactyly syndrome (GCPS), as the digital changes are similar. In 1990, Hendriks *et al.* found a case with increased birth weight and cerebellar hypoplasia, which could be considered as additional features of ACLS. They also found an extra bone within the anterior fontanel, suggesting similarity to the autosomal dominant mouse mutant *Xt* (*extra toes*), which is homologous to GCPS in humans (Green, 1981) (see Section V,A,1). Schinzel and Kaufman (1986) first suggested that the syndrome was recessive, as they observed four cases from a small region in Switzerland with two affected siblings from normal nonconsanguineous parents.

As more cases were described the picture became more complicated. Schinzel and Kaufman (1986) raised the possibility that the Finnish "Hydrolethalus" syndrome might be a lethal allele for ACLS, because the combination of agenesis of the corpus callosum and pre- or postaxial polydactyly are also features of this condition. Philip *et al.* (1988) postulated that preaxial polydactyly may not be a complete feature of ACLS, after reporting two cases in which it was absent. In one of these cases there were additional features of hypospadias, cryptorchidism, inguinal hernias, duplication, and syndactyly of the phalanges of the big toe and bipartite right clavicle. The other exhibited an arachnoidal cyst, calvarian defect, and digitalization of the thumbs. Dandy-Walker malformation was also associated, as reported by Moeschler *et al.* (1989). By 1992, 21 well-documented cases had been reported (Cataltepe and Tuncbilek, 1992).

Pfeiffer *et al.* (1992) found a child that had all the typical abnormal features of ACLS and a duplication of almost the whole p arm of chromosome 12. Interestingly, trisomies and tetrasomies of chromosome 12 do show features similar to those of ACLS. This suggested two possible chromosomal locations that may play a role in this syndrome: 7p13, the map location for Greig syndrome, and/or 12p.

b. Molecular Genetic Analysis of Acrocallosal Syndrome As a consequence of the positive localization of GCPS to 7p13-p12.3, Brueton *et al.* in 1992 carried out linkage analysis in a family with ACLS. Using the linked markers for GCPS they obtained data that appeared to exclude ACLS from the GCPS region.

D. Molecular Approaches to Craniofacial Dysmorphologies Displaying an X-Linked Inheritance Pattern

1. Amelogenesis Imperfecta

a. Clinical Features and Incidence of Amelogenesis Imperfecta Amelogenesis imperfecta is characterized by defective formation of tooth enamel in the primary and permanent dentition (Fig. 10). The prevalence of the disease is geographically variable, with 1 in 8000 found in Israel (Chosak

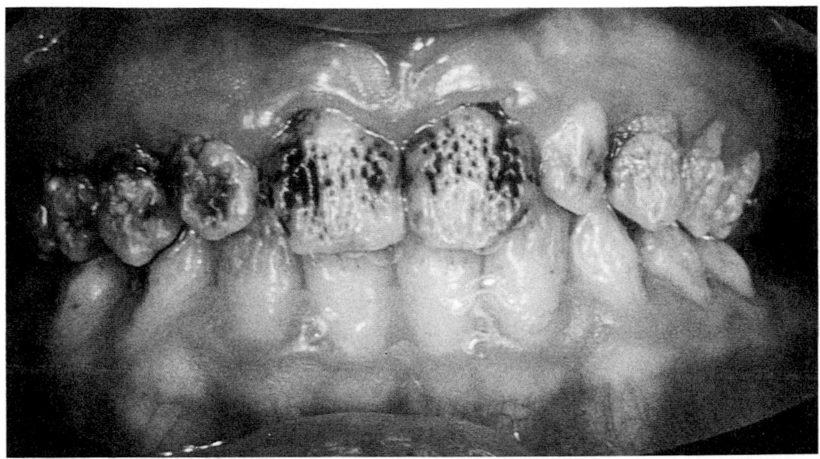

FIG. 10. Vertically pitted and grooved enamel found in XAI.

et al., 1979) and 1 in 700 in Sweden (Backman and Holm, 1986). Its inheritance is heterogeneous, with families demonstrating autosomal dominant, recessive, and X-linked patterns (Witkop and Sauk, 1976; Backman and Holmgren, 1988). This article focuses on the X-linked forms (XAI).

XAI can be broadly divided into two different clinical forms: one showing hypoplasia of the enamel and the other hypomaturation. In the first form the hypoplastic enamel is correctly mineralized but abnormally thin, showing macroscopic defects, whereas in the latter form the enamel is soft and insufficiently mineralized (Weinmann *et al.,* 1945). Since 1945, several classifications have evolved with around 10 subtypes, characterized by both mode of inheritance and clinical phenotype (Witkop, 1989). With respect to XAI two phenotypic variants have been described: a hypoplastic and a hypomineralization form. Affected males and females show different phenotypes. In the hypoplastic form, the males have a very thin layer of enamel and the females have enamel that is thicker in parts, giving rise to alternating grooves and ridges of abnormal and normal enamel. In the hypomineralization form males have teeth of normal size and shape but with irregular mottling; the females display vertical bands of mottling. The wide variation in the females is consistent with the Lyon phenomenon (Berkman and Singer, 1971).

The two most abundant proteins found in enamel are amelogenins and enamelins (Termine *et al.,* 1980). The amelogenins are a family of ectoderm-derived extracellular proteins, found in the enamel, that are abundantly and transiently expressed by the ameloblasts during tooth development. The molecular mass of the predominant enamelins is 70 kDa, making up around 10% of the matrix proteins in enamel. Any alterations in the genes for these two enamel proteins might lead to the defects seen in amelogenesis imperfecta.

b. Molecular Genetic Analysis of Amelogenesis Imperfecta The cloning of the human gene involved in this X-linked disorder is a good example of how a mouse model can help in identifying the molecular defect in humans. In 1989, Lau *et al.* mapped the human amelogenin gene to the X chromosome, using the murine homolog (Snead *et al.*, 1985). There was also thought to be a corresponding Y locus. Localization of the amelogenin gene was then further refined to p22.3-p22.1 on the human X chromosome (AMGX) and near to the centromere on the Y chromosome at Yp11 (AMGY) (Lau *et al.*, 1989). This was carried out using human–mouse hybrid cell lines containing various deletions of the X and Y chromosomes. The localization of the amelogenin gene to the X chromosome was consistent with its potential involvement in XAI, as was the locus on the Y chromosome with the possible regulation of tooth size and shape. In 1992, Salido *et al.* showed that both the AMGX and AMGY genes are expressed in developing male tooth buds albeit with the latter at a 90% lower level.

Using two large families with X-linked amelogenesis imperfecta, Lagerström *et al.* (1989, 1990) mapped the mutated locus to Xp22. They found a maximum lod score of 4.45 with $\theta = 0$ with DXS85. The amelogenin gene therefore clearly represents a good candidate for XAI. In 1991, Lagerström *et al.* reported a 5-kb deletion in the amelogenin gene including at least two exons that segregated with the disease in a family with XAI, termed AIH1. It segregated with 15 family members, all showing the hypomineralization form of the disease. Lagerström-Fermer *et al.* (1993) further examined the molecular basis of this large deletion. It removed the third to seventh exons of the coding region, starting at the end of the second exon and leaving only the first two codons of the mature protein. The extent of the deletion was consistent with the severe phenotype seen in this family.

In 1992, Aldred *et al.* (1992a) found a nonsense mutation, a deletion of one cytosine in exon 5, associated with the disorder in another XAI (AIH1) family. The mutation introduced a stop codon into the exon immediately 3′ of the deletion. This family displays combined phenotypes of enamel hypoplasia and hypomineralization. These authors proposed that the amelogenin gene could be implicated in both the formation of enamel thickness and in the process of biomineralization. There is also evidence for a second amelogenin locus on the X chromosome. Aldred *et al.* (1992b) found another XAI family that was excluded from the AMGX region but showed a small positive linkage with two probes, DXS144E and F9 to the q arm between Xq22-q28.

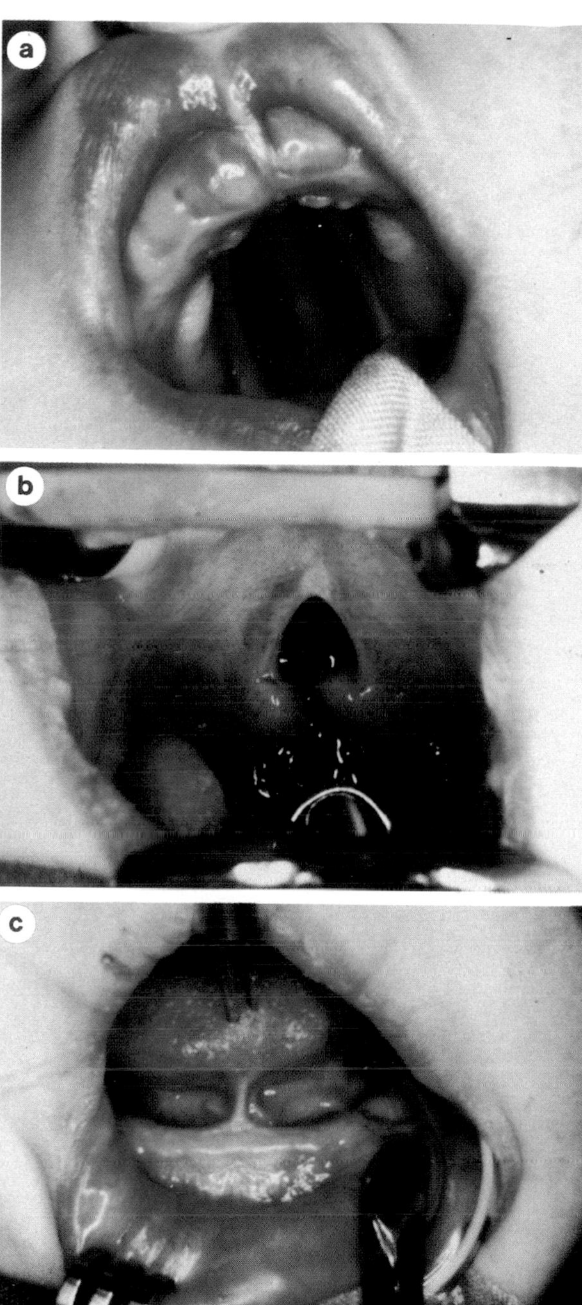

FIG. 11 The phenotype of the Icelandic CPX family: (a) complete secondary cleft palate; (b) bifid uvula; (c) ankyloglossia (tongue-tie).

2. X-Linked Cleft Palate and Ankyloglossia

a. Clinical Features and Incidence of X-Linked Cleft Palate and Ankyloglossia Facial clefting disorders are among the commonest dysmorphologies seen in humans. Isolated cleft palate (CP) occurs with a frequency of 1 in 1500 and is inherited most frequently as the result of interaction between genetic and environmental factors. Secondary cleft palate requires surgery and has associated medical and psychological problems, especially related to feeding and speech development.

The development and fusion of the secondary palate is a complex event, beginning at week 6 and continuing until week 12 of gestation. Mesenchymal cells originating from the neural crest migrate to the oral cavity to form the maxillary processes. Bilateral palatal shelves arise from these maxillary processes, initially growing vertically down the sides of the tongue. As the tongue begins to lower and flatten, the shelves rapidly elevate to a horizontal position until they meet. The medial edge epithelia of the palatal shelves are now able to fuse to form the midline seam. This rapidly degenerates, allowing continuity in the mesenchyme across the horizontal palate. At the same time, palatal epithelia differentiate regionally to form bone and cartilage that comprises the hard and soft palate. Cleft palate can be the result of disturbance to any of the above events, in particular, defective shelf growth, elevation, or fusion. Mechanisms such as a failure in directed cell signaling or programmed cell death (apoptosis) may also be involved (Sulik *et al.*, 1988; Ferguson, 1988).

Cleft palate is frequently associated with cleft lip (CLP) although this condition is etiologically distinct from cleft palate alone (Fogh-Andersen, 1942; Fraser, 1980). CLP varies in frequency in different ethnic groups: 1 : 1000 in whites, 1.7 : 1000 in Japanese, and 0.4 : 1000 in American blacks, with the majority of affected cases being males. Cleft palate alone, however, has a population incidence of 1 : 1500 with little ethnic variation and is more common in females than males (Thompson and Thompson, 1986). The recurrence risk of cleft palate in siblings is only about 2% and it has been concluded that the majority of cases are inherited as a consequence of interaction between genetic and environmental factors.

Although the multifactorial-threshold model may best explain the inheritance of CP (Fraser, 1980), there have been a small number of families in which CP has been shown to be inherited in an autosomal dominant fashion (Fogh-Andersen, 1942). In addition, a small number of families have been documented in which CP occurs through apparent semidominant, X-linked inheritance without the obvious intervention of nongenetic factors (Bjornsson *et al.*, 1989; Lowry, 1970; Rushton, 1979; Rollnick and Kaye, 1986; Bixler, 1987; Hall, 1987). The CP phenotype of these families includes clefting of the hard and soft palate or cleft uvula with or without

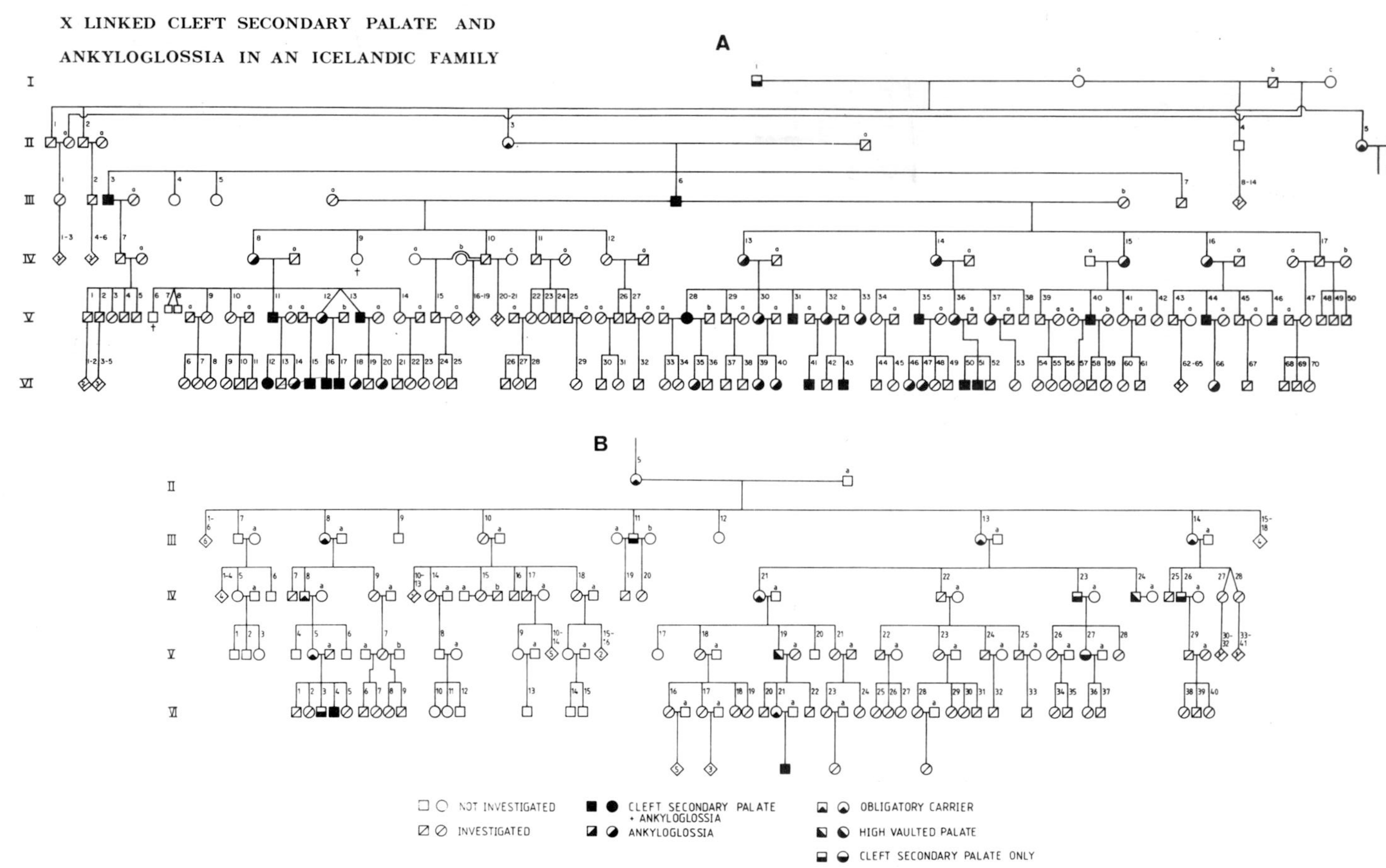
X LINKED CLEFT SECONDARY PALATE AND
ANKYLOGLOSSIA IN AN ICELANDIC FAMILY
A
B
NOT INVESTIGATED
INVESTIGATED
CLEFT SECONDARY PALATE + ANKYLOGLOSSIA
ANKYLOGLOSSIA
OBLIGATORY CARRIER
HIGH VAULTED PALATE
CLEFT SECONDARY PALATE ONLY

submucous clefting of the soft palate. Ankyloglossia (tongue-tie) was found to be associated with CP in an Icelandic family by Bjornsson *et al.* (1989) (Fig. 11) and was also described in an Illinois family (Rollnick and Kaye, 1986). This same phenotype was also found in a British Columbian family (Gorski *et al.*, 1992b). Ankyloglossia is found in the majority of affected males with CP and alone in 71% of the carrier females. This additional common feature adds credence to the theory that these families might share defects in the same gene. Although X-linked CP constitutes less than 1% of the total occurrence of CP, the study of these families could be essential in dissecting the genetic from nongenetic components in the commoner forms of the disorder.

b. Molecular Genetic Alkalysis of X-Linked Cleft Palate and Ankyloglossia The first positional data for cleft palate and ankyloglossia (CPX) resulted from genetic linage analysis using the Icelandic pedigree described by Bjournsson *et al.* (1989) (Fig. 12). The family was documented over 7 generations and contained more than 300 individuals. CPX was linked to the proximal half of Xq with a significant lod score of 3.07 ($\theta = 0$) using probe DXYS1X, which maps to the XY homology region in Xq21 (Moore *et al.*, 1987). Multipoint linkage analysis suggested two possible localizations for the CPX locus, either 5 cM proximal or 7 cM distal to DXYS1X (Ivens *et al.*, 1988). Recombination analysis indicated that CPX was most likely to be at the more distal location, between DXYS12X and DXS17. Another linkage study of the British Colombian family favored the more proximal location, suggesting that CPX is located in an interval of 8–9 cM, between the probes PGK1 and DXYS1X, with a maximum lod score of 7.44 ($\theta = 0$) at DXS72 (Gorski *et al.*, 1992a).

With the availability of new probes in the area, further analysis has been carried out on the Icelandic kindred. The analysis includes 116 individuals of which 22 are affected males, 13 affected females, 8 unaffected males, and 12 unaffected females that were all the children of either affected or obligatory carrier mothers. These have been typed with a further 14 informative polymorphic markers. The data obtained have excluded the area between DXYS12X and DXS17 and indicate that the region centromeric to DXYS1X, within the interval Xq21.1-q32.31, is now the most likely position of CPX. Two recombination events at DXYS1X form the distal boundary whereas one recombination event extends past DXS326, forming the proximal boundary (Stanier *et al.*, 1993). Multipoint linkage analysis is supportive of this interval with a peak lod score of 10.54, 21 times more likely than the next most favorable position. Recom-

FIG. 12. The Icelandic CPX family pedigree.

binations were detected with all of the close flanking markers. The diagram in Fig. 13 indicates the localization of markers around the CPX locus, showing the proximal and distal recombinant individuals flanking the region.

Yeast artificial chromosome clones are now being selected around the area with the aim being to create a YAC contig spanning the region between the flanking markers. Between the markers DXS72 and DXYS1X a consensus estimate of the genetic distance is 4–5 cM (Keats *et al.*, 1989; Huang *et al.*, 1992). From the known order of markers, the region must be smaller. Some data have also been published with regard to the physical size of several of the X chromosome deletions in the area. In patient XL45, for example, a deletion spanning from immediately within the distal side of DXS72 down to DXS95 has been estimated to be about 5 Mb (Cremers *et al.*, 1990). It is not yet known how the interval between DXS326 and DXYS1X compares.

The smallest physical interval containing the CPX locus is no more than several megabases with three crossovers remaining, indicating that the region can still be further reduced in size by genetic analysis. The next step is to complete the genetic mapping and to construct a physical map around the CPX locus to allow the gene responsible for X-linked CP to be isolated.

V. Mouse Models in the Study of Molecular Aspects of Craniofacial Dysmorphologies

Despite the great increase in the understanding of inherited human diseases through the advent of positional cloning, there are still many fields of

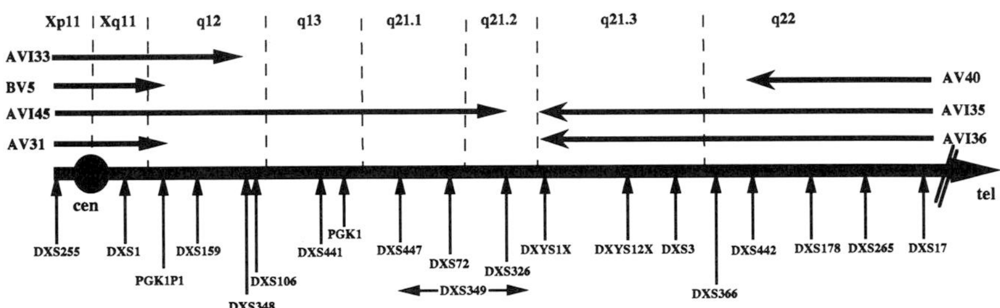

FIG. 13. A map of the markers surrounding the CPX locus. Bold horizontal arrows represent the chromosomes of family members with cross-overs. The direction of the arrows points to the region of the chromosome where the CPX locus must be. Therefore CPX must reside between DXS326/DXS349 and DXYS1X.

study that are particularly dependent on mouse or other animal models. Certainly the field of craniofacial malformations has been one, with the majority of understanding of the mechanisms involved being derived from experimental animals. This field, like many others, can greatly benefit from combining approaches from both mouse and humans. For medical and diagnostic reasons we are fundamentally interested in the mechanisms and developmental pathways that occur in the human, but our access to human tissue is often limited and controls are sometimes difficult to set up. Naturally occuring mouse models, or transgenic lines, can act as models for a parallel human disorder. The mode of action of teratogens that produce malformations that are phenocopies of genetic mutations can also give much insight into thc study of the causes of specific dysmorphologies. For this review, several applied human examples have been chosen in which molecular genetic analysis in the mouse has helped in the human. In some of the examples there are direct links [e.g., GCPS (and maybe ACLS) and AIH1], whereas in others there are potential future links [DiGeorge (CATCH 22), TCS, and CPX].

A. Genetic Mutants

There are hundreds of known mouse mutants that have had their phenotypes carefully studied (Green, 1989). About 30% have reasonable chromosomal localizations. The synteny between the mouse genome and human is now fairly well understood (Nadeau and Reiner, 1989). As phenotypes for many of the human syndromes are documented, some mouse mutations can be assigned their potential human homologs. If the mouse mutant has been precisely mapped, the syntenic region in the human can become a candidate region of interest. Finer mapping of the mouse mutant using breeding programs can be used to pinpoint the critical region on the human chromosome and often aid in the cloning of the gene itself.

1. *Xt* Mouse Mutant *extra toes:* A Mouse Model for GCPS?

In 1967, Johnson described a mouse mutant, *extra-toes* (*Xt*), which affects limb development and displays spontaneous semidominant inheritance. There is almost complete penetrance in heterozygotes but expressivity is variable. There are several *Xt* alleles, all mapping to mouse chromosome 13, which shares homologous synteny to human 7p13 (Lyon and Kirby, 1992). Heterozygotes for *Xt* exhibit predominantly preaxial polydactyly of the hindlimbs, postaxial nubbin on the forelimbs, enlarged interfrontal bone, with hydrocephaly in some cases. This phenotype parallells that seen in GCPS patients. All the *Xt* homozygotes display more extreme limb and facial abnormalities, including extra malformations such as exencephaly and edema. On the basis of both comparative mapping and phenotypic similarities, Winter and Huson (1988) proposed that *Xt* was a possible mouse model for GCPS.

In 1992, a deletion in the 5′ end of the murine *Gli3* gene was found in an *Xt* mutant by Vortkamp *et al.;* Schimmang *et al.* (1992) reported reduced expression of *Gli3* in this mutant. In 1993, Hui and Joyner showed that the *Xt* mutation contains a intragenic deletion of the *Gli3* gene. The murine *Gli, Gli2,* and *Gli3* genes were isolated by their sequence homology to the zinc finger domain of the human GLI genes (Kinzler *et al.,* 1988). These genes were showed to be expressed at day 7 of mouse gestation.

Gli3 was shown to be most prominently expressed in the developing limb buds and the facial primordia. In the *Xt* mutant mice, overgrowth of the limb buds and pharyngeal arches was therefore suspected to be due to an insult to the epitheliomesenchymal interaction. *Gli3* is highly expressed in the mesenchyme. The deficiency of *Gli3* expression found in *Xt* mutants could thus contribute significantly to the epitheliomesenchymal malformations observed. This mouse mutant provides an exciting model in which to study the effects of this gene family, and the ability to relate its expression to that seen in the human, especially with reference to GCPS. With respect to ACLS, the human gene localization has currently ruled out the possibility of it being allelic to GCPS. It may be that there is a phenocopy of the GLI gene family on another chromosome that is responsible for ACLS when mutated. By studying the mouse *Gli* gene family in transgenics or by screening mouse expression libraries for homologs, the genes involved in ACLS could also be elucidated.

In the case of XAI the work on the mouse amelogenin genes helped pinpoint the human homologs. Availability of the mouse genes will enable "knock-out" and mutated forms of the genes to be used in the production of transgenics lines. The effects of mutations seen in XAI in the human can then be studied at the corresponding developmental stages of tooth development in the mouse.

B. Transgenic Mutants

Transgenic insertional mutations that lead to congenital malformations can be useful in the study of genes involved in craniofacial development, as the transgene sequence may serve as a marker enabling localization and gene retrieval.

A Transgenic Mouse with X-Linked Cleft Palate: A Model for Human CPX?

During production of transgenic lines generated to express the latent membrane protein of Epstein–Barr virus, 1 line of 17 displayed a sex-linked lethal phenotype. Within this strain, the predominant phenotype was secondary cleft palate and neonatal lethality in males. Females did not suffer from clefting, but did have poorly formed mandibles and maxillae. The sequences *DXRib1* adjacent to the transgenic insertion have been mapped to the *Araf* locus, which has been localized to p11 on the human X chromosome. However, the CPX locus is at Xq21.2-q21.31, which is some 30 cM from Xp11. The disrupted locus in the transgenic line could be separated from the *Araf* locus on the human chromosome but this has yet to be determined. It is speculated that the disrupted locus on the mouse X chromosome may be the murine homolog of human X-linked cleft plate syndrome (CPX) (Wilson *et al.*, 1992). Any candidate genes found to be localized in the *DXRib1* region in the mouse could be analyzed in the human CPX critical region, to check for possible involvement in X-linked clefting in humans.

C. Teratogen-Induced Mutants

Despite the focus of this article on the genetic clues or molecular approaches to craniofacial dysmorphology, it is important not to discount those teratogens that cause congenital malformations. These are drugs, chemicals, infectious, and physical and metabolic agents that cause abnormalities to the fetus when present in the intrauterine environment. They can provide extremely important clues to the mechanisms of normal development, indirectly indicate genes involved, and provide useful mouse models for genetic analysis.

There are several issues to consider when studying the effect of a teratogen on fetal development. The dosage of the agent will be important, with some teratogens showing clear dosage-related threshold effects. The developmental timing of exposure to the drug can be critical. Each separate fetal tissue will have a critical susceptible period of development, and exposure before or after this may well be benign. There will be differences in susceptibility, depending on the genetic predisposition of both the fetus and mother. Finally, there will always be interaction of the causative agent with others. This may seriously complicate analysis of the effects seen. It is the complex combination of all these factors that has required the use of animal models for the study of the teratogenic effects of many natural and human-made substances.

There are many teratogens that cause craniofacial abnormalities and many reviews on the topic. The two examples mentioned here are linked with human craniofacial syndromes that have clear genetic components, and show the importance of information gained from the study of the action of teratogens.

1. Retinoic Acid Syndrome

The use of isotretinoin, etretinate, and large doses of retinol (vitamin A) for skin disorders such as psoriasis and acne has been clearly associated with serious congenital abnormalities in the fetus. Craniofacial malformations include microcephaly with central nervous system anomalies resulting in hydrocephaly and posterior fossa cysts. Facial features include asymmetry with midfacial hypoplasia, metopic synostosis, microphthalmia, oculomotor palsies, cleft palate and lip, and anomalies of the external auditory canal (Lammer *et al.*, 1985). Many of these malformations are similar to those of hemifacial microsomia, suggesting similar neural crest cell involvement. Malformations seen in human retinoic acid syndrome (RAS) can be induced in animals by the drug 13-*cis*-retinoic acid (Johnston and Bronsky, 1991).

2. Fetal Alcohol Syndrome

Fetal alcohol syndrome (FAS) is the name given to a spectrum of abnormalities that have been associated with maternal alcohol abuse. Clinical features include prenatal growth deficiency, characteristic facies, and abnormalities of the central nervous system. There are also many associated anomalies that are sometimes seen, including ptosis of the eyes, posterior rotation of the ears, and lateral palatine ridges with cleft lip and palate (Clarren and Smith, 1978) (Fig. 14). There is also a wide range of cardiac, renal, and skeletal defects reported.

It is currently estimated that 0.2% of children have FAS in the United States. It is documented that 1 in 30 pregnant women abuses alcohol and 6% of these will have affected offspring (Abel and Sokol, 1987). The exact amount of alcohol needed to cause malformations has not been clearly established and varies depending on the genotype and predisposition of both mother and fetus. An average of two or more drinks per day during pregnancy has been associated with adverse fetal outcome. However, more than five drinks per day is probably required to produce the more serious spectrum of associated abnormalities (Hanson *et al.*, 1978). The pathogenesis is unclear, alcohol having potentially many effects on cell metabolism and growth. It could be that alcohol itself or one of its metabo-

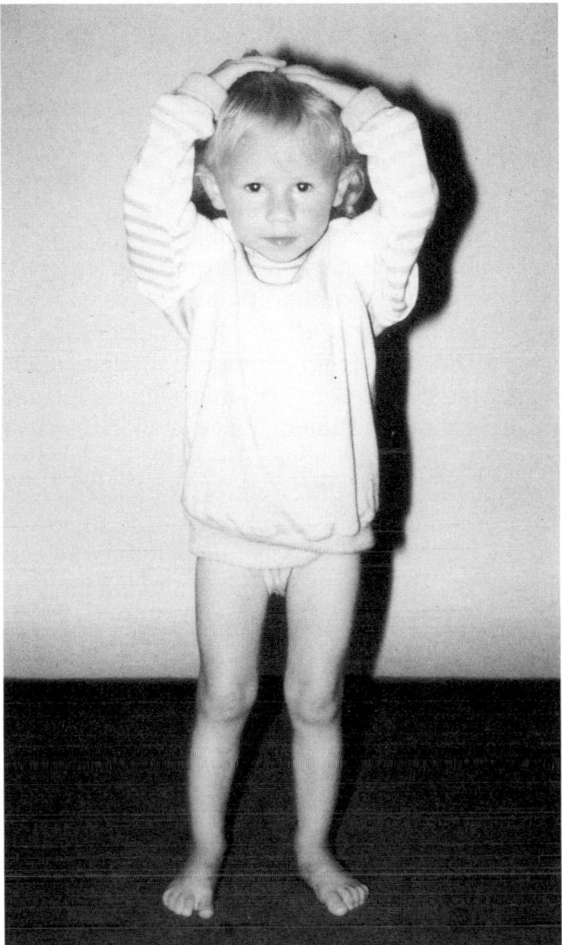

FIG. 14. A typical fetal alcohol syndrome (FAS) patient.

lites, such as acetaldehyde, causes the damage. It may also be that the indirect effects of alcoholism, such as malnutrition or vitamin deficiencies, could add to the level of malformation seen.

Studies in animal models have shown that fetal alcohol syndrome could be interpreted as a mild form of holoprosencephaly—a "single-cavity" forebrain (Sulik and Johnston, 1983). Use of animal models will allow manipulation of the amount of alcohol administered, and exact timing of the dose. This will give a continuous spectrum of the severity of the malformations observed.

3. Retinoic Acid Syndrome and Fetal Alcohol Syndrome: Links with Treacher Collins and DiGeorge Syndromes?

If 13-*cis*-retinoic acid is administered at the time of ganglion formation, excessive cell death occurs, limited to trigeminal ganglion neuroblasts of placodal origin. Secondary effects on neural crest cells in the region of the ganglion give rise to malformations identical to those of Treacher Collins sydrome (TCS) (Webster *et al.,* 1986).

Features of RAS also have striking similarities to the DiGeorge sequence. Features similar to DiGeorge syndrome can also be produced by ethanol administration in mice, and a link has been made with maternal alcoholism and FAS (Ammann *et al.,* 1982). Unlike RAS, the DiGeorge syndrome displays a short upper lip. In mouse models, ethanol administered before and during neural crest cell migration is lethal to migrating crest cells even where the migration distance is short, such as that to the frontonasal region that forms the upper lip (Sulik *et al.,* 1987). This may explain the differences between the effect of 13-*cis*-retinoic acid and that of ethanol, as 13-*cis*-retinoic acid is rarely lethal for migrating crest cells.

It may well be that the genes and developmental pathways involved in these syndromes and the effects of isoretinoin and ethanol are linked, and that information gained about the actions of these teratogens will give insight into the type of genes or secondary pathways involved in TCS and DiGeorge syndrome.

VI. Future Directions

A. Prenatal Diagnosis: Ultrasound and Fetal Surgery

As the genes involved in craniofacial dysmorphologies are elucidated, prenatal diagnosis will become a diagnostic reality for many affected families. Amniocentesis and chorionic villus sampling are now standard procedures in many hospitals. The increased risk of miscarriage for both procedures at 1 and 2 %, respectively, can be assessed by the individual families in comparison with feelings about caring for an affected child. Using the polymerase chain reaction diagnosis of a gene defect can take just a few days, and is also inexpensive as most cases require no radioactive labeling to visualize the bands.

In other cases in which the genes and causes are still unknown, high-resolution real-time ultrasound during pregnancy can often diagnose affected status. Using transvaginal ultrasonography, a profile of the face can be achieved at 12 weeks and the facial anatomy can be examined fairly easily at 14 weeks. Diagnosis of incomplete fusion of the palate and lips is straightforward at this time. Intraocular distance can be reliably measured at 11 to 12 weeks. Intraocular structures such as the hyoid artery and lens are clearly visible owing to interface with the intraocular fluid, and by the beginning of the second trimester congenital cataracts can be discerned. Prenatal ultrasonography can accurately detect many facial abnormalities, especially combined with prior knowledge of a specific family syndrome or a history of maternal teratogen exposure (Pilu *et al.,* 1986, Hegge *et al.,* 1986).

The cranium and brain are obviously dependent on one another for normal development, and disruption of one will cause problems for the other. Anencephaly (Campbell *et al.,* 1972), and other related brain and cranium defects such as hydrocephalus (Cardoza *et al.,* 1988), cephalocele (Chervenak *et al.,* 1984), holoprosencephaly (Blackwell *et al.,* 1982), and microcephaly (Avery *et al.,* 1972), can be fairly reliably diagnosed in the second and third trimesters. The majority of severe head abnormalities can be seen or measured by 20 weeks gestation. Malformations of the neck can also be seen using ultrasonography, with hypoplasia of the mandible, maxilla or orbit, microtia, and hemivertebrae all diagnosable (Tamas *et al.,* 1986).

Oncc thcsc malformations have been diagnosed by ultrasound, little can be done in terms of repair. At present *in utero* fetal medicine or surgery is really restricted to a few simple procedures such as blood transfusion, shunting, and aminiotic fluid addition or reduction. There have been attempts at more complex procedures such as open surgical reduction of diaphragmatic hernias and resection of fetal tumors, but these involve a risk to the mother and are performed only in a few experimental centers. When more is understood about the genes and pathways involved, then other therapies may become a possibility, for example, growth factor addition to the amniotic fluid at important developmental stages.

B. Gene Therapy

Gene therapy has become an exciting area of research that has rapidly accelerated during the past few years. With the successful cloning of the genes for several purely genetic disorders such as cystic fibrosis (CF) and Duchenne muscular dystrophy, and the better understanding of many cancers, this technique has become a reality.

Since the gene for CF was cloned 4 years ago, four clinical trials are underway on two continents, involving either modified viruses or liposomes as delivery vehicles. However, each individual disorder will pose separate problems with regard to the tissue needing therapy, age of onset, and type of delivery vehicle needed.

At the moment, viruses are probably the best candidates for the majority of accessible diseases, in particular cancer. The two types of virus currently in use are adenoviruses and retroviruses. The chief advantage of the adenovirus over the more frequently used retrovirus is its ability to infect quiescent cells. The retrovirus requires postmitotic dividing cells. Unlike the retrovirus, the adenovirus will not integrate into the host genome, minimizing risks associated with insertional mutagenesis. Retroviruses are, however, still the vector of choice for targeting stem cells, making them the most likely candidates for the treatment of hematological disorders. The safety of recombinant viruses for long-term use, and the need for repeated treatments, also needs careful assessment.

With respect to craniofacial malformations the problems are much more complex. It is most probable that in the majority of syndromes, we will need to correct many defective genes. Even for single-gene disorders, it will be hard to access the target tissues *in utero,* which is when most of the damage occurs. However, as the techniques and understanding improve it is entirely possible that the successful treatment of CF in children will be followed by the successful treatment of DiGeorge syndrome or CATCH 22 *in utero*. Diagnosis would have to be just after conception and treatment would need to begin at 4–5 weeks gestation.

C. Ethical Issues

There are many eithical issues to be considered as the genes for these disorders are elucidated. Should each individual know about his/her faulty genome and their potential risk for malformed offspring? What right to life has the unborn baby with cleft palate, which is not a life-threatening condition? Should affected DiGeorge and CATCH 22 fetuses be born? The decisions will be made by physicians, families, and ethical committees but one wonders if our quest to clone and understand the whole genome may not bring more problems than it solves.

VII. Concluding Remarks

This is an exciting era in the understanding of the causes of congenital malformations. The technology to clone genes that when mutated or deleted cause specific human craniofacial disorders is now available. Families with such syndromes have in most cases been documented and are ready for analysis.

This article does not include any discussion of which candidate genes are most likely to be involved in the various disorders. From the evidence collected to date, these will likely prove to encode a whole range of classes of protein products, from structural proteins such as amelogenin to transcriptional regulators such as GLI3 and TUPLE1. Such speculation forms a significant part of the excitement that is an essential component of the laborious daily quest for the genes involved in craniofacial malformations.

Acknowledgments

I thank all my colleagues from the Action Research Laboratory for the Molecular Biology of Fetal Development, with special thanks to Dr. Philip Stanier for assisting with the diagrams and text. My thanks to Professor John Burn, Professor Gerry Winter, and Dr. Louise Brueton for photographs and to Drs. Nick Lench, Mike Dixon, and Peter Scambler for critical reading of their relevant research areas. Finally, I acknowledge our funding bodies, The Dunhill Medical Trust, Wellbeing, Action Research, the Medical Research Council (UK), and the Special Health Authority and Trust for Hammersmith and Queen Charlotte's and Chelsea Hospitals.

References

Abel, E. L., and Sokol, R. J. (1987). Incidence of fetal alcohol syndrome and economic impact of FAS-related anomalies. *Drug Alcohol Depend.* **19,** 51–70.

Aldred, M. J., Crawford, P. J., Roberts, E., and Thomas, N. S. T. (1992a). Identification of a nonsense mutation in exon 5 of the amelogenin gene in a family with X-linked amelogenesis imperfecta. *Hum. Genet.* **90,** 413–416.

Aldred, M. J., Crawford, P. J., Roberts, E., Gillespie, C. M., Thomas, N. S. T., Fenton, I., Sandkuijl, L. A., and Harper, P. S. (1992b). Genetic heterogeneity in X-linked amelogenesis imperfecta. *Genomics* **14,** 567–573.

Allen, N. D., Cran, D. G., Barton, S. C., Hettle, S., Reik, T., and Surani, M. A. (1988). Transgenes as probes for active chromosomal domains in mouse development. *Nature (London)* **333,** 852–855.

Ammann, A. J., Wara, D. W., Cowan, M. J., Barrett, D. J., and Stiehm, E. R. (1982). The DiGeorge syndrome and the fetal alcohol syndrome. *Am. J. Dis. Child.* **136,** 906–908.

Aslanidis, C., Jansen, G., Amemiya, C., Shutler, G., Mahadevan, M., Tsifidia, C., Chen, C., Allerman, J., Wormskamp, N. G. M., Vooijs, M., Buxton, J., Johnson, K., Smeets, H. J. M., Lennon, G. G., Carrano, A. V., Korneluk, R. G., Wieringa, B., and deJong, P. J. (1992). Cloning of the essential myotonic dystrophy region and mapping of the putative defect. *Nature (London)* **355,** 548–551.

Augusseau, S., Jouk, S., Jalbert, P., and Prieur, M. (1986). DiGeorge syndrome and 22q11 rearrangements. *Hum. Genet. Lett.* **74,** 206.

Avery, G. B., Meneses, L., and Lodge, A. (1972). The clinical significance of "measurement microcephaly." *Am. J. Dis. Child.* **123,** 214–217.

Baccichetti, C., Artifoni, L., and Zanardo, V. (1982). Deletions of the short arm of chromosome 7 without craniosynostosis. (Letter) *Clin. Genet.* **21,** 348–349.

Backman, B., and Holm, A.-K. (1986). Amelogenesis imperfecta: Prevalence and incidence in a northern Swedish country. *Community Dentistry Oral Epidimiol.* **14,** 43–47.

Backman, B., and Holmgren, G. (1988). Amelogenesis imperfecta: A genetic study. *Hum. Hered.* **38,** 189–206.

Baldwin, C. T., Hoth, C. F., Amos, J. A., de Silva, E. O., and Milunsky, A. (1992). An exonic mutation in the HuP2 paired domain gene causes Waardenburg's syndrome. *Nature (London)* **355,** 637–638.

Balestrazzi, P., Baeteman, M. A., Mattei, M. G., and Mattei, J. F. (1983). Franceschetti sydrome in a child with a de novo balanced translocation (5;13)(q11;p11) and significant decrease of hexomidase. B. *Hum. Genet.* **64,** 305–308.

Bendon, R. W., Siddiqi, T., Soukup, S., and Srivasatava, A. (1988). Prenatal detection of triploidy. *J. Pediatr.* **112,** 149–153.

Berger, W., Meindl, A., van de Pol, T. J. R., Cremers, F. P. M., Ropers, H. H., Doerner, C., Monaco, A., Bergen, A. A. B., Lebo, R., Warburg, M., Zergollern, L., Lorenz, B., Gal, A., Biecker-Wagemakers, E. M., and Meitinger, T. (1992). Isolation of a candidate gene for Norrie disease by positional cloning. *Nat. Genet.* **1,** 199–203.

Berkman, M. D., and Singer, A. (1971). Demonstration of the Lyon hypothesis in X-linked dominant hypoplastic amelogenesis imperfecta. *Birth Defects. Orig. Artic. Ser.* **7,** 204–209.

Berry, G. A. (1889). Note on a congenital defect (coloboma?) of the lower lid. *R. London Ophthalmol. Hosp. Rep.* **12,** 255–257.

Bird, A. P. (1986). CpG-rich islands and the function of DNA methylation. *Nature (London)* **321,** 209–213.

Bixler, D. (1987). Letter to the editor: X-linked cleft palate. *Am. J. Med. Genet.* **28,** 503–505.

Bjornsson, A., Arnason, A., and Tippet, P. (1989). X-linked midline defect in an Icelandic family. *Cleft Palate J.* **26,** 3–8.

Blackwell, D. E., Spinnato, J. A., Hirsch, G., Giles, H. R., and Sackler, J. (1982). Antenatal ultrasound diagnosis of holoprosencephaly: A case report. *Am. J. Obstet. Gynecol.* **143,** 848–849.

Boue, J., Boue, A., and Lazar, P. (1975). Retrospective and prospective epidemiological studies of 1500 karyotpyed spontaneous human abortions. *Teratology* **12,** 11–26.

Brueton, L., Huson, S. M., Winter, R. B., and Williamson, R. (1988). Chromosomal localisation of a rare development gene in man: Direct DNA analysis demonstrates that Greig cephalopolysyndactyly maps to 7p13. *Am. J. Med. Genet.* **31,** 799–804.

Brueton, L. A., Chotai, K. A., van Herwerden, L., Shinzel, A., and Winter, R. M. (1992). The acrocallosal syndrome and Greig syndrome are not allelic disorders. *J. Med. Genet.* **29,** 635–637.

Burke, D. T., Carle, G. F., and Olson, M. V. (1987). Cloning of large segments of exogenous DNA into yeast by means of artificial chromosome vectors. *Science* **236,** 806–812.

Campbell, S., Johnstone, F. D., Holt, E. M., and May, P. (1972). Anencephaly: Early ultrasonic diagnosis and active managment. *Lancet* **2,** 1226–1227.

Cardoza, J. D., Goldstein, R. B., and Filly, R. A. (1988). Exclusion of fetal ventriculomegaly with a single measurement: The width of the lateral ventricular atrium. *Radiology* **169,** 711–714.

Carey, A. H., Kelly, D., Halford, S., Wadey, R., Wilson, D., Goodship, J., Burn, J., Paul, T., Sharkey, A., Dumanski, J., Nordenskjold, M., Williamson, R., and Scambler, P. J. (1992). Molecular genetic study of the frequency of monosomy 22q11 in DiGeorge syndrome. *Am. J. Hum. Genet.* **51,** 964–970.

Cataltepe, S., and Tuncbilek, E. (1992). A family with one child with acrocallosal syndrome one child with anencephaly-polydactyly and parental consanguinity. *Eur. J. Pediatr.* **151,** 288–290.

Chelly, J., Tumer, Z., Tonnesen, T., Petterson, A., Ishikawa-Brush, Y., Tommerup, N., Horn, N., and Monaco, A. (1993). Isolation of a candidate gene for Menkes disease that encodes a potential heavy metal binding protein. *Nat. Genet.* **3,** 14–19.

Chen, Z.-Y., Hendriks, R. W., Jobling, M. A., Powell, J. F., Breakefield, X. O., Sims, K. B., and Craig, I. W. (1992). Isolation and characterization of a candidate gene for Norrie disease. *Nat. Genet.* **1,** 204–208.

Chervenak, F. A., Isaacson, G., Mahoney, M. J., Berkowitz, R. L., Tortura, M., and Hobbins, J. C. (1984). Diagnosis and management of fetal cephalocele. *Obstet. Gynecol. (NY)* **64,** 86–91.

Chomczynski, P., and Sacchi, N. (1987). Single step method of RNA isolation by acid guanidinium thiocyanate-phenol-chloroform extraction. *Anal. Biochem.* **162,** 156–159.

Chosak, A., Eidelmann, E., Wisotski, I., and Cohen, T. (1979). Amelogenesis imperfecta among Israeli Jews and the description of a new type of local hypoplastic autosomal recessive amelogenesis imperfecta. *Oral Surg., Oral Med. Oral. Pathol.* **47,** 148–156.

Clarren, S. K., and Smith, D. W. (1978). The fetal alcohol syndrome. *N. Engl. J. Med.* **298,** 1063–1067.

Collins, F. S., and Weissman, S. M. (1984). Directional cloning of DNA fragments at a large distance from an initial probe: A circulization method. *Proc. Natl. Acad. Sci. U. S. A.* **81,** 6812–6816.

Conley, M. E., Beckwith, J. B., Mancer, J. F., and Tenckhoff, L. (1979). The spectrum of the DiGeorge syndrome. *J. Pediatr.* **94,** 883–890.

Cotton, R. G., Rodrigues, N. R., and Campbell, R. D. (1988). Reactivity of cytosine and thymine in single-base-repair mismatches with hydroxylamine and osmium tetroxide and its application to the study of mutations. *Proc. Natl. Acad. Sci. U. S. A.* **85,** 4397–4401.

Cremers, F. P. M., Sankila, E.-M., Brunsmann, F., Jay, M., Jay, B., Wright, A., Pinckers, A. J. L. G., Schwartz, M., van de Pol, D. J. R., Wieringa, B., de la Chapelle, A., Pawlowitzki, I. H., and Ropers, H.-H. (1990). Deletions in patients with classical choroideremia vary in size from 45 kb to several megabases. *Am. J. Hum. Genet.* **47,** 622–628.

Davies, A. A., Moss, S. E., Crompton, M. R., Jones, T. A., Spurr, N. K., Sheer, D., Kozak, C., and Crumpton, M. J. (1989). The gene coding for the p68 calcium-binding protein is localised to bands q32-q34 of human chromosome 5 and to mouse chromosome 11. *Hum. Genet.* **82,** 234–238.

DeCicco, F., Steele, M. W., Pan, S., and Park, S. C. (1973). Monosomy of chromosome no. 22: A case report. *J. Pediatr.* **83,** 836–838.

De la Chapelle, A., Herva, R., Koivisto, M., and Aula, P. (1981). A deletion of chromosome 22 can cause DiGeorge syndrome. *Hum. Genet.* **57,** 253–256.

Dietz, H. C., Cutting, G. R., Pyeritz, R. E., Maslen, C. L., Sakai, L. V., Corson, G. M., Puffenberger, E. G., Hamosh, A., Nanthakumar, E. J., Curristen, S. M., Stetten, G., Meyers, D. A., and Francomano, C. A. (1991). Marfan syndrome caused by a recurrent de novo missense mutation in the fibrillin gene. *Nature (London)* **352,** 337–339.

DiGeorge, A. M. (1965). Discussion on a new concept of the cellular basis of immunology. *J. Pediatr.* **67,** 907.

Dixon, M. J., Haan, E., Baker, E., David, D., McKensie, N., Williamson, R., Mulley, J., Farrall, M., and Callen, D. (1991a). Association of Treacher Collins syndrome and translocation 6p21.31/16p13.11: Exclusion of the locus from these candidate regions. *Am. J. Hum. Genet.* **48,** 274–280.

Dixon, M. J., Read, A. P., Donnai, D., Colley, A., Dixon, J., and Williamson, R. (1991b). The gene for Treacher Collins syndrome maps to the long arm of chromosome 5. *Am. J. Hum. Genet.* **49,** 17–22.

Dixon, M. J., Dixon, J., Houseal, T., Bhatt, M., Klinger, K., and Landes, G. M. (1993).

Narrowing the position of the Treacher Collins syndrome locus to a small interval between three new microsatellite markers at 5q32-33.1. *Am. J. Hum. Genet.* **52,** 907–914.

Dorin, J. R., Dickinson, P., Alton, E. W. F. W., Smith, S. N., Geddes, D. M., Stevenson, B. J., Kimber, W. L., Fleming, S., Clarke, A. R., Hooper, M. L., Anderson, L., Beddington, R. S. P., and Porteous, D. J. (1992). Cystic fibrosis in the mouse by targeted insertional mutagenesis. *Nature (London)* **359,** 211–215.

Drabkin, H., Sage, M., Helms, C., Green, P., Gemmill, R., Smith, D., Erickson, P., Hart, I., Ferguson-Smith, A., Ruddle, F., and Tommerup, N. (1989). Regional and physical mapping studies characterizing the Greig polysyndactyly 3;7 chromosome translocation, t(3;7)(p21.1;p13). *Genomics* **4,** 518–529.

Driscoll, D. A., Budarf, M. L., and Emanuel, B. S. (1992a). A genetic etiology for DiGeorge syndrome: Consistent deletions and microdeletions of 22q11. *Am. J. Hum. Genet.* **50,** 924–933.

Driscoll, D. A., Spinner, N. B., Budarf, M. L., McDonald-McGinn, D. M., Zackai, E. H., Goldberg, R. B., Shprintzen, R. J., Saal, H. M., Zonana, J., Jones, M. C., Mascarello, J. T., and Emanuel, B. S. (1992b). Deletions and microdeletions of 22q11.2 in velocardiofacial syndrome. *Am. J. Med. Genet.* **44,** 261–268.

Dryja, T. P., McGee, T. L., Reichel, E., Hahn, L. B., Cowley, G. S., Yandell, D. W., Sandberg, M. A., and Berson, E. L. (1990). A point mutation of the rhodopsin gene in one form of retinitis pigmentosa. *Nature (London)* **343,** 364–366.

Duyk, G. M., Kim, S., Myers, R. M., and Cox, D. R. (1990). Exon trapping: A genetic screen to identify candidate transcribed sequences in cloned mammalian genomic DNA. *Proc. Natl. Acad. Sci. U. S. A.* **87,** 8995–8999.

Estivill, X., Farrall, M., Scambler, P. J., Bell, G. M., Hawley, K. M. F., Lench, N. J., Bates, G. P., Kruyer, H. C., Frederick, P. A., Stanier, P., Watson, E. K., Williamson, R., and Wainwright, B. J. (1987). A candidate gene for the cystic fibrosis locus isolated by selection of methylation-free islands. *Nature (London)* **326,** 840–845.

Ferguson, M. W. (1988). Palate development. *Development (Cambridge, UK)* **103,** Suppl., 41–60.

Fibison, W. J., Budarf, M., McDermid, H., Greenberg, F., and Emmanuel, B. S. (1990). Molecular studies of DiGeorge syndrome. *Am. J. Hum. Genet.* **46,** 888–895.

Fogh-Andersen, P. (1942). "Inheritance of Cleft Lip and Palate." Arnold Busck, Copenhagen.

Franceschetti, A., and Klein, D. (1949). Mandibulo-facial dysostosis: New hereditary syndrome. *Acta Ophthalmol.* **27,** 143–224.

Franceschetti, A., Brocher, J. E. H., and Klein, D. (1949). Dysostose mandibulo-facial unilaterale avec deformations multiples du sequelette (Processus paramastoide, synostose des vertebres, sacralisation, etc.) et torticolis clonique. *Ophthalmologica* **118,** 796–814.

Fraser, F. C. (1980). The genetics of cleft lip and palate: Yet another look. *In* "Current Research Trends in Prenatal Craniofacial Development" (Pratt and Christiansen, eds). Elsevier/North-Holland, Amsterdam.

Frazen, L. E., Elmore, J., and Nadler, H. L. (1967). Mandibulo-facial dysostosis (Treacher Collins syndrome). *Am. J. Dis. Child.* **113,** 406–410.

Fu, Y. H., Pizzuti, A., Fenwick, R. G., Jr., King, J., Rajnarayan, S., Dunne, P. W., Dubel, P. W., Nasser, G. A., Ashizawa, T., de Jong, P., Wieringa, B., Korneluk, R., Perryman, M. B., Epstein, H. F., and Caskey, H. F. (1992). An unstable triplet repeat in a gene related to myotonic muscular dystrophy. *Science* **255,** 1256–1259.

Fukushima, Y., Ohashi, H., Wakui, K., Nishida, T., Nakamura, Y., Hoshino, K., Ogawa, K., and Oh-ishi, T. (1992). DiGeorge syndrome with del(4)(q21.3q25): Possibility of the fourth chromosome region responsible for DiGeorge syndrome. *Am. J. Hum. Genet.* **51,** A80.

Giuffra, L. A., Kennedy, J. L., Castiglione, C. M., Evans, R. M., Wasmuth, J. J., and Kidd, K. K. (1988). Glucocorticoid receptor maps to the distal long arm of chromosome 5. *Cytogenet. Cell Genet.* **49,** 313–314.

Gnamey, D., and Farriaux, J. P. (1971). Syndromes dominant associant polysyndactylie pouces en spatule, anomalies faciales et retard mental une forme paticulière de l' acrocephalo-polysyndactylie de type Noack. *J. Genet. Hum.* **19,** 299–316.

Goate, A., Chartier-Harlin, M. C., Mullan, M., Brown, J., Crawford, F., Fidani, L., Guiffra, L., Haynes, A., Irving, N., James, L., Mant, R., Newton, P., Rooke, K., Rogues, P., Talbot, C., Pericak-Vance, M., Roses, A., Williamson, R., Rossor, M., Owen, M., and Hardy, J. (1991). Segregation of a missense mutation in the amyloid precursor protein gene with familial Alzheimer's disease. *Nature (London)* **349,** 704–706.

Gorlin, R. J., Cohen, M. M., and Levin, L. S. (1990). "Syndromes of the Head and Neck," Oxford Monog. Med. Genet. No. 19, 3rd ed. Oxford Univ. Press, New York and Oxford.

Gorski, S. M., Adams, K. J., Birch, P. H., Friedman, J. M., and Goodfellow, P. J. (1992a). The gene responsible for X-linked cleft palate (CPX) in a British Columbian native kindred is localized between PGK1 and DXYS1. *Am. J. Med. Genet.* **50,** 1129–1136.

Gorski, S. M., Adams, K. J., Birch, P. H., Chodirker, B., Greenberg, C. R., and Goodfellow, P. J. (1992b). Linkage studies in two families with X-linked cleft palate and ankyloglossia (CPX). *Am. J. Med. Genet.* **51,** Suppl., A188.

Gosden, C. M., Wright, M. O., Paterson, W. G., and Grant, K. A. (1976). Clinical details, cytogenetic studies, and cellular physiology of a 69,XXX fetus, with comments on biological effect of triploidy in man. *J. Med. Genet.* **13,** 371–380.

Gray, M. R., Colot, H. V., Guarente, L., and Rosbash, M. (1982). Open reading frame cloning: Identification, cloning, and expression of open reading frame DNA. *Proc. Natl. Acad. Sci. U. S. A.* **79,** 6598–6602.

Green, M. C. (1981). Catalog of mutant genes and polymorphic loci. *In* "Genetic Variants and Strains of Laboratory Mouse" (M. C. Green, ed.), pp. 8–278, Fischer Verlag.

Green, M. C. (1989). *In* "Genetic Strains of the Laboratory Mouse" (M. F. Lyon and A. G. Searle, eds.), 2nd ed., Chapter 2, pp. 12–403. Oxford Univ. Press, Oxford.

Greenberg, F., Crowder, W. E., Paschall, V., Colon-Linares, J., Lubianski, B., and Ledbetter, D. H. (1984). Familial DiGeorge syndrome and associated partial monosomy of chromosome 22. *Hum. Genet.* **65,** 317–319.

Greenberg, F., Valdes, C., Rosenblatt, H. M., Kirkland, J. L., and Ledbetter, D. H. (1986). Hypoparathyroidism and T cell immune defect in a patient with 10p deletion syndrome. *J. Pediatr.* **109,** 489–492.

Greenberg, F., Elder, F. F. B., Haffner, P., Northrup, H., and Ledbetter, D. H. (1988). Cytogenetic findings in a prospective series of patients with DiGeorge anomaly. *Am. J. Hum. Genet.* **43,** 605–611.

Greig, D. M. (1926). *Oxycephaly. Edinburgh Med. J.* **33,** 189–218.

Haffner, R., and Willison, K. (1987). *In situ* hybridization to messenger RNA in tissue sections. *In* "Mammalian Development: A Practical Approach" (M. Monk, ed.), pp. 199–215, IRL Press; Oxford.

Halford, S., Wadey, R., Roberts, C., Daw, S. C. M., Whiting, J. A., O'Donnell, H., Dunham, I., Bentley, D., Lindsay, E., Baldini, A., Francis, F., Lehrach, H., Williamson, R., Wilson, D. I., Goodship, J., Cross, I., Burn, J., and Scambler, P. J. (1993). Isolation of a putative transcriptional regulator from the region of 22q11 deleted in DiGeorge syndrome, Shprintzen syndrome and familial congenital heart disease. *Hum. Mol. Genet.* **2,** 2099–2107.

Hall, B. D. (1987). Letter to the editor: A further X-linked isolated non-syndromic cleft palate family with a nonexpressing obligate affected male. *Am. J. Med. Genet.* **26,** 239–240.

Hamaguchi, M., Sakamoto, H., Tsuruta, H., Sasaki, H., Muto, T., Sugimura, T., and

Terada, M. (1992). Establishment of a highly sensitive and specific exon-trapping system. *Proc. Natl. Acad. Sci. U. S. A.* **89,** 9779–9783.

Hanson, J. W., Streissguth, A. P., and Smith, D. W. (1978). The effects of moderate alcohol consumption during pregnancy on fetal growth and morphogenesis. *J. Pediatr.* **92,** 457–460.

Hegge, F. N., Prescott, G. H., and Watson, P. T. (1986). Fetal facial abnormalities identified during obstetric sonography. *J. Ultrasound Med.* **5,** 679–684.

Hendriks, H. J. E., Brunner, H. G., Hagen, T. A. M., and Hamel, B. C. J. (1990). Acrocallosal syndrome. *Am. J. Med. Genet.* **35,** 443–446.

Hootnick, D., and Holmes, L. B. (1972). Familial polysyndactyly and craniofacial anomalies. *Clin. Genet.* **3,** 128–134.

Horowitz, S. L., and Morishima, A. (1974). Palatal abnormalities in the syndrome of gonadal dysgenesis and its variants and in Noonan's syndrome. *Oral Surg., Oral Med. Oral Pathol.* **38,** 839–844.

Huang, T. H. M., Cottingham, R. W., Jr., Ledbetter, D. H., and Zoghbi, H. Y. (1992). Genetic mapping of four dinucleotide repeat loci, DXS453, DXS458, DXS454, DXS424, on the X chromosome using multiplex polymerase chain reaction. *Genomics* **13,** 375–380.

Hui, H.-H., and Joyner, A. L. (1993). A mouse model of Greig cephalopolysyndactyly syndrome: The extra toes mutation contains an intragenic deletion of the gene. *Nat. Genet.* **3,** 241–246.

Huntington's Disease Collaborative Research Group (1993). A novel gene containing a trinucleotide repeat that is expanded and unstable on Huntington's disease chromosomes. *Cell (Cambridge, Mass.)* **72,** 971–983.

Ide, C., and Holt, J. E. (1975). Median cleft face syndrome assciated with orbital hypertelorism and polysyndactyly. *Ears Eyes Nose Throat Mon.* **54,** 150–151.

Ivens, A., Moore, G. E., Chambers, J., Arnason, A., Jensson, O., Bjornsson, A., and Williamson, R. (1988). X-linked cleft palate: The gene is localised between polymorphic DNA markers DXYS12 and DXS17. *Hum. Genet.* **78,** 356–358.

Jabs, E. W., Coss, C. A., Hayflick, S. J., Whitmore, T. E., Pauli, R. M., Kirkpatrick, S. J., Meyers, D. A., Goldberg, R., Day, D. W., and Rosenbaum, K. N. (1991a). Chromosomal deletion 4p15.32-p14 in a Treacher Collins syndrome patient: Exclusion of the disease locus from and mapping of anonymous DNA sequences to this region. *Genomics* **11,** 188–192.

Jabs, E. W., Xiang, L., Coss, C. A., Taylor, E. W., Meyers, D. A., and Weber, J. L. (1991b). Mapping the Treacher Collins syndrome locus to 5q31.3-q33.3. *Genomics* **11,** 193–198.

Jensen, B. L. (1985). Craniofacial morphology in Turner syndrome. *J. Craniofacial Genet. Dev. Biol.* **5,** 327–340.

Johnson, D. R. (1967). Extra-toes: A new mutant gene causing multiple abnormalities in the mouse. *J. Embryol. Exp. Morphol.* **3,** 543–581.

Johnston, M. C., and Bronsky, P. T. (1991). Animal models for human craniofacial malformations. *J. Craniofacial Genet. Dev. Biol.* **11,** 227–291.

Jones, K. L., Smith, D. W., Harvey, M. A. S., Hall, B. D., and Quan, L. (1975). Older paternal age and fresh gene mutations: Data on additional disorders. *J. Pediatr.* **86,** 84–88.

Keats, B., Ott, J., and Conneally, M. (1989). Report of the committee on linkage and gene order. *Cytogenet. Cell Genet.* **51,** 459–502.

Kelley, R. I., Zackai, E. H., Emanuel, B. S., Kistenmacher, M., Greenberg, F., and Punnett, H. H. (1982). The association of DiGeorge anomalad with partial monosomy of chromosome 22. *J. Pediatr.* **101,** 197–200.

Kelly, D., Goldberg, R., Wilson, D., Lindsay, E., Carey, A., Goodship, J., Burn, J., Cross, I., Shprintzen, R. J., and Scambler, P. J. (1993). Confirmation that the velo-cardio-facial syndrome is associated with haplo-insufficiency of genes at chromosome 22q11. *Am. J. Med. Genet.* **45,** 308–312.

Keppen, L. D., Fasules, J. W., Burks, A. W., Gollin, S. M., Sawyer, J. R., and Miller, C. H. (1988). Confirmation of autosomal dominant transmission of the DiGeorge malformation complex. *J. Pediatr.* **113,** 506–508.

Klein, D. (1953). Dysostose mandibulofaciale. *Prat. Odontol.-Stomatol.* **487-490,** 1–8.

Kinzler, K. W., Ruppert, J. M., Bigner, S. H., and Vogelstein, B. (1988). The GLI gene is a member of the Kruppel family of zinc finger proteins. *Nature (London)* **332,** 371–374.

Korting, G. W., and Ruther, H. (1954). Ichthyosis vulgaris und akrofaciale Dysostosis. *Arch. Dermatol. Syph.* **197,** 91–104.

Kruger, G., Gotz, J., Kvist, U., Dunker, H., Erfurth, F., Pelz, L., and Zech, L. (1989). Greig syndrome in a large kindred due to reciprocal chromosome translocation t(6;7)(q27;p13). *Am. J. Med. Genet.* **32,** 411–416.

Kwee, M. L., and Lindhout, D. (1983). Frontonasal dysplasia, coronal craniosynostosis, pre- and postaxial polysyndactyly and split nails: A new autosomal dominant mutant with reduced penetrance and variable expression. *Clin. Genet.* **24,** 200–205.

Lagerström, M., Dahl, N., Iselius, L., Backman, B., and Pettersson, U. (1989). Linkage analysis of X-linked amelogenesis imperfecta. *Cytogenet. Cell Genet.* **51,** 1028 (Abstr.).

Lagerström, M., Dahl, N., Iselius, L., Backman, B., and Pettersson, U. (1990). Mapping of the gene for X-linked amelogenesis imperfecta by linkage analysis. *Am. J. Hum. Genet.* **46,** 120–125.

Lagerström, M., Dahl, N., Nakahori, Y., Nakagome, Y., Backman, B., Landegren, U., and Pettersson, U. (1991). A deletion in the amelogenesis gene (AMG) causes X-linked amelogenesis imperfecta (AIH1). *Genomics* **10,** 971–975.

Lagerström,-Fermer, M., Pettersson, U., and Landegren, U. (1993). Molecular basis and consequences of a deletion in the amelogenin gene analyzed by capture PCR. *Genomcis* **17,** 89–92.

Lammer, E. J., Chen, D. T., Hoar, R. M., Aguish, N. D., Benke, P. J., Braun, J. T., Curry, C. J., Fernhoff, P. M., Grix, A. W., Jr., Lott, I. T., Richard, J. M., and Sun, S. C. (1985). Retinoic acid embryopathy. *N. Engl. J. Med.* **313,** 837–841.

Lau, E. C., Mohandras, T. K., Shapiro, L. J., Slavkin, H. C., and Snead, M. L. (1989). Human and mouse amelogenin gene loci are on the sex chromosomes. *Genomics* **4,** 162–168.

Lee, B., Godfrey, M., Vitale, E., Hori, H., Mattei, M.-G., Sarfarazi, M., Tsipouras, P., Ramirez, F., and Hollister, D. W. (1991). Linkage of Marfan syndrome and a phenotypically related disorder to two different fibrillin genes. *Nature (London)* **352,** 330–334.

Lischner, H. W. (1972). DiGeorge syndrome(s). *J. Pediatr.* **81,** 1042–1044.

Loftus, S. K., Edwards, S. J., Scherpbier-Heddema, T., Buetow, K. H., Wasmuth, J. J., and Dixon, M. J. (1993). A combined genetic and radiation hybrid map surrounding the Treacher Collins syndrome locus on chromosome 5q. *Hum. Mol. Genet.* **2,** 1785–1792.

Loughna, S., Bennett, P. R., Gau, G., Nicolaides, K., Blunt, S., and Moore, G. E. (1993). Over-expression of Esterase D in kidney from human trisomy 13. *Am. J. Hum. Genet.* **53,** 810–816.

Lovett, M., Kere, J., and Hinton, L. M. (1991). Direct selection: A method for the isolation of cDNAs encoded by large genomic regions. *Proc. Natl. Acad. Sci. U. S. A.* **88,** 9628–9632.

Lowry, R. B. (1970). Sex linked cleft palate in a British Columbian Indian family. *Pediatrics* **46,** 123–128.

Lyon, M. F., and Kirby, M. C. (1992). Mouse chromosome atlas. *Mamm. Genome* **90,** 22–43.

Marshall, R. E., and Smith, D. W. (1970). Frontodigital syndrome: A dominantly inherited disorder with normal intelligence. *J. Pediatr.* **77,** 129–133.

Marx, J. L. (1985). Hopping along the chromosome. *Science* **228,** 1080.

Maslen, C. L., Corson, G. M., Maddox, B. K., Glanville, R. W., and Sakai, L. Y. (1991).

Partial sequence of a candidate gene for the Marfan syndrome. *Nature (London)* **352,** 334–337.

Mercer, J. F. B., Livingston, J., Hall, B., Paynter, J. A., Berg, C., Chandrasekharappa, S., Lockhart, P., Grimes, A., Bhave, M., Siemieniak, D., and Glover, T. W. (1993). Isolation of a partial candidate gene for Menkes disease by positional cloning. *Nat. Genet.* **3,** 20–25.

Moeschler, J. B., Pober, B. R., Holmes, L. B., and Graham, J. M., Jr. (1989). Acrocallosal syndrome: New findings. *Am. J. Med. Genet.* **32,** 306–310.

Monaco, A. P., Neve, R. L., Colletti-Freener, C., Bertelson, C. J., Kurnit, D. M., and Kunkel, L. M. (1986). Isolation of candidate cDNAs for portions of the Duchenne muscular dystrophy gene. *Nature (London)* **323,** 646–650.

Moore, G. E., Ivens, A., Chambers, J., Farrall, M., Williamson, R., Page, D. C., Bjornsson, A., Arnason, A., and Jensson, O. (1987). Linkage of an X-chromosome cleft palate gene. *Nature (London)* **326,** 91–92.

Moore, K. L. (1988). The Developing Human, "4th ed. Saunders, Philadelphia.

Muller, W., Peter, H. H., Wilken, M., Juppner, H., Kallfelz, H. C., Krohn, H. P., Miller, K., and Rieger, C. H. L. (1988). The DiGeorge syndrome: I. Clinical evaluation and course of partial and complete forms of the syndrome. *Clin. Immunol. Immunopathol.* **147,** 496–502.

Nadeau, J. H., and Reiner, A. H. (1989). Linkage and synteny homologies in mouse and man. *In* "Genetic Strains of the Laboratory Mouse" (M. F. Lyon and A. G. Searle, eds.), 2nd ed. Chapter 7, pp. 506–536. Oxford Univ. Press, Oxford.

Nelson, M. M., and Thompson, A. J. (1982). The acrocallosal syndrome. *Am. J. Med. Genet.* **12,** 195–199.

Olsen, M., Hood, L., Cantor, C., and Botstein, D. (1989). A common language for physical mapping of the human genome. *Science* **245,** 1434–1435.

Opitz, J. M., and Lewin, S. O. (1987). The development field concept in pediatric pathology-especially with respect to fibrillar A hypoplasia and the DiGeorge anomaly. *Birth Defects Orig. Artic. Ser.* **23,** 277–292.

Orita, M., Iwahana, H., Kanazawa, H., Hayashi, K., and Sekiya, T. (1989). Detection of polymorphisms of human DNA by gel electrophoresis as single strand conformation polymorphisms. *Proc. Natl. Acad. Sci. U. S. A.* **86,** 2766–2770.

Pagon, R. A., Graham, J. M., Jr., Zonana, J., and Young, S. L. (1981). Coloboma, congenital heart disease, and choanal atresia with multiple anomalies: CHARGE association. *J. Pediatr.* **99,** 223–227.

Palacios, J., Gamallo, C., Garcia, M., and Rodriquez, J. I. (1993). Decrease in thyrocalcitonin-containing cells and analysis of other congenital anomalies in 11 patients with DiGeorge anomaly. *Am. J. Med. Genet.* **46,** 641–646.

Parimoo, S., Patanjali, S. R., Shukla, H., Chaplin, D. D., and Weissman, S. (1991). cDNA selection: Efficient PCR approach for the selection of cDNAs encoded in large chromosomal DNA fragments. *Proc. Natl. Acad. Sci. U. S. A.* **88,** 9623–9627.

Pettigrew, A. L., Greenberg, F., Ledbetter, D. H., and Caskey, C. T. (1989). Greig syndrome associated with an interstitial deletion of 7p: Confirmation of the localization of Greig syndrome to 7p13. *Am. J. Hum. Genet. Suppl.* **45,** A87.

Pettigrew, A. L., Greenberg, F., Caskey, C. T., and Ledbetter, D. H. (1991). Greig syndrome associated with an interstitial deletion of 7p: Confirmation of the localization of Greig syndrome to 7p13. *Hum. Genet.* **87,** 452–456.

Pfeiffer, R. A., Legat, G., and Trautman, U. (1992). Acrocallosal syndrome in a child with de novo inverted tandem duplication of 12p11.2-p13.3. *Ann. Genet.* **35,** 41–46.

Philip, N., Apicella, N., Lassman, I. Ayme, S., Mattei, J. F., and Giraud, F. (1988). The acrocallosal syndrome. *Eur. J. Pediatr.* **147,** 206–208.

Piatak, M., Jr., Luk, K.-C., Williams, B., and Lifson, J. D. (1993). Quantitative competitive polymerase chain reaction for accurate quantitation of HIV DNA and RNA species. *Bio Techniques* **14,** 70–81.

Pilu, G., Reece, E. A., Romero, R., Bovicelli, L., and Hobbins, J. C. (1986). Prenatal diagnosis of craniofacial malformations with ultrasonography. *Am. J. Obstet. Gynecol.* **155,** 45–50.

Raatika, M., Rapola, J., Tuuteri, L., Louhimo, I., and Savilahti, E. (1981). Familial third and fourth pharyngeal pouch syndrome with truncus arteriosus: DiGeorge syndrome. *Pediatrics* **67,** 173–175.

Rhoads, D. D., Dixit, A., and Roufa, D. J. (1986). Primary structure of human ribosomal protein S14 and the gene that encodes it. *Mol. Cell. Biol.* **6,** 2774–2783.

Riley, J., Butler, R., Ogilvie, D. J., Finniear, R., Jenner, D., Anand, R., Smith, J. C., and Markham, A. F. (1990). A novel rapid method for the isolation of terminal sequences from yeast artificial chromosome (YAC) clones. *Nucleic Acids Res.* **18,** 2887–2890.

Riordan, J. R., Rommens, J. M., Kerem, B., Alon, N., Rozmahel, R., Grzelczak, Z., Zielenski, J., Lok, S., Plavsic, N., Chou, J.-L., Drumm, M. L., Ianuzzi, M. C., Collins, F. S., and Tsui L.-C. (1989). Identification of the cystic fibrosis gene: Cloning and characterisation of complementary DNA. *Science* **245,** 1066–1073.

Robinson, H. B. (1975). Familial third-fourth pharyngeal pouch syndrome with apparent autosomal dominant transmission. *Perspect. Pediatr. Pathol.* **2,** 173–206.

Rohn, R. D., Leffell, M. S., Leadem, P., Johnson, D., Rubio, T., and Emanuel, B. S. (1984). Familial third-fourth pharyngeal pouch syndrome with apparent autosomal dominant transmission. *J. Pediatr.* **105,** 47–51.

Rollnick, B. R., and Kaye, C. I. (1986). Mendelian inheritance of isolated nonsyndromic cleft palate. *Am. J. Med. Genet.* **24,** 465–473.

Rose, E. A., Glaser, T., Jones, C., Smith, C. L., Lewis, W. H., Call, K. M., Minden, M., Champagne, E., Bonetta, L., Yeger, H., and Housman, D. E. (1990). Complete physical map of the WAGR region of 11p13 localizes a candidate Wilm's tumour gene. *Cell* **60,** 495–508.

Rosenkranz, W., Kroisel, P. M., and Wagner, K. (1989). Deletion of EGFR-gene in one of two patients with Greig cephalopolysyndactyly syndrome and microdeletion of chromosome 7p. *Cytogenet. Cell Genet.* **51,** 1069 (abstr.).

Rosenthal, I. M., Bocian, M., and Krmpotic, E. (1972). Multiple anomalies including thymic aplasia associated with monosomy 22. *Pediatr. Res.* **6,** 358A.

Rovin, S., Dachi, S. F., Borenstein, D. B., and Cotter, W. B. (1964). Mandibulofacial dysostosis, a familial study of five generations. *J. Pediatr.* 65, 215–221.

Royer-Pokora, B., Kunkel, L. M., Monaco, A. P., Goff, S. C., Newburger, P. E., Baehner, R. L., Cole, F. S., Curnutte, J. T., and Orkin, S. H. (1986). Cloning the gene for an inherited human disorder-chronic granulomatous disease- on the basis of its chromosomal location. *Nature (London)* **322,** 32–38.

Ruppert, J. M., Kinzler, K. W., Wong, A. J., Bigner, S. H., Kao, F. T., Law, M. L., Seuanez, H. N., O'Brien, S. J., and Vogelstein, B. (1988). The GLI-Kruppel family of human genes. *Mol. Cell. Biol.* **8,** 3104–3113.

Ruppert, J. M., Vogelstein, B., Arheden, K., and Kinzler, K. W. (1990). GLI3 encodes a 190-kilodalton protein with multiple regions of GLI similarity. *Mol. Cell. Biol.* **10,** 5408–5415.

Rushton, A. R. (1979). Sex linked inheritance of cleft palate. *Hum. Genet.* **48,** 179–181.

Sage, M., Hart, I., Green, P., Tommerup, N., and Drabkin, H. (1987). Isolation and analysis of the Greig polysyndactyly-craniofacial anomalies syndrome 3;7 translocation. *Am. J. Hum. Genet.* **41,** A106.

Salido, E. C., Yen, P. H., Koprivnikar, K., Yu, L.-C., and Shapiro, L. J. (1992). The human enamel protein gene amelogenin is expressed from both the X and Y chromosomes. *Am. J. Hum. Genet.* **50,** 303–316.

Scambler, P. J. (1993). A genetic aetiology for DiGeorge syndrome, velo-cardio-facial syndrome and familial congenital heart disease. *In* "Annual of Cardiac Surgery" (M. Yacoub and J. Pepper, eds.), pp. 5–12. Current Science, London.

Scambler, P. J., Dumanski, J. P., Nordenskjold, M., Williamson, R., and Carey, A. (1990). Molecular detection of 22q11 deletions in patients with DiGeorge syndrome and normal karyotpye. *Am. J. Hum. Genet.* **47,** A235.

Scambler, P. J., Kelly, D., Lindsay, E., Williamson, R., Goldberg, R., Shprintzen, R., Wilson, D, I., Goodship, J. A., Cross, I. E., and Burn, J. (1992). Velo-cardial-facial syndrome associated with chromosome 22 deletions encompassing the DiGeorge locus. *Lancet* **339,** 1138–1139.

Schimmang, T., Lemaistre, M., Vortkamp, A., and Ruther, U. (1992). Expression of the zinc finger gene Gli3 is affected in the morphogenetic mouse mutant extra-toes (Xt). *Development (Cambridge, UK)* **116,** 799–804.

Schinzel, A. (1982). Acrocallosal syndrome. *Am. J. Med. Genet.* **12,** 201–203.

Schinzel, A., and Kaufman, U. (1986). The acrocallosal syndrome in sisters. *Clin. Genet.* **30,** 399–405.

Schinzel, A., and Schmid, W. (1980). Hallux duplication, postaxial polydactyly, absence of the corpus callosum, severe mental retardation and additional anomalies in two unrelated patients: A new syndrome. *Am. J. Med. Genet.* **6,** 241–249.

Schwartz, D. C., and Cantor, C. R. (1984). Separation of yeast chromosome-sized DNAs by pulsed field gradient gel electrophoresis. *Cell (Cambridge, Mass.)* **37,** 67–75.

Sheffield, V. C., Cox, D. R., Lerman, L. S., and Myers, R. M. (1989). Attachment of a 40-base-pair G + C-rich sequence (GC-clamp) to genomic DNA fragments by polymerase chain reaction results in improved detection of single-base changes. *Proc. Natl. Acad. Sci. U. S. A.* **86,** 232–236.

Sinclair, A. H., Berta, P., Palmer, M. S., Hawkins, J. R., Griffiths, B. L., Smith, M. J., Foster, J. W., Frischauf, A. M., Lovell-Badge, R., and Goodfellow, P. N. (1990). A gene from the human sex-determining region encodes a protein with homology to a conserved DNA-binding motif. *Nature (London)* **346,** 240–244.

Snead, M. L., Lau, E. C., Zeichner-David, M., Finchman, A. G., Woo, S. L. C., and Slavkin, H. C. (1985). DNA sequence for cloned cDNA for murine amelogenin reveal the amino acid sequence for enamel-specific protein. *Biochem. Biophys. Res. Commun.* **129,** 812–818.

Snouwaert, J. N., Brigam, K. K., Latour, A. M., Malouf, N. N., Boucher, R. C., Smithies, O., and Koller, B. H. (1992). An animal model for cystic fibrosis made by gene targeting. *Science* **257,** 1083–1088.

Stanier, P., Forbes, S. A., Arnason, A., Bjornsson, A., Sveinbjornsdottir, E., Williamson, R., and Moore, G. E. (1993). The localization of a gene causing X-linked cleft palate and ankyloglossia (CPX) in an Icelandic kindred is between DXS326 and DXYS1X. *Genomics* **17,** 549–555.

Steele, R. W., Limas, C., Thurman, G. B., Schuelien, M., Bauer, H., and Bellanti, J. A. (1972). Familial thymic aplasia: Attempted reconstitution with fetal thymus in a Millipore diffusion chamber. *N. Engl. J. Med.* **287,** 787–791.

Stevens, C. A., Carey, J. C., and Shigeoka, A. O. (1990). DiGeorge anomaly and velocardio-facial syndrome. *Pediatrics* **85,** 526–530.

Sulik, K. K., and Johnston, M. C. (1983). Sequence of developmental changes following ethanol exposure in mice: Craniofacial features of fetal alcohol syndrome (FAS). *Am. J. Anat.* **166,** 257–269.

Sulik, K. K., Johnston, M. C., Smiley, S. J., Speight, H. S., and Jarvis, B. E. (1987). mandibulofacial dysostosis (Treacher Collins syndrome): A new proposal for its pathogenesis. *Am. J. Med. Genet.* **27,** 359–372.

Sulik, K, K., Cook, C. S., and Webster, W. S. (1988). Teratogens and craniofacial malformations: Relationships to cell death. *Development (Cambridge, UK)* **103,** Suppl. 213–232.

Swaroop, A., Hogan, B. L. M., and Francke, U. (1988). Molecular analysis of the cDNA for human SPARC/osteonectin/BM-40:sequence, expression, and localization of the gene to chromosome 5q31-q33. *Genomics* **2,** 37–47.

Taitz, L. S., Zarate-Salvador, C., and Schwartz, E. (1966). Congenital absence of the parathyroid and thymus glands in an infant (III and IV pharyngeal pouch syndrome). *Pediatrics* **38,** 412–418.

Tamas, D. E., Mahony, B. S., Bowie, J. D., Woodruff, W. W., and Kay, H. H. (1986). Prenatal sonographic diagnosis of hemifacial microsomia (Goldenhar-Gorlin syndrome). *J. Clin. Ultrasound* **5,** 461–463.

Tassabehji, M., Read, A. P., Newton, V. E., Harris, R., Ballings, R., Gruss, P., and Strachen, T. (1992). Mutations in the PAX3 gene causing Waardenburg syndrome type 1 and type 2. *Nature (London)* **355,** 635–637.

Taylor, M. J., and Josifek, K. (1981). Multiple congenital anomalies, thymic dysplasia, severe congenital heart disease, and oligosyndactyly, with a deletion of the short arm of chromosome 5. *Am. J. Med. Genet.* **9,** 5–11.

Termine, J. D., Belcourt, A. B., Christner, P. J., Conn, K. M., and Nylen, M. U. (1980). Properties of dissociatively extracted fetal tooth matrix proteins. I. Principal molecular species in developing bovine enamel. *J. Biol. Chem.* **255,** 9760–9768.

Thériault, A., Boyd, E., Harrap, S. B., Hollenberg, S. M., and Connor, J. M. (1989). Regional chromosomal assignment of the human glucocorticoid receptor gene to 5q31. *Hum. Genet.* **83,** 289–291.

Thompson, J. S., and Thompson, M. W. (1986). "Genetics in Medicine." Sanders, Philadelphia.

Thomson, A.(1846–1847). Notice of several cases of malformation of the external ear, together with experiments on the state of hearing in such persons. *Mon. J. Med. Sci.* **7,** 420; cited by Klein (1953).

Tommerup, N., and Nielsen, F. (1983). A familial reciprocal translocation t(3:7)(p21.1;p13) associated with Greig polysyndactyly–craniofacial anomalies syndorme. *Am. J. Med. Genet.* **16,** 313–321.

Ton, C. C., Hirvonen, H., Miwa, H., Weil, M. M., Monaghan, P., Jordan, T., van-Heyningen, V., Hastie, N. D., Meijers-Heijboer, H., Drechsler, M., Royer-Pokora, B., Collins, F., Swaroop, A., Strong, L. C., and Saunders, G. F. (1991). Positional cloning and characterization of a paired box- and homeobox-containing gene from the aniridia region. *Cell (Cambridge, Mass.)* **67,** 1059–1074.

Townes, P. L., and White, M. R. (1978). Inherited partial trisomy 8q(22-qter). *Am. J. Dis. Child.* **132,** 498–501.

Toynbee, J. (1847). Description of a congenital malformation in the ears of a child. *Mon. J. Med. Sci.* **1,** 738–739.

Treacher Collins, E. (1960). Cases with symmetrical congenital notches in the outer part of each lid and defective development of the malar bones. *Trans. Ophthalmol. Soc. U.K.* **20,** 190–192.

Turner, H. H. (1938). A syndrome of infantilism, congenital webbed neck and cubitus valgus. *Endocrinology (Baltimore)* **23,** 566–574.

Van den Berghe, H., Van Eygen, M., Fryns, J. P., Tanghe, W., and Verresen, H. (1973). Partial Trisomy 1, karyotype 46,XY,12-t(1q, 1wp)+.*Humangenetik* **18,** 225–230.

Verkerk, A. J., Pieretti, M., Sutcliffe, J. S., Fu, Y. H., Kuhl, D. P., Pizzuti, A., Reiner, O., Richards, S., Victoria, M. F., and Zhang, F. P. (1991). Identification of a gene (FMR-1) containing a CGG repeat coincident with a breakpoint cluster region exhibiting length variation in fragile X syndrome. *Cell (Cambridge, Mass.)* **65,** 905–914.

Vetrie, D., Vorechovsky, I., Sideras, P., Holland, J., Davies, A., Flinter, F., Hammarström, L., Kinnon, C., Levinsky, R., Bobrow, M., Smith, C. I. E., and Bentley, D. R. (1993). The gene involved in X-linked agammaglobulinaemia is a member of the src family of protein-tyrosine kinases. *Nature (London)* **361,** 226–233.

Vortkamp, A., Gessle, M., and Grzeschik, K. -H. (1991). GLI3 zinc-finger gene interrupted by translocations in Greig syndrome families. *Nature (London)* 352, 539–540.

Vortkamp, A., Franz, T., Gessler, M., and Grzeschik, K. H. (1992). Deletion of GLI3 supports the homology of the human Greig cephalopolysyndactyly syndrome (GCPS) and the mouse mutant extra-toes (Xt). *Mamm. Genome* **3,** 461–463.

Vulpe, C., Levinson, B., Whitney, S., Packman, S., and Gitshier, J. (1993). Isolation of a candidate gene for Menkes disease and evidence that it encodes a copper-transporting ATPase. *Nat. Genet.* **3,** 7–13.

Wagner, K., Kroisel, P. M., and Rosenkranz, W. (1990). Molecular and cytogenetic analysis in two patients with microdeletions of 7p and Greig syndrome: Hemizygosity for PGAM2 and TCRG genes. *Genomics* **8,** 487–491.

Weber, F., deVilliers, J., and Schaffner, W. (1984). An SV40 "enhancer trap" incorporates exogenous enhancers or generates enhancers from its own sequences. *Cell (Cambridge, Mass.)* **36,** 983–992.

Weber, J. L., and May, P. E. (1989). Abundant class of human DNA polymorphisms which can be typed using the polymerase chain reaction. *Am. J. Hum. Genet.* **44,** 388–396.

Webster, W. S., Johnston, M. C., Lammer, E. J., and Sulik, K. K. (1986). Isotretinoin embryopathy. *J. Craniofacial Genet. Dev. Biol.* **6,** 211–217.

Weinmann, J. P., Svoboda, J. F., and Woods, R. W. (1945). Hereditary disturbances of enamel formation and calcification. *J. Am. Dent. Assoc.* **32,** 397–418.

Weissenbach, J., Gyapay, G., Dib, C., Vignal, A., Morissette, J., Millasseau, P., Vayssiex, G., and Lathrope, M. (1992). A second-generation linkage map of the human genome. *Nature (London)* **359,** 794–801.

White, M. B., Carvalho, M., Derse, D., O'Brien, S. J., and Dean, M. (1992). Detecting single base substitution as heteroduplex polymorphisms. *Genomics* **12,** 301–306.

Williams, F. E., and Trumbly, R. J. (1990). Characterization of TUP1, a mediator of glucose repression in Saccharomyces cerevisiae. *Mol. Cell. Biol.* **10,** 6500–6511.

Wilson, D. I., Cross, I. E., Goodship, J. A., Coulthard, S., Carey, A. H., Scambler, P. J., Bain, H. H., Hunter, A. S., Carter, P. E., and Burn, J. (1991). DiGeorge syndrome with isolated aortic coarctation and isolated ventricular septal defect in three sibs with a 22q11 deletion of maternal origin. *Br. Heart J.* 66, 308–312.

Wilson, D. I., Cross, I. E., Goodship, J. A., Brown, J., Scambler, P. J., Bain, H. H., Taylor, J. F. N., Walsh, K., Bankier, A., Burn, J., and Wolstenholme, J. (1992a). A prospective cytogenetic study of 36 cases of DiGeorge syndrome. *Am. J. Hum. Genet.* **51,** 957–963.

Wilson, D. I., Goodship, J. A., Burn, J., Cross, I. E., and Scambler, P. J. (1992b). Deletions within chromosome 22q11 in familial congenital heart disease. *Lancet* **340,** 573–575.

Wilson, D. I., Burn, J., Scambler, P. J., and Goodship, J. A. (1993). DiGeorge syndrome, part of CATCH 22. *J. Med. Genet.* **30,** 852–856.

Wilson, J. B., Ferguson, M. W. J., Jenkins, N. A., Lock, L. F., Copeland, N. G., and Levine, A. J. (1992). Transgenic mouse model of X-linked cleft palate. *Cell Growth Differ.* **4,** 67–76.

Winter, R. M., and Huson, S. M. (1988). Greig cephalopolysyndactyly syndrome: A possible mouse homologue (Xt-Extra-toes). *Am. J. Med. Genet.* **31,** 793–798.

Witkop, C. J. (1989). Amelogenesis imperfecta, dentinogenesis imperfecta and dentin dysplasia revisited: Problems in classification. *J. Oral Pathol. Med.* **17,** 547–553.

Witkop, C. J., and Sauk, J. J. (1976). Heritable defects of enamel. *In* "Oral Facial Genetics" (R. E. Stewart and G. H. Prescott, eds.), pp. 151–226. Mosby, St. Louis, MO.

Young, R. S., Reed, T., Hodes, M. E., and Palmer, C. G. (1982). The dermatoglyphic and clinical features of the 9p trisomy and partial 9p monosomy syndromes. *Hum. Genet.* **62,** 31–39.

Zinn, K., DiMaio, D., and Maniatis, T. (1983). Identification of two distinct regulatory regions adjacent to the human beta-interferon gene. *Cell (Cambridge, Mass.)* **34,** 865–879.

Effects of Electromagnetic Fields on Molecules and Cells

Eugene M. Goodman*, Ben Greenebaum*, and Michael T. Marron†
*Biomedical Research Institute, University of Wisconsin–Parkside, Kenosha, Wisconsin 53141 and †Office of Naval Research, Arlington, Virginia 22217

Evidence suggests that cell processes can be influenced by weak electromagnetic fields (EMFs). EMFs appear to represent a global interference or stress to which a cell can adapt without catastrophic consequences. There may be exceptions to this observation, however, such as the putative role of EMFs as promoters in the presence of a primary tumor initiator. The nature of the response suggests that the cell is viewing EMFs as it would another subtle environmental change. The age and state of the cell can profoundly affect the EMF bioresponse. There is no evidence that direct posttranscription effects occur as a result of EMF exposure. Although transcription alterations occur, no apparent disruption in routine physiological processes such as growth and division is immediately evident. What is usually observed is a transient perturbation followed by an adjustment by the normal homeostatic machinery of the cells. DNA does not appear to be significantly altered by EMF. If EMF exposure is associated with an increased risk of cancer, the paucity of genotoxic effects would support the suggestion that the fields act in tumor promotion rather than initiation. The site(s) and mechanisms of interaction remain to be eleborated. Although there are numerous studies and hypotheses that suggest the membrane represents the primary site of interaction, there are also several different studies showing that *in vitro* systems, including cell-free systems, are responsive to EMFs. The debate about potential hazards or therapeutic value of weak electromagnetic fields will continue until the mechanism of interaction has been clarified.

KEY WORDS: Electromagnetic fields, Electric fields, Magnetic fields, EMF.

I. Introduction

Beginning with the mid-18th century experiments of Galvani (1737–1798), the bioeffect(s) of electromagnetic fields (EMFs) have been a continuing source of skepticism and interest to the scientific community. In the past 20–25 years, data obtained from experiments using both cells and organisms have shown a variety of EMF responses. More recently, a great deal of attention has been generated by epidemiological studies associating weak EMFs with small increases in the odds ratios for various types of cancers (Savitz *et al.,* 1988; London *et al.,* 1991; Feychting and Ahlbom, 1993). Still more recently, weak higher-frequency fields associated with mobile telephones have been the subject of popular concern, although without the support of any epidemiology. These public discussions of potential hazards associated with the weak EMFs generated by our use of electric power and various communications and broadcast systems occur against a background of therapeutic use of stronger, low- and intermediate-frequency fields in bone fracture repair. Another medical application of EMFs is the well-known heating effect of higher-frequency fields that has uses ranging from diathermic warming of painful tissues to inducing hyperthermia in tumors. Time-varying magnetic fields of both low and high frequency are also used in medical diagnostic procedures such as magnetic resonance imaging (MRI).

Experiments showing biological effects of weak EMFs (<100 Hz) have been challenged on two fronts. First, there are a number of negative findings as well as positive ones, and there have been some unsuccessful attempts to reproduce prior positive effects. Second, there is no widely accepted biophysical explanation for the way weak EMFs produce noticeable changes in biological systems. Both of these challenges are themselves challenged. Many apparent replications have departed from the original experiments in various ways; and many of the assumptions of the theoretical arguments have been questioned. Both areas are addressed in the body of this article. In addition, the phenomena that have been observed do not have a simple dose–response relationship. Essentially none shows a linear increase with exposure length or intensity; many seem to saturate quickly once a threshold has been passed. In some experiments, "windows" of frequency or intensity have been observed: effects are visible at one frequency and/or intensity that disappear for both higher and lower values of the parameters (Blackman *et al.,* 1985a,b).

It is our opinion that although there have been a large number of negative experimental findings concerning the biological effects of EMF, there exists a persistent pattern of evidence, both *in vitro* and *in vivo,* showing that fields can produce significant biological responses even at low levels. We present a discussion of the experiments on cells that form an important part of the basic scientific background of these debates. The scientific background underlying public policy debates or development of a new medical treatment must also include whole-organism results and theoretical discussions. The scientific community can then determine not only whether the fields have an effect, but also the physiological consequences of EMF exposure. Such a synthesis is beyond the scope of this article. Our purpose in this review is to provide an introduction to the current body of knowledge and the literature about EMF effects on cells. We are convinced that the mechanisms of interaction between EMFs and cells will be understood only by careful experimentation employing the powerful tools of molecular biology.

To set the stage for a detailed discussion of experiments on specific types of biological end points, we outline theories proposed to account for EMF effects. We then discuss the choice and design of EMF exposures and exposure apparatus and some features of experimental design particularly relevant to the study of EMF effects. The bulk of the article discusses experiments showing biological responses to EMF exposure, grouped roughly according to the type of response. We assume the reader is familiar with basic cell biology but may have little knowledge of electromagnetic field effect experiments.

II. Theories Concerned with Electromagnetic Field Mechanism(s)

In an attempt to place the question of mechanism in perspective, we first describe some of the models that have been proposed to explain EMF effects and examine the consistency of these models with available data. It is important to note that application of an oscillating magnetic field always involves both a magnetic and at least a very small induced electric field component, regardless of the intensity of the field being applied. Thus in experiments involving a magnetic field alone, induced electric fields cannot be totally dismissed. Although the converse also holds, that is, when an alternating electric field alone is applied a small magnetic field will also be induced, this induced field is truly minuscule (see Barnes, 1992).

The mystery of how EMFs exert their influence can be traced to the fact that a typical cell membrane of ~90-Å thickness has a resting membrane potential of about 100 mV, which creates an electric field across the membrane of approximately 10^5 V/cm. How can an applied field gradient five to six orders of magnitude weaker than the existing membrane gradient perturb a cell (Adair, 1991)?

Much of the skepticism concerning the ability of weak electromagnetic fields to induce bioeffects is based on computations showing that an effect would be a violation of classic equilibrium thermodynamic principles (Adair, 1991, 1992a,b, 1994): The energy being applied is several orders of magnitude less than the random thermal energy fluctuations (kT) in the cell. Adey and colleagues (Lin-Liu and Adey, 1982; Adey, 1989, 1990, 1993) have addressed this criticism by suggesting that the cellular responses to weak EMFs can be explained by employing nonequilibrium, rather than equilibrium, thermodynamic theory.

An underlying tenet of many models is that the plasma membrane is the primary site of interaction. Figure 1 shows a schematic representation of one mechanism by which an extracellular, non-EMF signal can be transduced. Although the diagram is not an inclusive representation of all membrane signaling mechanisms, it identifies most of the transductive mechanisms we discuss in this article. In Adey's three-step model, the transduction process involves (1) binding of a stimulating molecule at a receptor site, (2) amplification and transduction of the signal to the interior by a transmembrane portion of integral membrane proteins, and (3) internal coupling to an intracellular enzyme system. It is proposed that EMFs amplify binding at membrane glycoproteins (step 1). In each of the three steps changes in the Ca^{2+} flux have also been observed. It is noteworthy that alterations in the intracellular and extracellular Ca^{2+} environment have also been observed by several investigators studying EMF effects (Bawin and Adey, 1976; Lin-Liu and Adey, 1982; Blackman *et al.*, 1985a); see Section V for further discussion.

Luben (1991) developed a model to explain the response of a cell to pulsed magnetic fields (PMFs; described below) based on the response of bone cells to hormones. Like Adey, Luben suggests that EMFs modify the normal signal processes in the membrane. Luben *et al.* (1982) and Cain *et al.* (1987) found that when bone cells were exposed to PMFs for 10 min, adenylate cyclase (an enzyme that converts ATP to cyclic AMP) showed a diminished response to parathyroid hormone (PTH). Normally, PTH increases bone resorption and decreases bone formation; these processes appear to be intracellularly modulated by cyclic AMP and GTP. Cain *et al.* (1987) also showed that PMFs decreased the activation of G protein (Fig. 1); the net result of PMF exposure was an increase in collagen synthesis and a decrease in bone resorption. Although the fields apparently did not interfere with the binding of PTH to its membrane receptor, Luben

FIG. 1 The phosphatidylinositol (PIP_2) pathway. Binding of certain hormones and some other ligands to a variety of membrane receptors (R) leads to activation of GTP-binding proteins, which then activate phospholipase C, which cleaves the inositol triphosphate (IP_3) moiety from PIP_2 and leaves diacyl glycerol (DG) in the membrane. Diacylglycerol, in conjunction with phosphotidylserine (PS), activates a protein kinase (C-kinase) that phosphorylates enzymes associated with key metabolic pathways, thereby activating or inactivating them. Meanwhile the polar IP_3 moiety binds to intracellular receptors on the endoplasmic reticulum, resulting in the liberation of calcium into the cytosol. The calcium binds to calmodulin, which then activates another group of protein kinases (the Ca^{2+}/CaM-dependent protein kinases). Hormonal stimulation can be short-lived because the IP_3 and DG are rapidly degraded to inactive forms that are ultimately recycled to PIP_2; $Ca^{2+}$is pumped back into the cytoplasmic reticulum, where it is sequestered. [From Zubay, G. (1993). "Biochemistry," 3rd ed. Copyright© Wm. C. Brown Communications, Inc., Dubuque, IA. All rights reserved. Reprinted by permission.]

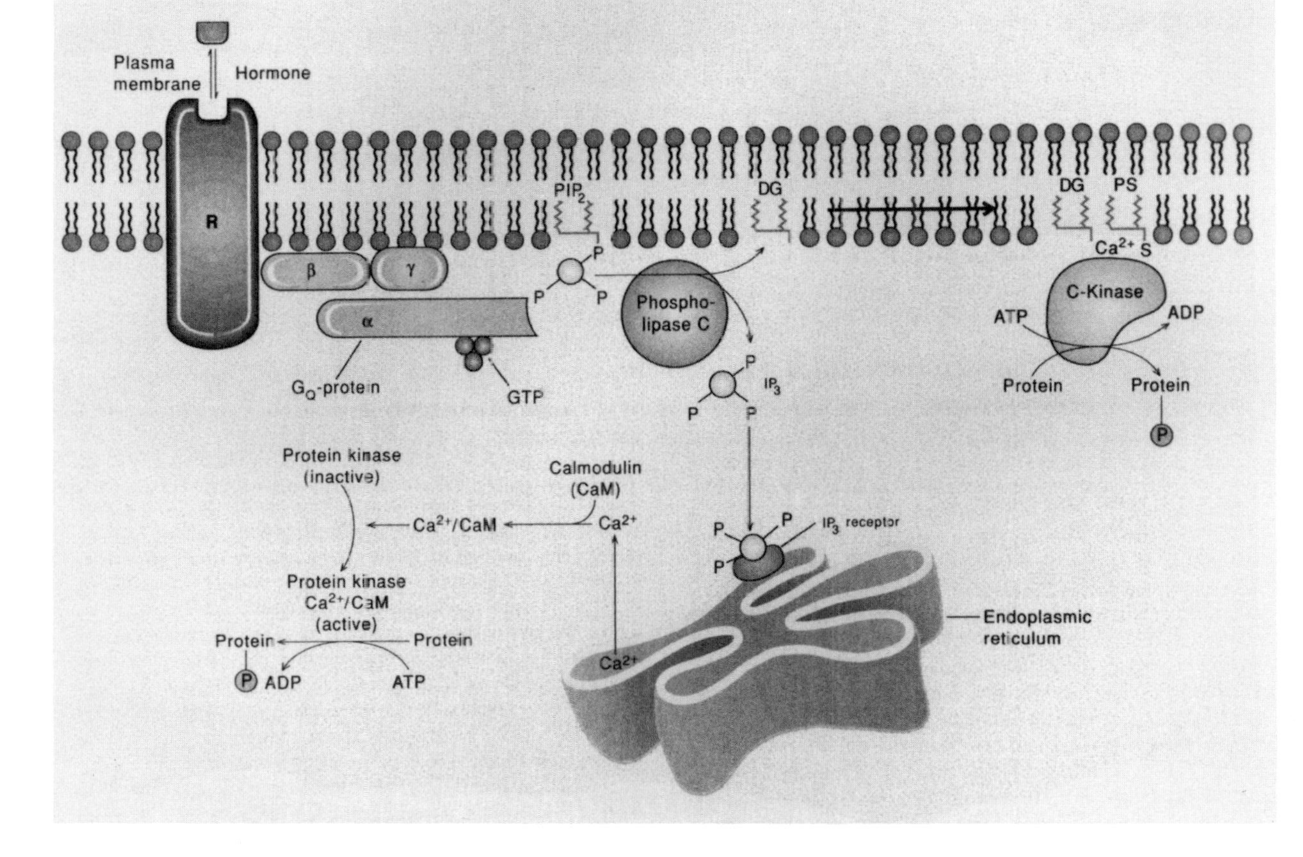

Plasma membrane
Hormone
R
β
γ
α
G_Q-protein
GTP
PIP_2
P
P
P
Phospho-lipase C
DG
IP_3
P
P
P
DG
PS
Ca^{2+} S
C-Kinase
ATP
ADP
Protein
Protein
P
Protein kinase (inactive)
Calmodulin (CaM)
Ca^{2+}/CaM
Ca^{2+}
Protein kinase Ca^{2+}/CaM (active)
Protein
Protein
P
ADP
ATP
IP_3 receptor
P
P
P
P
Ca^{2+}
Endoplasmic reticulum

concludes that a change has occurred in some critical molecule and possibly a signal amplification may be involved in the EMF transduction process. An example of a membrane transductive process that might be related to the system discussed by Luben is shown in Fig. 1.

An electrochemical model involving EMF-induced changes in the activity of enzymes in the membrane has been proposed by Blank and colleagues (Blank and Goodman, 1988; Blank and Soo, 1990, 1992; Blank, 1992). They assume that the applied or induced electric field component is the primary stimulus. Experiments supporting this model involved applying an electric field to a solution containing the integral membrane enzyme, Na^+,K^+-ATPase. Because the field-induced ionic current must move around particles in an enzyme suspension, Blank and Soo (1989) suggest that under the influence of an alternating (AC) field the motion of an ion is not fully symmetric, unlike its motion in pure solvent. As a result, at the end of a full AC cycle there will be some displacement of the ions. Several experiments support this model. When Na^+,K^+-ATPase in suspension was exposed to oscillating E fields (10 Hz–100 kHz) with current densities in the range of 0.05–50 mA/cm^2 (rms), enzyme activity was found to decrease in a frequency-dependent manner (Blank and Soo, 1990). They estimate that the threshold E field across the enzyme suspension was about 5 μV/cm at a current density of 8 nA/cm^2 and 100 Hz. In contrast, enzyme activity was enhanced in the presence of an inhibitor such as ouabain (Blank and Soo, 1989, 1992; Blank, 1992). They conclude that AC fields can either interfere with or enhance enzyme activation by affecting the ion concentration immediately available to the enzyme. Markov *et al.* (1993) also report that EMFs can modulate the kinetic response of an enzyme system. An important question relating to the model proposed by Blank and colleagues is whether similar changes occur *in vivo*. A related four-step model was proposed by Westerhoff *et al.* (1986) to explain how transmembrane enzymes gain free energy from an oscillating electric field. They present calculations for sinusoidal, square-wave, and positive-only time-varying electric fields.

In addition to those described above, models have also been proposed for EMF interactions that do not explicitly require a membrane interaction. Weaver and Astumian (1990), for example, discuss a mechanism by which EMFs could alter reaction rates. Weaver (1992) and Weaver and Astumian (1992) discuss the role of theory in selecting experiments to compare potential field effects to background noise. Litovitz *et al.* (1991) take the position that regardless of how the initial interaction occurs, the applied field will eventually influence both anabolic and catabolic synthetic rates, thereby exerting an effect subtly over time. A chemical reaction rate-based explanation for some of the "window" phenomena that have been observed has been proposed by Litovitz *et al.* (1992). They note that in some experiments an effect will appear for a specific field intensity or exposure time without similar effects occurring at either higher or lower field intensity or exposure times. Because different intensities and frequencies might differentially affect synthetic rates, the level of metabolites would be expected to peak at different times. Thus the absence of a response might be attributed to the fact that the investigator(s) either waited too long (or not long enough); a second implication of the Litovitz model is that larger bioeffects would occur with shorter exposure times.

In a brief review Grundler *et al.* (1992) examine several nonequilibrium mechanisms for EMF effects that would not be constrained by the thermodynamic argument. Because free radicals and triplet-state molecules both have large unpaired electron spins they are susceptible to fields that could change the relative polarization of spins and effect their lifetimes or chemical reaction rates. McLauchlan (1992) also presents a model showing how low-intensity static fields induce free radical triplet pairs to form singlets. At an intensity of 8 mT, two of the three triplet states become completely decoupled from the singlet state. This has the net effect of inducing a large chemical shift in response to a very low magnetic perturbation.

Kirschvink (1992a) observed single-domain ferromagnetic inclusions in insect and human tissue where none had been expected; if these inclusions prove to be universal in nature, they would provide a site for direct magnetic field interactions. A model utilizing these observations has been proposed by Kirschvink *et al.* (1992c). Using a novel approach to the question of mechanism, Lednev (1991, 1993) proposed a resonance model involving both an alternating magnetic field and the earth's steady field; he proposes that $Ca2+$-binding protein activity might be affected by a change in vibrational level populations at resonance. The possibility that nonmembrane transduction processes exist is also supported by experiments in our laboratory that employed a cell-free expression system (E. M. Goodman *et al.*, 1993a) and a totally synthetic transcription system (Frederiksen *et al.*, 1993). In both systems enhanced levels of both protein and mRNA were observed following exposure to a 45- or 72-Hz, 1.1-mT (rms) sinusoidal field.

In the proceedings of a June 1991 symposium on dosimetry of extremely low-frequency (ELF) fields, mechanisms similar to those discussed above have been summarized by Tenforde (1992). Participants also examined the patterns of field and current distribution at the cellular or subcellular level, particularly for cells in tissue, and the consequences of the high degree of local nonuniformities that exist at this level (McLeod, 1992; Wachtel, 1992b). Barnes (1992) and Wachtel (1992a,b) both use an engineering approach to model signal-to-noise enhancement due to local nonuniformities, and McLeod (1992) and Loew (1992) discuss some possible experimental approaches to the same problem. Barnes (1992) also considers several other mechanisms, including the coherent locking together of biological oscillators, parametic amplification processes, and neural networks.

Litovitz and colleagues (1993a,b,c, 1994; Farrell *et al.,* 1993; J. M. Mullins *et al.,* 1993) have suggested that cells impose stringent temporal and spatial coherence conditions at each potential receptor site. In examining systems that respond to EMF signals they found that the effect could be eliminated by adding incoherence or noise. An intriguing aspect of this finding is their suggestion that this approach offers a method for examining the validity of an EMF response.

In addition to the question of how an applied electromagnetic field might induce a bioeffect, there are also divergent views as to whether the active component is the electric field, the magnetic field, or both. This is especially true in experiments involving pulsed magnetic fields in which the magnetic field component is capable of inducing an electric field that ranges from 0.05 V/m to about 0.5 V/m. A variety of waveforms and frequencies have been applied to cells, using a pulsed-field format. For example, one waveform involves a train of 22 sawtooth magnetic field pulses with a 200-μsec rise and a 20-μsec fall from a peak intensity of 0.6 mT; pulse bursts are repeated at frequencies of 5 to 25 Hz. Investigators employing this type of waveform must eventually address the question of whether the observed effects occur as a result of the applied magnetic field, the induced electric field, or a combination of both. It is worth reiterating that with the exception of Kirschvink (1992c), most theories provide no direct role for the magnetic field component. Nevertheless, there are data suggesting that weak magnetic fields alone can produce a bioeffect (Goodman *et al.,* 1988; Marron *et al.,* 1988; Blackman *et al.,* 1993a,b).

In vitro experiments have been reported that employ a wide variation in the intensities, frequencies, and spatial relationships of applied electric and/or magnetic field, in the electric fields induced by time-varying magnetic fields, and in the current densities accompanying an electric field (Berg, 1993). Some experimenters report electromagnetic effects only in frequency and/or intensity windows, outside of which the fields are not effective (Postow and Swicord, 1989). Others, usually for experiments employing stronger fields, report a broad-band response that is not particularly sensitive to the frequency or waveform being applied.

Critics of reports of electromagnetic field effects point to the lack of a simple dose–response relationship with respect to frequency, intensity, or duration of exposure (Foster, 1988), in addition to the previously noted thermodynamic problem (Adair, 1991, 1992a; Sandweiss, 1990). Several authors have suggested possible explanations that remove these contradictions, at least in some cases (Weaver and Astumian, 1990, 1992; Wachtel, 1992a).

The majority of the EMF experiments have been done without reference to the constant magnetic field of the earth. However, some investigators have found sharp, resonance-like features in response to specific combinations of time-varying and constant fields. In some cases the fields were parallel (Thomas *et al.,* 1986; Liboff *et al.,* 1987; Smith *et al.,* 1987) and in others, perpendicular (Blackman *et al.,* 1990). Theoretical models embodying both constant and alternating fields have been proposed. The first of these, motivated by the numerical coincidence of observed frequency windows with the cyclotron resonance frequency of free ions in a magnetic field, was the cyclotron resonance model (Liboff, 1985; Liboff and McLeod, 1988). A number of experimental tests have supported this model, but others have reported an inability to reproduce some of the experimental data (Liboff and Parkinson, 1991; Parkinson and Sulik, 1992). This model has also been challenged on theoretical grounds having to do with ion hydration, time between collisions with other molecules, orbit size, and other considerations (Halle, 1988; Durney *et al.,* 1988; Sandweiss, 1990; Galt *et al.,* 1993a,b).

Other articles have proposed mechanisms that preserve the successes of the cyclotron resonance model in predicting windows in some situations, but avoid the problems of interpreting it literally. Liboff (1985) suggests effects due to unhydrated ions moving in a membrane channel. Lednev (1991) calculates frequency modulation of ion quantum-mechanical levels in an isolated structure, such as calcium in calmodulin. Blanchard and Blackman (1994) have made some corrections in the Lednev calculation and extended it to other ions, supporting their predictions with experimental data from PC-12 cell neurite outgrowth in experiments similar to those discussed below (Blackman *et al.,* 1994). Edmonds (1993a,b) discusses the uneven Larmour precession rate of an ion in a combined static and alternating magnetic field, speculating that this could produce periodically altered activity in the case of an isolated ion such as calcium in calmodulin, or a net motion of ions such as sodium in the Debye layer near a charged surface.

Except for models incorporating the earth's field, most hypotheses about the mechanism of electromagnetic interaction with cells have invoked an interaction between an electric field or current and a charge or electric polarization in all or part of a cell (see discussion of various models in Blackman *et al.,* 1988, 1989; Weaver and Astumian, 1992). Models of magnetic field interactions invoke the idea of an interaction between the electric field or current induced by a changing magnetic field and a charge or elecrtric polarization in all or part of a cell. An exception is the Kirschvink model (1992a), which invokes Kirschvink's recent finding of magnetite inclusions in different types of cells, including brain cells.

It may well be that there are several distinct types of interactions involved. Alternatively, the same basic interaction may be effective at different sites or have somewhat different manifestations, depending on the type of field (electric, magnetic, or both) or the range of field intensities or frequencies involved. Table I summarizes the range of fields and intensities generally found in various environments.

III. Electromagnetic Field Exposure Systems

Investigators conducting EMF experiments have usually attempted to establish well-defined field conditions. For the most part, investigators have avoided applying fields with spatial nonuniformities, more than one frequency, or multiple pulsed waveforms. There have also been conscious efforts to reduce background fields and transients (asynchronous, usually very fast changes in field due to random events) from building power systems and laboratory apparatus.

These attempts at uniformity may in themselves reduce the probability of observing an EMF effect if a bioresponse depends on nonuniformities,

TABLE I
Representative Exposures[a]

Exposure	Frequency	Duty cycle	Internal H	Internal E
Power lines	50- or 60-Hz^{+} harmonics for 3 phase	Continuous	0.1–0.5 mT under line	1 mV/m in tissue
Household (away from appliances)	50 or 60 Hz	Continuous but intensity varies	0.5–0.05 mT	10–100 μV/m in tissue
Household appliances	50- or 60-Hz^{+} harmonics	Short	0.5–1 μT or more	10–1000 μV/m in tissue
Bone-healing therapies	Pulses	1–8 hr/day	1.5–2.5 mT	0.5–1 V/m in tissue

[a] The general range in which the indicated characteristics will fall for most ordinary exposures to EMF. Many exceptions will occur, especially near electrical wiring or apparatus. Electric fields induced in tissue by bone-healing therapy magnetic pulses are estimates calculated using 1- to 2-mT field amplitudes, tissue features of 1- to 10-cm radius, and 10^{-3}- to 10^{-5}-sec pulse rise times.

transients, or interactions between more than one type of field. Further, within organisms (and even in cell clusters, confluent sheets, or organ explants) the applied electric fields are distorted so that the local field at an individual cell can be different from the average field in tissue. Physical properties of tissue do not distort applied magnetic fields, although electric fields and currents induced by time-varying magnetic fields are dependent on the local geometry, conductivities, and so on. However, both electric and magnetic fields experienced in the environment, as well as fields applied therapeutically, are nonuniform. Environmental fields also vary considerably in time and often contain transients and harmonics (Kaune, 1992; Bracken, 1992; Hart, 1992).

Many, although not all, experiments are conducted by applying the fields continuously for times that range from a few minutes to hours, days, or weeks. The exceptions are experiments simulating applications of waveforms presumed to be therapeutic because clinical requirements demand spaced periods of application. Some experimental evidence indicates that the greatest effects may occur with onset or removal of exposure, both in cells (Russell and Webb, 1981; R. Goodman *et al.*, 1992d) and in humans (Cook *et al.*, 1992).

Near high-voltage power transmission lines and other three-phase equipment the instantaneous direction of the electric or magnetic field is qualitatively different from that produced by single-phase wiring. Instead of alternating back and forth through zero intensities along a single direction (so-called "linear polarization"), fields from a three-phase source may change orientation as well as intensities and may never be precisely zero ("elliptical or circular polarization"; see Deno, 1976). This distinction has been ignored by most investigators. Some have seen differences between cultures exposed to circularly polarized fields and unexposed cells at 60 Hz, but not at 70 Hz (Gundersen *et al.*, 1986). Differences have also been reported by Phillips *et al.* (1986a,b), but their methodology has been challenged; Cohen *et al.* (1986a,b) also used circularly polarized fields and reported negative results (see also Sakamoto *et al.*, 1993; discussed in Section V,D,6). Again, primarily in a three-phase situation, there can be components of the field at the small third harmonic (180 Hz) and higher harmonics.

Experimenters have striven to subject both field-exposed and sham-exposed cultures to the same temperature, atmosphere, lighting levels and cycles, vibration, and noise emanating from the field-generating apparatus. Exposed and sham-exposed cells are generally taken to be similar when they are derived from the same stock culture flask at the same time and handled identically and simultaneously. Thermal regulation is a critical experimental parameter, and increasingly investigators have taken extra precautions to mitigate any field-induced temperature increases. Thermal and field effects do not seem to be additive, as demonstrated by Iannacone *et al.* (1988) using rabbit osteochondral junctions to compare changes due to EMFs that increased temperature by 1.3°C to those induced by a water bath producing the same temperature rise. In all experiments they used unexposed cells at a stable temperature as a control.

The basic physical, engineering, and biological considerations involved in designing an *in vitro* experiment have been extensively reviewed by Misakian *et al.* (1993); therefore we do not attempt to duplicate their presentation. Experimental apparatus for exposure of cells to electric and magnetic fields must be constructed with attention paid to two important criteria: first, the fields must be well characterized, stable, and of the desired intensity, frequency, and so on; second, the apparatus (and the fields themselves) must exert no additional influence on the cells. Artifacts can be introduced, for example, through heating or vibration in magnetic field coils or shielding of light by support structures, as well as by heating in the sample by applied or induced electric currents. Finally, the coupling between magnetic and electric field requires careful thought in creating apparatus that allows varying the two types of field in an independent manner.

Magnetic fields are usually applied by placing culture dishes in a coil or system of coils through which a current is passed; the magnetic field is parallel to the coil axis and is uniform near the axis and the center of the coil or coil system. Many investigators use an approximation to the Helmholtz pair of coils: few actually use the true configuration. A Helmholtz pair consists of two parallel, circular coils separated by a distance equal to their radius (Kraus, 1984). They produce a field uniform within 10% in a region that extends radially about halfway from the axis to the cylinder that includes the coils and roughly halfway from the plane midway between the coils to the coils themselves. In a number of experiments, square or rectangular coils or groupings with more than two coils are used; several of these configurations offer better field uniformities (Kirschvink, 1992b). Less often employed techniques include using a sheet of current in place of a coil, in which case the magnetic field is parallel to the plane of the sheet (Gundersen *et al.,* 1986; Miller *et al.,* 1989), or using a steel-cored electromagnet (R. D. Mullins *et al.,* 1993).

In magnetic field applications, care must be taken to reduce the stray field of the exposure apparatus at the location of the sham-exposed cultures. Placing the two at some distance from each other is an obvious answer, but in cell culture experiments this approach may involve using more than one incubator with the accompanying complications of possibly nonidentical temperature, CO_2 and humidity control, environmental or incubator-induced fields, and so on. Interchanging the roles of control and exposed incubators, using a single water bath with a split flow path to establish similar temperatures, and other techniques can be used to overcome these problems. An alternative approach is to use a single incubator and use placement, coil configuration, and/or high permeability and/or high-conductivity shielding to lower the stray field levels (Misakian *et al.*, 1993; R. D. Mullins *et al.*, 1993). The shielding in these configurations must allow for free airflow and uniform temperature and atmospheric conditions, and investigators must confirm by measurements that conditions are similar.

Exposure and sham-exposure apparatus should be identical, incorporating full sets of coils and/or shields, and their roles should be interchanged from time to time. Sham-exposure apparatus has usually not been energized, but sham magnetic field exposure incorporating exposure-level input power loads (and accompanying heat dissipation, etc.) has been achieved by winding each coil in two equal parts and running current through each half in opposite directions (Kirschvink, 1992b). If energized sham control coils are employed, the wires leading to them must be wound and placed carefully to allow full cancellation; ideally the sham field should be on the other of 1/1000 or less than the exposure field.

Electric fields have been applied to cell cultures either through electrodes in direct or indirect contact with the growth medium or through electromagnetic induction. Misakian *et al.* (1993) discuss in some detail many of the methods that have been used, as well as some of the specific ways electrical contacts, dish and substrate materials, and so on have been known to affect cell growth. It is generally not satisfactory to place the dish between two electrodes in air, essentially placing it inside an air-gap capacitor. This method requires too high a voltage on the plates and involves too many uncertainties in determining the actual field in the medium. Relatively high electric fields in the medium can be generated using a method in which the electrode is in contact with the medium. High electric fields mean high currents, and care must be taken to prevent electrolysis at the electrodes from introducing unwanted ion species into the culture and to prevent contact resistance, bubbles, or other features of electrolysis from dropping the field level below that intended (Gundersen and Greenebaum, 1985). Electrodes and their electrolysis products are often isolated from the culture medium by the use of agar bridges, which should be changed often enough to prevent diffusion of products across the bridge (McLeod *et al.*, 1987; Gundersen and Greenebaum, 1985; Cohen *et al.*, 1986b). Electrodes should be matched to their environment, for example, using Ag–AgCl electrodes in solutions rich in chloride ions, and run at as low a voltage as possible.

Electromagnetic induction is another method for producing electric fields in culture medium. As noted above, any changing magnetic field will induce an electric field in conducting materials. Because this phenomenon depends on the rapidity of the change in magnetic flux, it is more marked for higher frequencies and for pulsed fields that transients, both of which usually embody rapid changes. The pulsed magnetic field devices used to treat fractures are thought to operate through the electric fields and/or currents induced in the tissue (Pilla *et al.,* 1993). The coupling between magnetic and electric fields requires careful thought in designing apparatus that allows varying the two types of fields in an independent manner. McLeod *et al.* (1983) and more recently Bassen *et al.* (1992) have calculated the electric field and current intensity patterns induced under common combinations of magnetic field directions and types of culture vessel. The same principle of electromagnetic induction has been used to produce strong sinusoidal electric fields in ring-shaped vessels (Kaune *et al.,* 1984; Blank and Soo, 1992) that enclose a magnetic field and employ iron core electromagnets to intensify the magnetic fields.

In any experiment invoking subtle effects, blind experimental techniques are advisable; however, most experiments discussed in this article were not done blinded. Blind technique is especially required for experiments that depend on scoring by human judgment, as contrasted to those that employ quantifiable, instrumentally based measurement techniques. Even in the latter type of experiments, whether blinded or not, care should be taken to handle exposed and sham-exposed cultures similarly, simultaneously, and in a random order. Despite these precautions, subtle differences in handling that can affect the data are possible.

Coupled with the necessity for careful handling of the cells and methods for data acquisition is the need to ensure selection of proper controls and control conditions. Although normally an important concern in any comparative investigation, this issue is particularly relevant in EMF research. This aspect is becoming increasingly complex as more investigators employ the techniques of molecular biology. For example, in *Escherichia coli* it is not immediately obvious what gene could be considered a constitutive housekeeping gene useful for internal control, analogous to β-macroglobulin in eukaryotic systems.

IV. Characteristics of an Effective Electromagnetic Field Experiment

An effective experimental design for examining EMF bioeffects should produce a clear biological response that is reproducible, both within the original laboratory and elsewhere. An ideal experimental series will follow the initial demonstration of an effect by locating the site where the field influences the cell, tracing the field influence to the biological effect through the appropriate biological/biochemical pathway(s), and identifying the effective components, frequencies, intensities, and other characteristics of the field and the target site.

The controversy related to the effects of electromagnetic fields is due to the subtle response of cells to EMFs and the conflicting results sometimes obtained when experimental replications are attempted. This situation is exacerbated by the absence of a generally accepted mechanism to explain how weak electromagnetic fields interact with cells and produce their effect. In this section we attempt to examine some of the problems affecting experimental reproducibility and to indicate how they might be addressed.

Critics questioning bioeffects associated with exposure to weak electromagnetic fields inevitably point to the inability of investigators to reproduce published experimental data. On closer examination of some of these attempts, one finds that the original experimental protocols were not carefully followed. For example, Brayman *et al.* (1985) report on their inability to replicate the decreased rate of respiration in the slime mold *Physarum polycephalum* originally reported by Greenebaum *et al.* (1975), Marron *et al.* (1978), and E. M. Goodman *et al.* (1979). In the original papers, the onset of an effect was measured as a decrease in the rate of respiration. This decrease was observed approximately 120 days following the start of exposure at 75 Hz. Further, when an alteration in respiration was observed, experiments were continued for another 60- to 90-day period before a new series of experiments was begun to confirm the effect. The exposure protocols described by E. M. Goodman *et al.* (1979) combined sinusoidal electric and magnetic fields of 0.7 V/m, 0.2 mT at 75 Hz; in addition, individual electric (0.7 V/m) and magnetic (0.2 mT) fields were also applied in separate experiments. Greenebaum *et al.* (1975) reported preliminary data suggesting that an effect could be induced faster when electric and magnetic fields were applied at 45 and 60 Hz (~30 and 90 days, respectively).

In the experiments of Brayman *et al.* (1985), the cells were exposed to 45- and 60-Hz electric fields of 0.7 or 70 V/m. In the original studies the *E*-only experiments involved exposure to a 75-Hz field and not the 45- and 60-Hz exposure reported by Brayman *et al.* (1985). Further, in the "replicate" experiments the total length of exposure never exceeded 100 days. An error analysis of the data reported by the Brayman group shows that the scatter in their data would have prevented them from observing the 7 to 10% differences reported by the original authors. In summary, the laboratory attempting to perform a replicate experiment used different fields, length of exposure, and level of precision to conclude that the original result could not be replicated. The problem with this type of quasi replication is that unless the reader goes back to the original article and performs a critical comparison of experimental protocols, one might accept the conclusion that this was another unreplicated EMF effect. Any investigator attempting to replicate experiments in a subject area as controversial as EMF effects must assiduously follow protocols reported in the literature or at least state unambiguously which aspects of their experiment are not exact replicates.

More recent studies reinforce the validity of this statement as evidence amasses showing that the state and age of a culture can profoundly affect its response to EMFs (Cadossi *et al.*, 1992). Similarly, many of the cell-related studies that show EMFs can induce transcriptional alterations (discussed in Section V,C,2 below) have been performed on transformed rather than primary cells. Investigators working with primary cultures use a given cell passage for only a few weeks before the line is discarded and a new passage restarted from frozen stocks. In contrast, the length of time a transformed cell is used before being discarded varies from laboratory to laboratory. This raises the question of whether genetic drift might have occurred in the transformed lines, maintained in different laboratories. Because of the potential problem of genetic drift in transformed cell lines experimental replication might be facilitated if the cell line to be used were obtained directly from the laboratory of the original investigator and if the number of passages from the initial source were limited. The importance of the latter suggestion is best exemplified in publications from Blackman *et al.* (1993a), in which the outgrowth of neurites in a 50-Hz magnetic field was examined in PC-12D cells (pheochromocytoma cells originally derived from rat adrenal medulla). In these experiments, Blackman and colleagues report that the magnetic field enhanced neurite outgrowth. In contrast, using another PC-12 line (obtained from the Tissue Culture Facility at the University of North Carolina, Chapel Hill, NC) and the same exposure conditions, a reduction in neurite outgrowth was observed (Blackman *et al.*, 1993b). Thus the same experiment using slightly different cell types resulted in opposite EMF responses. These authors have extended their experiments and claim to have explained their results with a theory involving modulated quantum-mechanical energy levels (Blackman *et al.*, 1994; Blanchard and Blackman, 1994).

One of the best examples of a concerted attempt to reproduce an EMF bioeffect involved six independent laboratories attempting to replicate the developmental alterations on chick development originally reported by Delgado *et al.* (1982). In the Delgado experiments chicken eggs that were exposed to pulsed magnetic fields (10, 100, and 1000 Hz, 0.12 to 1.2 μT) exhibited signficant developmental malformations. In the replicate studies all investigators used identical equipment constructed in one laboratory and followed similar but not identical protocols. Fertilized eggs were evaluated for fertility, development, morphology, and stage of maturity. Ten eggs were sham exposed and 10 eggs were exposed to a unipolar pulsed magnetic field (500-μsec pulse, 100 pulses/sec, 1-μT peak). Because of the need to obtain fertilized eggs locally, the genetic characteristics of the eggs were not controlled. In five of six laboratories more embryos exhibited structural anomalies than did the sham-exposed controls. However, statistically signficant differences were observed only in two laboratories; a third laboratory had marginally signficant differences ($p = 0.08$) (Berman *et al.*, 1990). If data from all laboratories were pooled, approximately 25% of exposed embryos showed abnormalities compared to 19% for sham-exposed controls.

The disconcerting lesson from this study is that subtle factors yet to be identified may play a dominant role in bioeffect studies of EMFs. In this study an effort was made by all of the research groups to reproduce the original experiment as closely as possible and yet the results were still ambiguous. As mentioned above, one possible explanation for the overall lack of agreement might be the genetic variability in the eggs. Support for this suggestion was obtained by Litovitz *et al.* (1993a), who, using conditions and end points similar to those in the Berman study, report evidence that changes in the breeding flock may be a strong determinant as to whether or not an EMF-related defect is observed. In some experiments involving multiple replications with eggs from one breeding flock, a strong effect was obtained each time the experiment was performed. In other experiments, using eggs from the same supplier but a different breeding flock, no defects were ever obtained. These data strongly suggest that a yet to be defined factor in a supposedly homogeneous breeding stock may be involved in susceptibility to EMFs.

V. Cellular Responses to Electromagentic Fields

This section discusses experiments on cells that have demonstrated positive responses, arranged according to the type of biological response involved. The reader should note that different types of cells and EMF exposures are often present in each subsection. Cell-level experiments are intended to detect and characterize an effect in a system simpler than a multicelled organism. Such experiments are by their nature free from endogenous homeostatic mechanisms encountered in whole organisms and thus may be more sensitive to applied fields.

A. Ion Effects

1. Calcium

a. Radio-Frequency Fields Because of the important physiological role(s) of Ca^{2+} and its relationship to membrane changes, the EMF effect on the flux of Ca^{2+} was a subject of early scrutiny. Studies were conducted using fields applied at both low frequencies (0–300 Hz) and frequencies in the megahertz range. In general, a radio-frequency (rf) carrier modulated at low frequencies appears to be the most effective in inducing a bioresponse. In rf experiments cells experience a coupled electric and magnetic field. Adey and colleagues (Bawin and Adey, 1976; Bawin *et al.*, 1978; Lin-Liu and Adey, 1982), using isolated cerebral hemispheres or brain synaptosomes, reported that EMFs enhanced the efflux of $^{45}Ca^{2+}$. In these experiments radio-frequency carriers (147 or 450 MHz) were modulated by sinusoidal signals between 6 and 20 Hz. Efflux was insensitive to the Ca^{2+} concentration in the bathing medium, but it was enhanced if the pH was lowered and decreased in the presence of La^{3+}, a Ca^{2+} antagonist. An unmodulated rf field or a field modulated at 60 Hz had no effect on Ca^{2+} efflux. It should be noted that these experiments could not distinguish between the release of membrane-bound Ca^{2+} and the Ca^{2+} from intracellular stores. Blackman *et al.* (1980) reported enhanced efflux of Ca^{2+} in brain tissue slices exposed to a 147-MHz signal amplitude modulated by a 16-Hz sine wave. They also reported a power-density window similar to the one at higher frequency (450-MHz amplitude modulated at 16 Hz) reported by Bawin *et al.* (1978).

Dutta *et al.* (1984) exposed cultured neuroblastoma cells for 30 min to a 915-MHz field amplitude modulated by a 16-Hz sinusoid prior to examining the rate of $^{45}Ca^{2+}$ efflux. In these experiments, signficant increases in $^{45}Ca^{2+}$ efflux were observed at two specific absorption rates (SARs, defined as the rate at which energy is being deposited in the tissue or medium): 0.05 and 1 mW/g. They observed that enhanced $^{45}Ca^{2+}$ efflux was dependent on the presence of 16-Hz amplitude modulation at the lower SAR (0.05 mW/g); at the higher intensity (1.0 mW/g) enhanced efflux also occurred in the absence of amplitude modulation. Pollack *et al.* (1992) exposed isolated bone calvaria cells for 20 min to frequencies ranging from 6 to 600 kHz at an E field intensity of 10 mV/cm and assessed intracellular levels Ca^{2+} fluorometrically using the dye Fura-2. Their data indicate transient intracellular increases in cytosolic Ca^{2+} at 20, 30, 60, 100, and 300 kHz.

b. Lower Frequency Fields Lyle *et al.* (1992) applied 60-Hz electric fields (1 mV/cm) to Jurkat E6 human T leukemic cells and examined their effect on intracellular Ca^{2+}. Changes in intracellular Ca^{2+} were determined with a FACScan cytometer (Becton Dickinson, Mountain View, CA); no significant differences were detected following EMF exposure. In a related experiment at a higher field intensity, Walleczek *et al.* (1992) applied a 50-Hz sinusoidal 50-mV/cm (calculated) electric field to the same cell type (Jurkat) and assessed the levels of intracellular Ca^{2+} and Mn^{2+}, using the fluorescent dye Fura-2. Their data showed a significant enhancement in intracellular Mn^{2+} following a 4-min exposure, but intracellular Ca^{2+} levels were unaffected. The calcium signaling role of the cell membrane on cells of the immune system exposed to EMFs has been briefly reviewed by Walleczak (1992).

Lindström *et al.* (1993) exposed Jurkat cells to 50-Hz, 0.1-mT magnetic fields and examined free intracellular Ca^{2+} in individual cells, using Fura-2. Within 15 to 200 sec after the Helmholtz coils were energized, intracellular Ca^{2+} increased to 200–400 n*M* from baseline levels of 50 to 100 n*M*. When similar experiments were performed with mitogen-activated peripheral lymphocytes, a similar enhancement in intracellular Ca^{2+} response was observed. However, only 10% of the lymphocytes tested responded to the magnetic fields, whereas about 85% of the Jurkat cells showed enhanced intracellular Ca^{2+} in response to an applied magnetic field. Magnetic fields failed to enhance intracellular Ca^{2+} levels in nondividing endocrine pancreatic cells from mice or rats or when Ca^{2+} was absent from the cell perfusion medium. These studies have the additional unique feature that single cells were observed continuously before and after MFs were activated and the change in Ca^{2+} level was perfectly correlated with MFs.

In contrast to the leukemic cell studies, Walleczek and Liburdy (1990) exposed thymocytes that had been prelabeled with $^{45}Ca^{2+}$ and activated with the mitogen concanavalin A (ConA) for 60 min, to a sinusoidal 60-Hz, 22.0-mT (rms) magnetic field. The calculated induced electric field was 1.0 V/cm. Their data show that EMFs enhanced $^{45}Ca^{2+}$ uptake about 170% in ConA-activated cells; no differenecs in Ca^{2+} influx were observed in cells that were exposed to EMF but not stimulated with ConA. The level of ConA activation also affected the response of the cells to EMFs; suboptimal mitogen-activated cells were most responsive to the applied EMFs. Walleczek and Budinger (1992), using mitogen ConA-treated rat thymocytes, applied a 3-Hz monopolar pulsed magnetic field for 30 min and examined the uptake of $^{45}Ca^{2+}$. The field intensities in this experiment were 1.6, 6.5, and 28 mT. They found that thymocytes exposed to pulsed magnetic fields showed a 45% reduction in $^{45}Ca^{2+}$ influx, rather than the enhanced uptake found with exposure to sinusoidal magnetic fields. They concluded that cells that were less responsive to mitogen stimulation showed enhanced uptake of Ca^{2+} when exposed to EMFs, whereas cells that responded to mitogen stimulation exhibited reduced Ca^{2+} uptake.

To examine the effectiveness of the magnetic or electric field component in altering Ca^{2+} flux, Liburdy (1992) employed both real-time fluorescence measurements of Ca^{2+}, using the fluorescent probe Fura-2, and multiring, annular culture dishes designed to provide a more careful control of the induced electric field component. In other experiments thymocytes were directly exposed in the presence or absence of ConA to a 60-Hz, 1.7-mV/cm electric field inside a spectrofluorometer. The applied field was similar to the induced *E* field calculated for the outer annular ring of the culture dish. The combined data from these studies strongly support the conclusion that the electric field component is responsible for altered calcium flux. Further, the Liburdy data suggest that the electric field operates by inducing an opening of the calcium channel in the membrane rather than by increasing Ca^{2+} mobilization from the endoplasmic reticulum. Yost and Liburdy (1992) also examined the effect of static and time-varying magnetic fields on Ca^{2+} influx. In these studies mitogen-activated lymphocytes were exposed to 16-Hz, 42.1-μT magnetic fields with the simultaneous application of a 23.4-μT static DC field. The combined fields inhibited Ca^{2+} influx in mitogen-activated lymphocytes; exposure to the individual fields or resting lymphocytes showed no field responses.

Carson *et al.* (1990) exposed HL-60 cells to a time-varying magnetic field and assessed the uptake of Ca^{2+} using the fluorescent dye Indo-1. The EMF source in these experiments was a magnetic resonance imaging (MRI) system. During imaging, this instrument produces a static magnetic field (0.15 T), a modulated rf field with a carrier frequency of 6.25 MHz, and three gradient orthogonal magnetic fields (G_x, 1.4 mT/m; G_y, 1.4 mT/m, G_z, 1.9 mT/m). The gradient fields are repetitively switched on and off, resulting in a time-varying magnetic field. Using the Indo-1 probe, the rf carrier alone (6.25 MHz) had no effect on Ca^{2+} uptake; however, with a 23-min exposure to time-varying magnetic fields, Ca^{2+} influx was enhanced about 25% over basal levels. The length of time betwen exposure and assay (10 to 60 min) did not affect the outcome of the experiment. These data are in partial agreement with experiments described above, in which a frequency-modulated rf carrier was used (Adey, 1981; Dutta *et al.*, 1992).

Stagg *et al.* (1992) used ^{45}Ca-loaded microsomes as a model system and applied 60-Hz, 0.1-mT magnetic fields to measure ATP-dependent uptake of $^{45}Ca^{2+}$. In these experiments the uptake of $^{45}Ca^{2+}$ decreased about 20% relative to controls following 3-, 5-, and 10-min EMF exposures. Because the experiments involved measuring $^{45}Ca^{2+}$ in the medium, the decrease may have resulted from either an inhibition of the membrane-bound calcium ATPase or leakage from the microsomal vesicle.

Jolley *et al.* (1983) applied a pulsed magnetic field (5-msec burst of 4-kHz triangular pulses repeated at 15 Hz) to isolated islet of Langerhans cells. They report that $^{45}Ca^{2+}$ efflux decreased about 25% relative to nonexposed controls. In comparison to other Ca^{2+} efflux experiments, these experiments used a significantly longer field exposure (18 hr, compared to a few minutes). These differences may thus represent a longer term, rather than a more immediate transient, cell response to EMF perturbations. In contrast, in a study using an intermediate exposure, Colacicco and Pilla (1983, 1984) exposed chick tibia cells to a pulsed magnetic field (similar to Jolley *et al.*, 1983) for 60 min and found that $^{45}Ca^{2+}$ uptake was stimulated. An important consideration in interpreting these experiments, as discussed in Sections IV and VI, is the age of the culture, a factor that can significantly affect the response of a cell to EMFs.

2. Sodium and Potassium

Compared to calcium there have been fewer studies on the effects of EMFs on Na^+ and K^+ movement across an intact cell membrane. Although Tsong and colleagues (Teissie and Tsong, 1981; Serpersu and Tsong, 1983, 1984) examined the effects of EMFs on erythrocytes, the applied fields were high, generally in the range of 15 to 32 V/cm. Erythrocytes exposed to these fields showed enhanced transport against a concentration gradient of Rb^+, a K^+ analog; ATP hydrolysis was not detected. The frequency dependence of the enhanced transport (effective range is between 0.1 kHz and 0.1 MHz) supports the conclusion that the response is not thermal in nature. Tsong and colleagues suggest that the enzyme (Na^+,K^+-ATPase) extracts energy from the oscillating field and transduces it to transport Rb^+. This mechanism is similar to the one proposed by Blank and Soo (1989, 1990) and discussed above.

B. Membrane Alterations

Most of the theories addressing the mechanism of interaction between biological systems and EMFs suggest that the primary site of interaction is the plasma membrane. However, direct experimental data showing EMF-induced membrane perturbations are sparse. Grandolfo *et al.* (1991) exposed chick myoblasts to a 50-Hz sinusoidal magnetic field at intensities between 1 and 10 mT. No field effects were observed on the fusion of myoblasts although decreases in the conductivity and permittivity of the cells were observed. The conductivity changes were interpreted as evidence for a decrease in transport of various ions, but the site of inhibition (channel, energy-driven pump, etc.) was not determined. The decrease in permittivity was interpreted as a change in surface charge owing to alterations in lipids, proteins, or both. An interesting aspect of this study was the observation that the maximum response of the cells was found at 5 mT rather than at the highest applied intensity of 10 mT, suggesting that a linear dose response is not involved. It is noteworthy that other investigators, for example, Goodman and Henderson (see below) report that even lower intensity fields in the microtesla range are more effective in inducing bioeffects than the more intense fields.

Paradisi *et al.* (1993) exposed K562 cells, an erythroleukemic line, to 2.5-mT, 50-Hz fields for 24–72 hr and compared the surface appearance of control and exposed cells by scanning electron microscopy. Exposed cells had many fewer microvilli and more blister-like blebs. They also examined the degree of membrane order, using electron paramagnetic resonance spectroscopy. Spectra were similar, but the exposed cells had a somewhat longer time constant, which they interpret as showing a more ordered membrane.

Data from our laboratory support the suggestion that both surface charge and membrane components change as a result of EMF exposure. We used a technique known as aqueous two-phase thin-layer countercurrent distribution (TLCCD) to assess EMF effects. In a two-phase system containing polyethylene glycol (PEG, $M_r = 8000$) and dextran T-500 ($M_r \sim 500{,}000$), cells distribute themselves between the interface and the PEG-rich upper phase according to their affinity for the PEG-rich layer. A potential difference of up to 2 mV between the upper phase and dextran-rich lower phase can be obtained by adjusting the concentration of salts in the phosphate buffer. Phosphate favors the dextran-rich phase and its asymmetric distribution results in a net positive upper phase. In separating cells by TLCCD, one performs a type of cell chromatography that partitions a cell population.

In these experiments amoebas from the slime mold *P. polycephalum* were exposed to sinusoidal 60-Hz, 0.1-mT, 1.0-V/m EMFs for 24 hr (Marron *et al.*, 1983, 1988); other groups of amoebas were exposed only to the electric or magnetic fields. Following EMF exposure the cells were partitioned. Two partitioning systems were used; one system utilized a 2-mV potential difference between the two phases and is referred to as a charged system. In the second system, NaCl replaced the phosphate to maintain osmolarity, producing no net potential difference between the two phases; this system is referred to as an uncharged system. Electromagnetic field-exposed cells were analyzed using both systems. Cells exposed to a combined electric and magnetic field for 24 hr showed an apparent increase in their net negative surface charge; that is, they preferred the upper phase in the charged system more than control cells. When exposed cells were placed in the uncharged phase system, there was a diminished affinity for the upper phase compared to the controls, indicating that the cell membranes were more hydrophobic.

If a magnetic field alone was applied, there was no difference in the profiles between the control and exposed cells in the charged system; but the uncharged system again indicated an increase in the hydrophobic nature of the membrane. If only an electric field was applied (60 Hz, 1.0 V/m) an increase in the net negative charge was observed in the charged phase system and there was no difference observed in the uncharged system. The data are summarized in Fig. 2. We interpret these data as indicating that magnetic field exposure decreases or alters the hydrophobic nature of the membrane, whereas the electric field increases the net negative surface charge on the cell. They are consistent with the data of Grandolfo *et al.* (1991) discussed above.

We have also exposed *Physarum* to pulsed magnetic fields similar to those used to facilitate bone repair and analyzed the cells by TLCCD. The response was similar to that observed with a sinusoidal $E + H$ field. Applying a pulsed electric field similar to the magnetically induced electric field produced changes in the profiles that were identical to those seen in cells exposed to a sinusoidal electric field (E. M. Goodman *et al.*, 1986). In a related study, Smith *et al.* (1991) exposed log-phase U-937 cells, a nonadherent, human histiocytic, monocyte-like cell line, to a pulsed magnetic field for 48 hr. The cells were then partitioned in a charged and uncharged TLCCD system. The data in the charged system show an increase in negative surface charge; no differences were detected in the uncharged system.

We may conclude from these data that field-induced surface changes are independent of applied waveform and are similar across a broad range of cell types. An important conclusion from these experiments is that electric and magnetic fields can differentially alter the cell surface. However, the long exposure period employed in these experiments (24–48 hr) makes it likely that we are observing a secondary rather than primary effect. Although few of the current mechanistic theories ascribe a role to the magnetic field component, our experiments clearly show that it can affect the cell surface.

Using murine bone marrow cells, Stegemann *et al.* (1993) examined the effect of a 30-min exposure to a 1.4-T stationary magnetic field on acetylcholinesterase activity. Static magnetic field exposure at 37°C decreased enzymatic activity by 80% in 2 hr; at 27°C it took 3.5 hr to achieve a similar reduction in activity. Cells recovered to about 93% of control levels at 37°C in about 4 hr; at 27°C the cells required about 15 hr to reach a similar level of recovery.

Lin-Liu *et al.* (1984) studied embryonic *Xenopus* myoblasts for the effect of a pulsed electric field (square pulses of 50 and 5 msec repeated

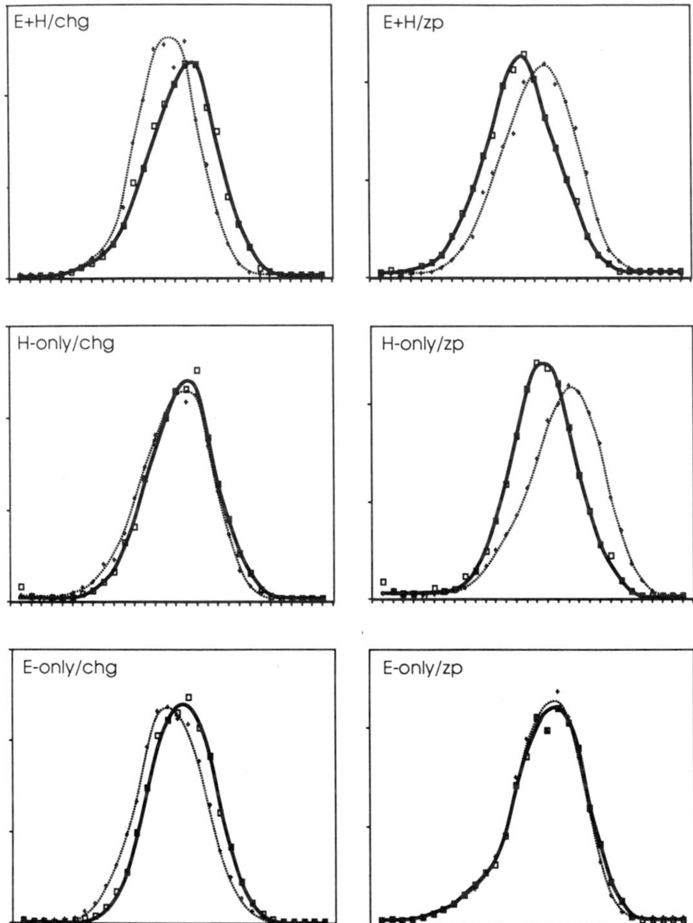

FIG. 2 Thin-layer countercurrent distributions (TLCCDs) plotted as cell count versus tube number. Cells exposed to fields (—); control cell populations (---). Field condition is indicated in each plot with $E + H$ indicating the presence of both electric and magnetic fields. Panels are paired to display TLCCDs in both charged (chg) and uncharged (zp) partitioning systems. Distributions have been normalized so that the total cell count summed over all chambers is constant. Note that a shift in the distribution peak from the center chamber (No. 15) by one chamber to No. 16 corresponds to a 14% increase in partition coefficient from 1.00 to 1.14, in which the partition coefficient is defined as the ratio of cells in the upper phase to those at the interface. Repeated analysis of control cells against control cells produced superimposable distributions similar to those shown in the bottom right-hand panel. [From Marron, M. T., *et al.*, (1988). *FEBS Lett.* **230,** 13–11. Reproduced by permission of Elsevier Science Publishers.]

at 10 and 100 Hz at intensities of 0.3 to 3.0 V/cm) on the distribution of fluorescently labeled ConA receptors. These experiments were designed to determine the minimum field required to induce an asymmetric distribution of ConA receptors and to compare the response of receptors to a pulsed field and to a DC field. The asymmetry in receptor distribution appears to be a function of both pulse width and frequency of the applied field. The asymmetric movement of ConA receptors toward the cathode was similar when either a pulsed electric or DC field was applied. Low duty cycles (<25%) extended the time (>50 min compared to <30 min for higher duty cycles) required for receptor distribution to reach a steady state.

Hamada *et al.* (1989) applied a 10-V/cm DC field to embryonic chick fibroblasts and examined the morphological changes that occurred over a 60-min exposure. They found that an applied field will induce a realignment of cell surface charges that ultimately produces an altered morphology. The results agreed with the observations of Lin-Liu *et al.* (1984), that cells realign toward the cathode in the presence of an electric field. With continued exposure, the cells became bipolar through a complex set of changes, losing the polygonal configuration usually seen *in vitro*. McCaig and Dover (1989) exposed cultured *Xenopus* myoblasts to a 120-mV/mm DC field and observed elongation of cells within 1 hr of exposure; the orientation appeared to be a Ca^{2+}-dependent process with influx occurring on the cathodal side. Unexpectedly, intracellular organelles such as the mitochondria, actin filaments, and ribosomes appeared to be asymmetrically distributed toward the anodal side. Additional experiments with Ca^{2+} channel agonists show Ca^{2+} enters at the cathode-facing side, the side opposite from where some of the organelles have redistributed. In another study Dover and McCaig (1989) report that differentiating myoblasts exposed to similar DC fields for 18 hr had more microfilaments and myofibrils than comparable, nonexposed control cultures. In addition, exposed myoblasts had more and thicker Z-lines. However, unlike fibroblasts and neural crest cells, which orient and migrate toward the cathode in an applied field (Erickson and Nuccitelli, 1984; Stump and Robinson, 1983), the myoblasts reoriented but remained in a fixed position.

Bedlack *et al.* (1992) examined the response of neurites to 0.1 to 1.0-mV/μm uniform DC fields; these intensities are within an order of magnitude of the endogenous electric fields in the central nervous system (Borgens, 1982). The galvanotropic response of cells had been described previously (Jaffe and Poo, 1979; Patel and Poo, 1982); this study examined whether localized calcium influx and membrane depolarization in response to an applied field are inducers of the galvanotropic response. They found that the number of filopodia facing the cathode increased within 5 min of field exposure; in contrast, filpodia on the anode side proportionally decreased. Within 30 sec after the onset of a field (0.38 mV/μm), an increase in intracellular Ca^{2+} (from 0.3 to 0.7 $\times$ 10^{-6} μM) was observed in the regions nearest the cathode; the intracellular Ca^{2+} levels fell almost to baseline within 20 min. Membrane depolarizations were examined using a fluorescent potentiometric dye (Di-8-ANEPPS); a cathode-localized depolarization occurred less than 1 min after application of a field (0.98 mV/μm). The investigators conclude that membrane depolarization results in a localized activation of a voltage-dependent calcium channel, producing a transient influx of Ca^{2+}. The alteration in the intracellular Ca^{2+} concentration leads to cytoskeletal reorganization and redirection of filopodia toward the cathode.

In a study with HeLa cells, Yamaguchi *et al.* (1992) rapidly switched the field intensity in an electromagnet between 0.35 and 1.77 T, inducing an electric eddy current in the medium. If the switching period was less than 10 sec, ouabain-sensitive Rb^+ influx (active K^+ influx) was significantly reduced; there was no apparent effect on passive or ouabain-insensitive influx. Over time, an increase in K^+ efflux was noted. They concluded that the changes occurred as a result of alterations in the electrical properties of the cell surface.

Fisher *et al.* (1986) examined the effect of pulsed magnetic fields on [^{3}H]glucosamine incorporation in plasma membranes of human fibroblasts and rat sarcoma cells. Their data show about a 100% increase in labeled glucosamine incorporation into existing cell surface glycoproteins following a 48-hr exposure to PMFs. Chromatographic data suggest that anionic residues (*N*-acetylglucosamine, *N*-acetylgalactosamine) are also being synthesized and incorporated. These results have been interpreted as showing that pulsed fields stimulate incorporation of anionic charges on the cell surface. They are also consistent with the previously discussed data from our laboratory showing an increase in the negative surface charge following *E*-field exposure.

When A431 cells (a human carcinoma line) were exposed to sinusoidal magnetic fields (50 Hz; 2.5, 5.0, and 10 mT), an alteration in cell morphology was observed at 5 mT (Malorni *et al.*, 1992). No changes were observed in the growth or proliferation of these cells for 96 hr at any of the field intensities examined. On the basis of data from fluorescent and scanning electron microscopy, these authors suggest that the microvilli and actin microfilament network have been altered. The net result of these changes is a decrease in the intracellular contacts and a retraction of the cells.

Cain *et al.* (1993) examined the formation of foci of cocultured cells exposed to intermittent (1 hr, four times per day), 60-Hz, 0.1-mT fields for 28 days in the presence of the tumor promoter 12-*O*-tetradecanoylphorbol-13-acetate (TPA). Magnetic fields alone had no effect on the formation of foci. In contrast, magnetic fields in the presence of TPA (10, 20, 40, 50, and 100 ng/ml) induced a 1.9-fold increase in foci compared to TPA sham controls. In addition, the total area of the foci increased 2.2-fold and the number of cells stained increased 2.3-fold.

Blackman *et al.* (1993a,b) have reported that magnetic fields can directly affect neurite outgrowth. In these experiments two lines of pheochromocytoma cells (PC-12 and PC-12D) obtained from different sources were exposed to EMFs. When PC-12 cells are used to study neurite regrowth, they require priming with nerve growth factor (NGF) at a concentration of 50 ng/ml every other day for 6 days prior to initiating the experiment. PC-12D cells do not require priming prior to initiating an outgrowth experiment. Both cell lines were exposed to 50-Hz, 7.9-μT magnetic fields for 22 hr, using annular ringed dishes. In this configuration, cells in the center would be exposed to an E field of < 3.6 μV/m. In the PC-12 cell line with sufficient but not excess NGF present (5 ng/ml), EMFs significantly reduced neurite outgrowth (27%) relative to nonexposed controls. In contrast, experiments with the PC-12D cells (in the absence of any NGF) showed enhanced neurite regrowth. They further conclude that the magnetic field is the effective component, because a 2-fold increase in the magnetic intensity resulted in an enhanced regrowth response whereas a 12-fold increase in the induced electric field had no effect of regrowth.

In an attempt to address the question of whether or not weak magnetic fields induce major changes in cells, Parola *et al.* (1993) exposed chick embryo fibroblasts to 100-Hz, 700-μT fields for 24 hr and examined their effect on the rotational dynamics of the enzyme adenosine deaminase (ADA). Adenosine deaminase, a malignancy marker in the plasma membrane of virally transformed cells, was monitored using phase-modulation spectrofluorometry. The data show that the magnetic fields induced differential phase profiles indicative of a more fluid lipid core. Further, the phase profiles were similar to those observed when cells are virally transformed. These investigators also report that magnetic fields increased the rate of cell proliferation and reduced the specific activity of ADA.

C. Nucleic Acids and Gene Expression

1. DNA

Relatively few direct effects have been reported on the influence of EMFs on DNA. In one of the earliest reports, Liboff *et al.* (1984) exposed fibroblasts to frequencies ranging from 15 Hz to 4 kHz and amplitudes that ranged from 2.3 μT to 56 mT. The maximum response was an ~60%

enhancement in incorporation of [^{3}H]thymidine when fibroblasts were exposed to a sinusoidal magnetic field of 76 Hz, 0.16 μT (rms). In a similar experiment, Takahashi *et al.* (1986) measured the uptake of [^{3}H]thymidine in Chinese hamster ovary cells exposed to a pulsed magnetic field. In these experiments, a significant enhancement (13%) in the uptake of [^{3}H]thymidine was found in cells exposed to a 10-Hz, 20-μT pulse 25 μsec wide; an approximate 30% enhancement in [^{3}H]thymidine incorporation was observed at a repetition rate of 100 Hz. If the intensity of the pulsed field was increased to 200 μT or greater, the incorporation of [^{3}H]thymidine was inhibited by up to 80%. In our laboratory human umbilical vein endothelial cells were transferred to medium containing [^{3}H]thymidine and immediately exposed to a pulsed magnetic field. The field consisted of a train of 22 sawtooth magnetic field pulses; each pulse had a 200-μsec rise to a peak intensity of approximately 2 mT and a 20-μsec fall. The pulse bursts were repeated at 25 Hz. No significant increase in the uptake of [^{3}H]thymidine was observed following 24 hr of exposure (E. M. Goodman *et al.*, 1993b). These data are in agreement with those reported earlier by Yen-Patton *et al.* (1988) in a similar experiment with endothelial cells.

Enhanced incorporation of [^{3}H]thymidine into DNA has been reported when a high-voltage electric field was used to stimulate human fibroblasts (Bourguignon and Bourguignon, 1987). In this study, either a 50 or 75-V unipolar pulsed field was applied at 100 pulses/sec for 20 min; the time-averaged current was about 50 μA. Armstrong *et al.* (1988) and Brighton *et al.* (1989) observed enhanced proliferation as determined by the uptake of [^{3}H]thymidine in chondrocytes exposed to capacitively coupled fields between 500m and 1000 V peak to peak, corresponding to electric fields between 15 and 30 mV/cm. Applied fields of 1500 V, peak to peak, inhibited the uptake of [^{3}H]thymidine. A study examining the duty cycle required to induce enhanced [^{3}H]thymidine suggests that an extremely short exposure is required to induce an effect; in this study a 2.5- and 5-min exposure in a 24-hr period was as effective as a 24-hr continuous exposure regimen (Brighton *et al.*, 1989).

In an earlier study Rodan *et al.* (1978) applied a 1.166-kV/cm unipolar pulse to chondrocytes from 10-day chick embryos and reported an approximate 30% enhancement in the uptake of [^{3}H]thymidine. In these experiments, the presence of antagonists for the transport of Na^+ (tetrodotoxin) or Ca^{2+} (verapamil) completely blocked the electric field effect on [^{3}H]thymidine incorporation. Enhanced uptake was seen in [^{3}H]thymidine uptake in chondrocytes and bone cells from the calvaria of 19 rat embryos, but not in skin fibroblasts from 10-day chick embryos or rat spleen lymphocytes. The investigators suggest that the mechanism may be tissue specific. Although one is initially inclined to dismiss such an explanation, if correct it might explain why some laboratories observe EMF effects whereas others are unable to detect any responses.

In a somewhat different experiment Fitzsimmons *et al.* (1986) exposed embryonic chick tibia and embryonic chick calvarial cells in monolayer culture to a 10-Hz sinusoidal electric field (10 μV/m). The cells were exposed for 72 hr, either continuously or for 30 min/day. The shorter daily exposure produced a greater response in cell proliferation compared to continuous exposure in both cell types as judged by the uptake of [^{3}H]thymidine. In a later study Fitzsimmons *et al.* (1989) applied a capacitively coupled electric field of 0.1 μV/cm to chick calvarial cells, using frequencies ranging from 8 to 24 Hz. The cells were exposed for 30 min, and the incorporation of [^{3}H]thymidine and cell number were assessed 18 hr later. They report a frequency-dependent effect on growth and proliferation with a maximum response at 16 Hz. Enhanced cell growth was also observed when cells were placed in "preconditioned" medium, obtained from cultures previously exposed to an electric field, for 30 min. They interpret these results to mean that cell proliferation is enhanced by an electric field-stimulated mitogen release. Subsequent experiments suggest that the mitogenic factor is an insulin-like growth factor II (IGF-II) (Fitzsimmons *et al.*, 1992). On the basis of these data, they propose that electric field exposure releases presynthesized growth factors into the medium, resulting in enhanced proliferation. When a capacitively coupled 14-Hz, 0.1-μV/m electric field is applied for 30 min to cultured osteosarcoma bone cells (TE-85), a 25% increase in the rate of proliferation was observed (Fitzsimmons *et al.*, 1992); enhanced proliferation was correlated with an increase in levels of IGF-II mRNA.

Mouse fibroblasts were exposed to a 50-Hz electric field (current density, 0.25 mA/cm^2) or a 2-mT magnetic field for 1 hr and then incubated for an additional 6 hr before determining the field effects on cell proliferation (Schimmelpfeng and Dertinger, 1993). The magnetic fields were applied using an iron magnet with an air gap placed in a box maintained at 37°C but without cooling. The electric field was applied using two electrodes immersed in the culture medium below and above the filter insert that carried the cell monolayer. Magnetic field exposure alone induced a small but significant decrease in the mean DNA content, whereas an increase was observed when an electric field was applied. An analysis of the cell cycle using a cell sorter indicated that the magnetic field enhanced the number of cells in G_1; in contrast, the electric field enhanced accumulation of cells in S, G_2, and M. Exposure to both electric and magnetic fields, using a solenoid, showed no statistically significant differences in either the mean DNA content or cell cycle distribution. In the latter experiment, the voltage drop between the windings generated the E field, which was estimated to be about 40% of the directly applied *E* field. The authors suggest that this finding is the result of simultaneous but opposite actions of the two field components. Schimmelpfeng and Dertinger (1993) also examined the effect of time and fields on the level of the secondary messenger, cAMP. Their data show that short-term exposure (5 min) to either an electric or magnetic field enhanced the intracellular cAMP concentration by about 15–20%. In contrast, exposure to the fields for 60 min resulted in a decreased cAMP concentration.

Guzelsu *et al.* (1994) examined both bidirectional and unidirectional time-dependent waveforms on primary cultures of chick tendon fibroblasts. The fields were energized 24 hr after cells had been transferred; the data based on total DNA synthesis indicate that both waveforms stimulate growth between days 2 and 3. On average, the computed doubling time of the field-exposed cells was about 16 hr compared to 21 hr for the nonexposed controls. At 3 to 4 days, the doubling time decreased in field-exposed cultures relative to nonexposed controls. They interpret these data as a slowing of growth in the magnetic field (MF)-exposed cultures as they approach confluence. In contrast to Murray and Farndale (1985), cells maintained in culture for 8 days showed no significant difference in the cAMP level.

2. RNA

Beginning with the work of R. Goodman *et al.* (1983), data have continued to accumulate that suggest that the level of some mRNAs are altered by EMFs. In early experiments, total RNA was labeled with [^{3}H]uridine and then separated by sucrose gradients or oligo(dT) cellulose chromatography. Analysis of fractionated RNA from the sucrose gradients showed new synthesis in the region generally indicative of mRNA; corroborative data were obtained from the chromatography experiments. The cells all showed enhanced incorporation of [^{3}H]uridine into the putative mRNA regardless of the applied waveform; however, some waveforms were more effective than others in inducing a stimulatory effect. A variety of signals (both pulsed magnetic fields and sine waves; see Fig. 3) were applied to the salivary chromosomes of *Sciara coprophila,* and incorporation of [^{3}H]uridine was followed using autoradiography (R. Goodman and Henderson, 1986, 1987). The autoradiograms revealed augmented banding patterns that were dependent on the signal being applied. Perhaps the most intriguing conclusion from these studies is that transcription was enhanced at loci already active at the time of EMF exposure.

To indicate more clearly what transcripts have been enhanced following EMF exposure, R. Goodman and Henderson (1991) and R. Goodman *et al.* (1992a,b) examined the effect of EMFs on *Drosophila melanogaster.*

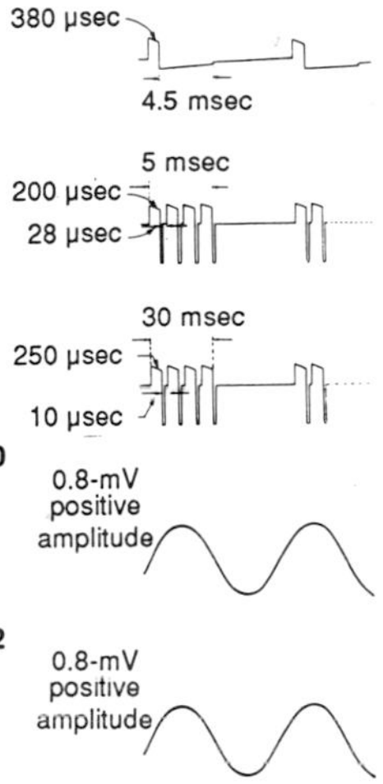

Signal	Waveform	Rate (Hz)	Positive Induced Amplitude (mV)	Positive Duration (μsec)	Burst Width (msec)	Negative Space (μsec)	Negative Spike (μsec)	Peak Magnetic Field (mT)
SP	380 μsec; 4.5 msec	72	15	380			4500	3.5
PT	5 msec; 200 μsec; 28 μsec	15	14.5	200	5	28	24	1.9
E-33	30 msec; 250 μsec; 10 μsec	1.5	2.5	250	30	10	4	0.38
CW60	0.8-mV positive amplitude	60	0.8					1.5
CW72	0.8-mV positive amplitude	72	0.8					1.1

FIG. 3 Major waveform characteristics of electromagnetic signals. In addition to the above-described characteristics, the frequency content of the three asymmetric pulses derived by discrete Fourier transforms differs significantly. The repetition rates are within the ELF range, but the frequency content of the asymmetric signals can range to 10 MHz. For example, the negative spike, and burst width of the SP signal, contains a 166-Hz component; the positive width contains a 2.631-kHz component. The positive width, negative spike, and burst width of the PT and E-33 signals also contain high-frequency components. The positive width in each case has a 4-kHz component. The negative spike of the E-33 signal contains a 166.6-kHz component; in the PT, a 41.6-kHz component. The burst width of the E-33 signal contains a 33.3-Hz component; in the PT, this is a 222.2-Hz component. [From R. Goodman *et al.* (1989) *Biochim. Biophys. Acta* **1009,** 216–220; reproduced by permission of Elsevier Science Publishers.]

Following a 20 min exposure to different waveforms, 13 defined regions of salivary gland chromosome 3R were identified by autoradiography as either being induced or showing enhanced transcript activity. The most interesting finding in this study was the elevated levels of transcripts for heat shock or stress proteins in *Drosophila* without any apparent increase in temperature. These data form the basis for the suggestion that EMFs may be perceived by cells as environmental stressors (Blank *et al.,* 1993). Using the same exposure regimens chromosomes 2R, 2L, 3R, 3L, and X from *Drosophila* were also examined (R. Goodman *et al.,* 1992a,b; Weisbrot *et al.,* 1993a). In all, 17 chromosome regions were identified as showing elevated transcript levels; many of the enhanced regions were associated with genes that control or regulate cell structure and development.

Some chromosomal regions responded in a signal-specific manner, whereas others showed a general transcriptional elevation following exposure.

Weisbrot *et al.* (1993b) examined the transcriptional response of three genes in *Saccharomyces cerevisiae* exposed to a 60-Hz sine wave at various field intensities (0.8, 8.0, and 80 μT) and lengths of exposure (10 to 60 min). They report that genes involved in uridine metabolism and a heat shock response showed elevated levels following 15 to 30 min of exposure; the lowest applied field generated the largest and earliest response. A third gene examined, a meiotic regulatory gene, did not respond to EMF exposure; however, if the cells were subjected to a heat shock (37°C) the transcript levels of all three genes increased. Collectively, these data support the conclusion that EMFs generate a differential gene response and may be recognized by the regulatory machinery of the cell as an exogenous stressor. They also underscore earlier observations that the EMF does not produce a linear dose–response relationship.

In another series of experiments, human HL-60 cells were exposed to 60-Hz fields at field intensities ranging from 0.57 to 570 μT for 10, 20, and 40 min (R. Goodman *et al.,* 1989, 1992c,d; Wei *et al.,* 1990; Blank *et al.,* 1992). In most experiments, the transcripts for actin, histone H2B, and c-Myc were elevated after a 20-min exposure to 5.7 μT, but not at higher or lower field intensities. Collectively, these data led R. Goodman *et al.* (1993) to the following conclusions: (1) direct transcriptional changes occur within 4 min when cells are exposed to any of several types of signals in the EMF range; (2) the transcripts that respond to the applied fields are those already being expressed in the cell; and (3) cells may perceive EMFs in a manner analogous to other environmental stressors (e.g., heat shock, chemicals). In a quasireplication attempt of the Goodman *et al.,* experiments, Krause *et al.* (1991) exposed HL-60 cells for 0.5–2 hr to a 60-Hz, 1.0-mT field. Although the fields in this experiment were considerably stronger, differences in the transcripts described by Goodman and Henderson were not observed; an enhancement in the transcription rate determined by [^{3}H]uridine incorporation was observed. Parker and Winters (1992) exposed mouse and tumor cells to a 60-Hz, 0.1-mT rotating magnetic field for 24, 48, and 72 hr. These fields approximate the conditions found under a high-power transmission line. Total RNA was extracted from control and EMF-exposed and probed for changes in transcription of v-*myc,* v-*fos,* v-*raf,* and the heat shock protein hsp-70. No significant alterations in RNA levels were detected for any of the probes or exposures.

Employing a single magnetic field pulse repeated at 72 Hz with a peak intensity of 3.5 mT, Phillips and McChesney (1991) exposed T lymphoblastoid CCRF-CEM cells for periods ranging from 30 min to 24 hr while following [^{3}H]uridine incorporation into RNA. They observed an increase in total RNA after 30 min; these levels remained elevated relative to controls over the next 22 hr. In contrast, mRNA peaked at 2 hr and then declined to control levels over the ensuing 14-hr period. No attempt was made to identify the enhanced mRNAs. In another study, Phillips *et al.* (1992) exposed CEM-CM3 T lymphoblastoid cells at different cell densities to a 60-Hz, 0.1-mT field for 15 to 120 min. Using a nuclear run-off assay, they report both time- and cell density-dependent changes in the transcription of c-*fos*, c-*jun*, c-*myc*, and the gene encoding protein kinase C. In the case of c-*fos*, the transcript levels appear to be independent of cell density (10^5 and 10^6 cell/ml), exhibiting a 2.5-fold enhancement following a 30-min exposure. These experiments and the Goodman RNA experiments have been reviewed by Phillips (1993).

Lin *et al.* (1994) transfected HeLa and mouse myeloma cells (PX3) with a plasmid containing the c-*myc* gene and promoter, including a 2329-bp DNA regulatory segment upstream from the P1 promoter. Both cell types showed elevated transcript or chloramphenicol acetyltransferase (CAT) levels following a 20-min exposure to 60 Hz, 8 and 80 μT. By systematically deleting specific regions of the DNA upstream from the promoter they conclude that the EMF-responsive region is located between -353 and -1257 bp from P1.

Greene *et al.* (1991) examined the effects of 60-Hz, 1.0-mT fields on transcription in HL-60 cells, using Petri dishes containing interior walls shaped as concentric rings. This configuration enabled them to separate the magnetic and electric field components, because the induced electric field component would be highest in the outer rings of the dish, while the magnetic field is constant throughout the dish. Their data show that cells exposed in the outer ring had a significant enhancement (about 50%) in the incorporation of [^{3}H]uridine, compared to cells in the inner ring. If the electric field component was held constant and the magnetic fields were altered, the kinetics of [^{3}H]uridine incorporation was similar for cells in both the inner and outer rings. Collectively, these data suggest that the induced electric field component is the primary agent stimulating elevated RNA transcription.

In experiments performed in our laboratory, *E. coli* were exposed to a sinusoidal 72-Hz, 1.1-mT (rms) magnetic field for periods ranging from 5 to 20 min. Following exposure, total RNA was isolated and probed by Northern blot analysis to determine the transcript levels for the α subunit of DNA-dependent RNA polymerase. An increase in the level of the α polymerase transcript occurred at 15 min but no significant differences in the level of this transcript were discernible at 10, 20, or 30 min of field exposure (Sustachek, 1992). These data suggest that weak EMFs have a transient effect on the level of mRNA for α polymerase, in agreement with other studies discussed above.

We conducted a series of *in vitro* experiments using a totally synthetic system to examine weak magnetic fields (45 Hz, 1.1 mT) on the transcription of *E. coli nusA*, an antitermination/termination factor. A 1-kb segment containing the promoter region for *nusA* was cut and amplified using the polymerase chain reaction. A reaction mix containing the amplified probe as a primer, RNA polymerase, and four ribonucleotides was placed into two tubes; one tube was placed in the control environment (no applied field), and the second tube was placed in a 45-Hz, 1.1-mT field. Following exposure for 1, 3, 5, and 7 min, 50-μl aliquots were removed from both the control and the experimental tubes and the reactions stopped. The data show that transcription in the EMF-exposed system was significantly enhanced relative to nonexposed controls at 5 and 7 min (Frederiksen *et al.*, 1993).

An *in vivo* study involving regenerating rat liver showed a small (about twofold) increase in the expression of the oncogenes c-*myc* and v-*ras* following PMF exposure (Battini *et al.*, 1991). In this report, the level of c-*myc* mRNA peaked at 3 hr and then decreased to control levels; in contrast, v-*ras* levels increased throughout the 40 hr of exposure period.

3. Proteins

Salivary glands from *S. coprophila* were subjected to a variety of pulsed and sinusoidal magnetic fields and then separated on two-dimensional (2-D) gels (R. Goodman and Henderson, 1988). The data obtained in these studies show both qualitative and quantitative differences in a number of proteins. Of interest is the fact that the investigators report both signal-specific and non-signal-specific changes in protein profiles following exposure. Using these data, Blank and Goodman (1989) conclude that EMF exposure results in an increase in lower molecular weight and an decrease in higher molecular weight proteins. More recently they have identified some of the changed proteins as similar to those stimulated by heat shock (Blank *et al.*, 1993). They speculate that the change in protein profiles might occur as a result of EMFs interfering with the translation process or by premature termination. They further suggest that both the frequency of the field and the charge on a protein will determine the extent of interference in the synthesis and ultimately the profiles of proteins affected by EMFs.

We have examined the effect of pulsed magnetic fields on the protein profiles of *E. coli*, using essentially the same approach as R. Goodman and Henderson (1988). Using an *E. coli* protein database (based on Neidhardt *et al.*, 1983) we were able to identify field-induced changes in the levels of 55 proteins from equilibrium gels and in 36 proteins from nonequilibrium gels. After closer examination of the data (Protein Data

Base, Inc., software, Huntington Station, NY) we decided to take a more conservative view of the actual differences. This decision was based in part on ambiguity in the software method of determining quantitative differences in regions of spot overlap. Of the proteins we did identify as being altered in the equilibrium gels, several are of physiological importance. These include the outer membrane proteins C and F (OMP-C and OMP-F), the α subunit of RNA polymerase, topoisomerase II (gyrase), and NusA, a termination/antitermination protein (E. M. Goodman *et al.*, 1994). Although other protein differences exist, these have attracted our attention because changes in their levels could produce the type of global effects on gene expression that have been observed when cells are exposed to EMFs. For example, a change in the level of gyrase would affect DNA supercoiling and subsequently enhance or depress transcription by making the promoter region more or less accessible to RNA polymerase. Further, genes that are expressing themselves in a cell would be more susceptible to this type of perturbation.

Having identified some of the proteins that change following field exposure, we examined the linkage map of *E. coli* to determine whether or not altered proteins were members of specific operons or regulons. We also examined the possibility that a specific replication time on the linkage map might be more susceptible to pulsed magnetic fields. No specific linkage periods, operons, or regulons appear to be uniquely susceptible to weak field exposure.

To a limited extent, we also attempted to confirm by independent techniques that proteins that appeared to be differentially expressed in the 2-D gel system were in fact altered. In these experiments, the same growth and exposure conditions were followed; *E. coli* were exposed before proteins were extracted and separated by polyacrylamide gel electrophoresis. Monoclonal antibodies to the α subunit of RNA polymerase and NusA were used to perform a Western blot for these proteins. The ratio of counts for both NusA and α polymerase from the 2-D gel were similar to the counts ratio obtained by excising the Western spots and counting them. Although counting the area defined by the Western blot leaves open the possibility that another protein may have comigrated with the protein of interest, the data do provide additional support for the suggestion that these specific proteins have been altered (see E. M. Goodman *et al.*, 1994).

Additional confirmation for an alteration in the expression of the α subunit has also been obtained with a cell-free transcription/translation system. In these experiments a cell-free expression system containing plasmids for the α_2, the $\beta\beta'$, or both subunits of DNA-dependent RNA polymerase from *E. coli* was exposed to 72-Hz sinusoidal magnetic fields of 1.1, 0.21, or 0.07 mT for periods ranging from 5 to 60 min (E. M.

Goodman *et al.*, 1993a). On the basis of incorporation of [^{35}S]methionine, we found evidence of elevated protein expression at each field intensity examined. All field intensities altered expression; however, a longer exposure was required to produce statistically significant alterations in protein levels if the field intensity was reduced to 0.07 mT. It is not immediately evident whether the increased levels of protein observed are a result of alterations in transcriptions, translation, or both. Alternatively, the stability of the mRNA might also be affected by field exposure.

Of added significance is the fact that increased expression for these proteins occurred in the absence of an intact cell membrane. These experiments demonstrate that an intact membrane is not an absolute requirement for magnetic bioeffects. It should be noted, however, that although an intact membrane is not present, membrane fragments may exist in the system. In spite of some inherent problems associated with the use of cell-free systems they do provide the tightly controlled environment required to address the question of where and how weak fields interact with biological systems. Although these systems offer a new approach for dissecting the mechanism of interaction between EMFs, additional *in vivo* experiments are required to establish the physiological relevance of these findings.

In a study designed to examine the effect of longer term exposure to EMFs, cells were allowed to proceed from a mitotic or cycling state to a noncycling (G_0) state (Rodemann *et al.*, 1989). In these experiments, normal human skin fibroblasts and transformed lung fibroblasts were subjected to a sinusoidal field of 20 Hz, 6 mT for 21 days. The cell medium was changed once a week; cells were exposed twice a day for 6 hr. Their data show that both the normal and transformed fibroblasts were stimulated to shift into nondividing postmitotic cells by the end of the 21-day exposure period. A concomitant 5- to 13-fold increase in the synthesis of total protein and collagen relative to nonexposed controls was also evident at the end of the exposure period. The effects of EMF exposure in transformed cells were more ambiguous: some subpopulations differentiated to nondividing postmitotic cells, whereas other populations were unaffected by field exposure.

D. Enzyme–Substrate Effects

1. Ornithine Decarboxylase

In general, few enzymes appear to respond immediately to an electromagnetic field. In light of the emerging data involving transcription and translation discussed above, one may speculate that the absence of an effect may be more related to selecting the wrong enzyme than to an inherent requirement for a latent response period. Of the enzymes investigated to date, studies of the response of ornithine decarboxylase (ODC) to EMFs are the most extensive (Byus *et al.*, 1987, 1988; Litovitz *et al.*, 1991). Byus *et al.* (1987) exposed human lymphoma (CEM) and mouse myeloma (P3) cells for 1 hr to 60-Hz, 10-mV/cm EMFs and induced a fivefold increase in activity in the former and a two- to threefold increase in the latter cell line. Hiraki *et al.* (1987) exposed rabbit costal chondrocytes to a pulsed magnetic field (0.2-mT bursts repeated at 15.4 Hz) for 48 hr and noted a 67% increase in ODC activity following parathyroid hormone treatment for 4 hr. When cells were exposed to pulsed magnetic fields but not treated with PTH, baseline levels of ODC were seen. If a Reuber H35 hepatoma was exposed for 1 hr to a lower field (0.1 mV/cm), a 30% decrease in ODC activity occurred (Byus *et al.*, 1988). In both experimental regimens, extending the exposure periods did not alter the response.

Similar results were obtained when Reuber H35, CHO, and 294T melanoma cells were exposed to a microwave field of 450 MHz, 1.0-mW/cm^2 amplitude modulated at 16 Hz (Byus *et al.*, 1988). In experiments exposing the H35 cells to the same carrier fields but amplitude modulated at 60 or 100 Hz, no effects were observed. The addition of the tumor promoter 12-*O*-tetradecanoylphorbol-13-acetate (TPA), a phorbal ester, coincidentally with field exposure gave a twofold enhancement in ODC activity beyond that seen with fields alone. In general, ODC activity increased following exposure, remained elevated for a period of 1–2 hr, and then returned to baseline levels. It is interesting that the response varies will cell type. The CEM and H35 hepatoma cells responded at the end of the 60-min exposure period, whereas the P3 cells showed enhanced ODC activity 1–2 hr after the exposure period. On the basis of the relatively long exposure period, one can probably conclude that the effects on ODC probably are not primary events in the response of cells to EMFs.

2. Protein Kinase C

The role of protein kinase C (PKC) as a possible intermediary in the EMF transductive process has also been receiving increasing attention. This interest is in part based on changes in the activity of protein kinase C with altered intracellular Ca^{2+} (Wolf *et al.*, 1985). As discussed above, one of the proposed mechanisms by which weak fields interact with cells is by altering the free Ca^{2+} concentration. Byus *et al.* (1984) reported a transient decrease in the activity of protein kinase C in lymphocytes after about 60 min of exposure to a 450-MHz, 1.0-mW/cm^2 field, amplitude modulated at 15, 40, and 60 Hz. In contrast, Monti *et al.* (1991) report enhanced activity of protein kinase C in HL-60 cells following exposure to a triangular magnetic field pulse (8-mT peak, 9-msec rise and fall followed by a 2-msec pause) for periods ranging from 10 to 20 min. The increased activity of protein kinase C remained constant for at least 60 min, in contrast to the transient decrease noted by Byus. Monti *et al.* also found that the field response was diminished in the presence of 2 m*M* ethylene glycol-bis(β-aminoethyl ether)-*N,N,N′,N′*-tetraacetic acid (EGTA), providing additional circumstantial evidence for a Ca^{2+} involvement. Although both studies used TPA to activate the system, only the latter measured ester-binding activity. Kavaliers *et al.* (1991) investigated the effect of a 60-Hz, 100-μT (rms) field on morphine-induced analgesia in the snail *Cepaea nemoralis*. In the presence of EMFs the level of morphine-induced analgesia was reduced; however, if a protein kinase C inhibitor was present the field effect decreased. One may infer that EMFs interfere with morphine analgesia through a protein kinase C mechanism: if PKC activity is inhibited, EMF effects are diminished.

3. cAMP

Luben *et al.* (1982; Luben, 1991) applied pulsed magnetic fields (induced *E* field, ~1.0 mV/cm) to cultured osteoblasts and examined their effect on the parathyroid hormone-induced level of cAMP. Although the parathyroid hormone-induced level of cAMP was reduced, fields did not directly affect the enzyme adenylate cyclase. They suggest that the fields reduced the cAMP level either by interfering with hormone–receptor interactions or through receptor–adenylate cyclase interactions in the plasma membrane. Brighton and McCluskey (1987) exposed fetal rat bone cells to a capacitively coupled electrical field [233 V (peak to peak), 60 kHz, calculated *E* field = 2.62 mV/cm] for 2.5 to 30 min and observed a 60% enhancement in cAMP levels. If parathyroid hormone was administered in the absence of an exogenous field, an approximate 36-fold enhancement in cAMP was detected. However, if parathyroid hormone was administered immediately after electrical stimulation the level of cAMP remained at control levels. Some variations in the level of cAMP were found, depending on the length of exposure and time of hormone administration.

An enhanced level of cAMP has also been reported in growth plate chondrocytes exposed for short time periods (2.5 and 5 min) to a capacitively coupled EMF [500 V (peak to peak) at 60 kHz] (Brighton and Townsend, 1988). If the exposure period was increased to 10 to 20 min, no significant differences were detected in the cAMP concentration. There was no attempt to determine if the increase was the result of increased adenylate cyclase activity or other factors. Hiraki *et al.* (1987) report a significant enhancement in the level of cAMP following administration of PTH (following a 96-hr exposure to PMF) to rabbit costal chondrocytes. In the absence of PTH, a slight decrease in cAMP occurred. An examination of electric (0.25 mA/cm^2) and magnetic field (2 mT) effects on cell proliferation by Schimmelpfeng and Dertinger (1993), discussed above, showed that both fields enhanced cAMP levels by 15–20% following a 5-min exposure, but not at 60 min. Exposing cells to a 4000-Hz, 0.1- to 0.6-mA/cm^2 field resulted in significant decreases of about 23% in cAMP levels (Knedlitschek *et al.*, 1993).

4. Myosin Light Chain Kinase

Markov *et al.* (1993) used a cell-free system to examine the effects of applied magnetic fields on the level of Ca^{2+} and calmodulin-dependent light chain phosphorylation. In their studies either a steady DC (0–200 μT, vertical and/or horizontal) or an atlernating (16 Hz, 20.9 μT) field was applied to isolated myosin light chains for periods ranging from 2 to 15 min. In these studies, the ambient magnetic field of the earth, as well as stray laboratory fields, were screened from the experimental system. When the vertical DC magnetic field was reduced below ambient (50 μT), using Mumetal shielding (Magnetic Shield Corp, Bensenville, IL), the rate of phosphorylation decreased 35–50% compared to control values. In contrast, when the vertical field was set to about two times the ambient field (110–140 μT) phosphorylation increased about 125%; increasing the fields to 150 μT or higher substantially enhanced the rate of phosphorylation to about 180% compared to controls. In addition to showing a distinct response in a cell-free environment, the data also provide additional evidence for the effectiveness of an applied magnetic field with an extremely low induced electric field. Furthermore, it is one of the few reports showing that DC fields are effective in inducing a bioresponse.

5. Acetylcholinesterase

Dutta *et al.* (1992) exposed growing neuroblastoma cells to 147-MHz rf fields modulated sinusoidally at 16Hz using specific absorption rates (SARs) of 0.001, 0.005, 0.01, 0.05, or 0.10 W/kg for 30 min. The cells were isolated, homogenized, and assayed for acetylcholinesterase (AChE) activity. Significantly enhanced AChE activity was observed at what appears to be a power density window between 0.02 and 0.05 W/kg. The results and field intensities of the latter experiment are consistent with the Ca^{2+} efflux experiments in which brain tissue was exposed to modulated radio-frequency fields (Bawin and Adey, 1976; Blackman *et al.*, 1980). Again the state of the cell appears to influence strongly the response to EMFs. For example, in a related study using the same fields and cell type, Verma *et al.* (1986) found that cells exposed to modulated rf fields for 1–2 hr after plating showed no change in AChE activity; the period of peak response appears to be 5.5–7.5 hr after plating.

6. Choline *O*-Acetyltransferase

In this study (Sakamoto *et al.*, 1993), parental rats were exposed to a 60-Hz, 50-μT, circularly polarized magnetic field and the levels of choline *O*-acetyltransferase were examined in embryos and neonates to determine potential developmental effects. No significant alterations in enzyme activity were found with either group as a result of EMF exposure.

7. ATP Synthesis and Related Enzymes

Using the yeast *S. cerevisiae*, Bolognani *et al.* (1992a) applied a pulsed magnetic field (1.8 mT) at 40, 80, and 120 Hz for 15 min and examined the effect on the adenine nucleotide pool. Their data show a significant increase in ATP levels at all frequencies, although 80 Hz was most effective. The levels of ADP and AMP were both elevated at 80 Hz; at 40 Hz no significant differences in either nucleotide were found, whereas only ADP showed elevated levels at 120 Hz. In a related study, Bolognani *et al.* (1992b) applied the same 80-Hz pulsed field to yeast and determined the amount of CO_2 evolved. Yeast exposed to EMFs showed a 27% enhancement in the microliters of CO_2 evolved per hour per 100 mg of yeast, compared to control cells.

In contrast, experiments on the haploid amoeboid state of the slime mold *Physarum* exposed long term (>120 days) to sinusoidal fields of 60 Hz, 0.1 mT, 1.0 V/m showed a depression of about 8% in ATP and approximately 20% in their Qo_2 (microliters of O_2 consumed per minute per milligram of protein) (Marron *et al.*, 1986). A comparison of the two studies is difficult because of the difference in both applied field intensities and length of exposure. In the slime mold study, cells were continuously exposed for about 4 months before significant differences in respiration or ATP levels were observed, as compared to 15 min in the yeast experiments.

Dachà *et al.* (1993) used human erythrocytes from donors 25–45 years of age and exposed them to a 50-Hz electric (0.1–20 kV/m), magnetic (20–200 μT), and mutually orthogonal electric and magnetic fields for up to 4 hr. The maximum electric field in the cell suspension was calculated to be ~2 mV/m (current density, 2 mA/m^2). Several key glycolytic enzymes were examined including hexokinase, glucose-6-phosphate 1-dehydrogenase, 6-phosphogluconic dehydrogenase, and phosphofructokinase in addition to adenine nucleotide levels and glucose consumption. No significant differences were observed in any of the indices examined.

Under low ionic strength conditions the cytochrome-*c* oxidase complex shows nonhyperbolic kinetics (non-Michaelis–Menten). This has led to the suggestion that the complex has two catalytic sites, low- and high-affinity sites that are involved in the binding and oxidation of cytochrome *c* (see Cooper, 1990). In the EMF experiments cytochrome-*c* oxidase was isolated from beef heart and the redox activity examined in the presence of either a DC or 50-Hz magnetic field at field intensities between 50 μT and 100 mT (Nossol *et al.*, 1993). A significant enhancement (90%) in activity in the high-affinity range was observed when the enzyme complex was exposed to static magnetic fields at 300 μT. Exposure to 50-Hz fields at 10 and 50 mT increased activity at the low-affinity site; at 300 μT (DC), a weak enhancement in activity in both the high- and low-affinity sites was seen.

8. Na^+, K^+-ATPase

Blank and Soo (1990) exposed Na^+, K^+-ATPase in suspension to an oscillating E field (10 Hz to 100 kHz) and showed that enzyme activity was diminished by the fields in the range of 0.05–50 mA/cm^2 (rms) in a frequency-dependent manner. By extrapolation, they suggest that the threshold E field across the enzyme suspension was about 5 μV/cm at a current density of 8 nA/cm^2 and 100 Hz. In contrast, enzyme activity was enhanced by fields in the presence of an inhibitor such as ouabain (Blank and Soo, 1989, 1992, 1993; Blank, 1992). They conclude that AC fields can either interfere with or enhance enzyme activation by affecting the ion concentrations immediately available to the enzyme.

E. Biorhythms and Hormones

The pineal gland in mammals is part of the visual system; pinealocytes of the gland are responsible for the synthesis of melatonin from serotonin (5-hydroxytryptamine). There are a number of reports suggesting that synthesis of melatonin and its precursors is altered in the presence of an electromagnetic field, although some other experiments have shown no effects (Wilson *et al.*, 1986; Welker *et al.*, 1983; Olcese *et al.*, 1985; Rudolph *et al.*, 1988; Lerchl *et al.*, 1990; Reiter and Richardson, 1992; Reiter, 1992). Melatonin has natural oncostatic activity on human estrogen receptor-positive cells (Cos *et al.*, 1991).

Liburdy *et al.* (1993) examined the effect of 60-Hz, 0.19- to 1.19-μT magnetic fields on cell proliferation of the human breast cancer cell line MCF-7 in the presence or absence of melatonin. When cells were grown without melatonin, 60-Hz magnetic fields had not effect on cell growth.

In contrast, cells grown in medium containing $10^{-9}M$ melatonin showed an 18% inhibition in growth when cells were exposed to a 24-μT field; this growth inhibition did not occur when cells were exposed to a magnetic field of 1.19 μT. These data are interpreted as showing that 60-Hz, 1.19-μT magnetic fields can interfere with the natural oncostatic activity of melatonin.

F. Genotoxic Effects

The question of whether genotoxic effects occur as a result of exposure to electromagnetic field has been of increased interest in view of the epidemiological data pointing to potential health hazards from exposure (Savitz *et al.*, 1988). Using mitogen-stimulated adult peripheral lympocytes, Rosenthal and Obe (1989) examined the effect of 50-Hz, 0.1- to 7.5-mT sinusoidal fields on sister chromatid exchange. The frequency of both chromosome breaks and chromosome exchange were not affected by field exposure. A similar study using a pulsed magnetic field at 1.0, 2.0, and 4.0 mT, applied as a quasirectangular pulse of 26 μsec, in 5-msec trains, repeated at 14 Hz, also showed no effect of sister chromatid exchange (Garcia-Sagredo *et al.*, 1990). A genotoxic study by Livingston *et al.* (1991) involved the application of 60-Hz electric fields (current densities of 3, 30, 300, and 3000 $\mu A/cm^2$ with a fixed magnetic field of 0.22 mT) to both CHO fibroblasts and phytohemagglutinin (PHA)-stimulated adult and newborn lymphocytes. No significant effects on sister chromatid exchange were found in any cell type or field examined. Cohen *et al.* (1986a,b) exposed human peripheral lymphocytes to 60-Hz, circularly polarized fields at an electric current density of 30 $\mu A/cm^2$ and/or magnetic fields at 0.1 and 0.2 mT. No statistically significant differences were observed in either the mitotic rate or chromosome breakage of lymphocytes exposed to electric, magnetic, or combined fields. Finally, micronuclei in lymphocytes were examined (Livingston *et al.*, 1991; Scarfi *et al.*, 1991, 1994) to determine if either electric or pulsed magnetic fields affect chromosomal breakage or induce missegregation. The micronuclear test is a sensitive genotoxic marker that "evaluates the frequency of small round bodies derived from chromosome fragments or whole chromosomes whose centromeres have lost their affinity for the mitotic spindle" (Scarfi *et al.*, 1991). Again, no effects were found in either study.

Frazier *et al.* (1990) examined the ability of lymphocytes exposed to ^{60}Co (5 Gy) to repair single-strand breaks in the presence of either combined or individual 60-Hz EMFs (0.05 to 1.0 mT and/or 0.2 to 20 V/m). The data indicate that the fields do not interfere or assist with the repair of single-strand breaks. A similar conclusion was reached by Cossarizza *et al.* (1989c) in a related study of chromosome repair.

In contrast to most of the studies described above, Garcia-Sagredo and Monteagudo (1991) report that lymphocytes from healthy donors subjected to 0.4-mT pulsed fields showed a statistically significant increase in chromosome breaks. However, in another study at the same intensities Garcia-Sagredo *et al.* (1990) reported no effect on sister chromosome exchanges. Khalil and Qassem (1991) exposed PHA-stimulted lymphocytes to a pulsed field (50 Hz, 1.05-mT peak; pulse duration, 10 msec) for 24, 48, and 72 hr. They found no significant alterations in the baseline frequency of sister chromatid exchange in the cultures exposed for 24 or 48 hr; in contrast, a significant increase was observed in cultures continuously exposed for 72 hr. In addition they also report a decrease in the rate of proliferation of cells exposed for 72 hr.

Cossarizza *et al.* (1989c) examined the question of whether electromagnetic fields interfere with DNA repair following γ irradiation. In these experiments lymphocytes were stimulated with PHA for 66 hr, subjected to 100 Gy of ^{60}Co radiation, pulsed with [^{3}H]thymidine, and subjected to pulsed EMFs for 6 hr. The latter 6-hr period corresponds to the normal period of radiation repair. Their data indicate that postexposure to EMFs did not interfere with the normal repair processes induced by γ irradiation.

To examine the mutagenicity of sinusoidal 50-Hz, 1- or 10-μT fields, Nafzinger *et al.* (1993) used the Ames test with *Salmonella typhimurium*. Mutations were determined by assaying for hypoxanthine–guanine phosphoribosyltransferase (HGPRT) activity in embryonic lung fibroblasts. In addition, Southern blot analysis was used to detect mutations in Epstein–Barr virus-transformed lymphoblastoid cells. All three tests failed to show any genotoxic effects of field exposure.

G. Generalized Cell Responses

The lymphocyte has been employed by numerous investigators seeking to understand weak field–cell interactions. As a result, both provocative and somewhat insightful results have emerged that might explain some of the disparate data and difficulty in reproducing effects. One of the most intriguing observations is that the age and state of the cells is a crucial determinant in their response to a field perturbation. Using the uptake of [^{3}H]thymidine as a marker, PHA-stimulated lymphocytes from young donors (ages 1–19 years) respond to pulsed magnetic field to a lesser extent than do cells from older donors (ages 46–55 years) (Cossarizza *et al.*, 1989a, 1991a,b). In another study using older donors and interleukin 2 (IL-2), this group reported data suggesting that the pulsed magnetic field enhanced utilization of IL-2 in older donor cells and enhanced their proliferation (Cossarizza *et al.*, 1989a). It remains to be determined whether the enhanced proliferative response also involves an increase in the plasma membrane IL-2 receptor. An enhanced response of PHA-stimulated lymphocytes to pulsed magnetic fields has been observed in cells obtained from individuals with lymphocytic leukemia (Emilia *et al.*, 1985). Similarly, enhanced field responses have also been observed in patients with Down's syndrome; these individuals possess an aging immune system as a result of a genetic dysfunction (Cossarizza *et al.*, 1991a). Although a lymphocyte fault or defect appears to enhance the field response to PHA stimulation, not all lymphocytes are responsive. For example, lymphocytes from patients with AIDS or testing seropositive did not show enhanced field responses (Cossarizza *et al.*, 1989b).

It is worth reiterating that Ca^{2+} influx is enhanced in lymphocytes and thymocytes exposed to pulsed magnetic fields, providing that the cells have first been stimulated with a mitogen (Emilia *et al.*, 1985; Walleczek and Liburdy, 1990; Lyle *et al.*, 1991). It also appears that aged lymphocytes or those defective in transducing exogenous chemical signals are more responsive to applied electromagnetic fields. Further, to observe a proliferative change in mitogen-stimulated cells, a relatively long field exposure (~6 hr) appears to be a necessary requirement (see review by Cadossi *et al.*, 1992). Ramoni *et al.* (1992) examined the effect of pulsed fields (50 Hz, 0.1 to 10 mT) on the cytotoxic activity of human natural killer cells and found no effect on the activity of these cells following exposure of the killer cells, target cells, or both to pulsed fields.

In a series of experiments in our laboratory we fused plasmodia from the slime mold *P. polycephalum* using mixtures of nonexposed controls and cells that had been exposed to EMFs of 72 Hz, 1.0 G, 1.0 V/m and which exhibited a lengthened mitotic cycle (E. M. Goodman *et al.*, 1988). In these fusion experiments the timing of mitosis was dependent on the relative volume of cytoplasm from the exposed cells. For example, if the fusion was a 50/50 mixture of nonexposed controls and exposed cells, the onset of metaphase was somewhat earlier than the time predicated by computing a simple average of the cell cycle lengths of the two unmixed cultures. If the mix was predominantly control cells (3:1) the length of the cell cycle was not significantly different from that of the control. In contrast, if the mix contained predominantly exposed cytoplasm (1:3) the length of the cell cycle was significantly shorter when compared to exposed plasmodia. At the time we speculated that the control may have contained some mitogenic factor in higher concentration than the EMF-exposed plasmodia. On the basis of advances in our understanding of cell division cycle (CDC) proteins (Pines, 1993), one may speculate that applied EMFs affect the *Physarum* cell cycle by interfering with the accumulation of CDC proteins.

Russell and Webb (1981) examined respiration in the insect larvae *Danaus archippus* before and after exposure to a 10- to 16-Hz magnetic field (0.5 μT). Similar experiments were also done on the yeast *S. cerevisiae* at higher frequencies (100–200 Hz). In both cases, a depression in the respiration rate (7.5% in larvae and 30% in yeast) was observed immediately following application of the field. It is noteworthy that the return to baseline levels occurred with 30 min regardless of whether or not the fields were present during the recovery period.

VI. Concluding Remarks

McLeod *et al.* (1992) examined the role of the biological state of an organism as a factor involved in its response to EMFs and put forward the hypothesis that a system that is quiescent or in a physiological steady state is less likely to respond to EMFs than one that is metabolically changing with time. They examined the state of the organism in several previously published experiments and found data supporting their hypothesis. Although the concept of the "physiological state" of an organism is difficult to define, these ideas are consistent with several of the points we have attempted to make regarding the need to place more emphasis on controlling the state and age of the cell culture being used for experimental purposes.

Current evidence suggests that cell processes can be influenced by weak electromagnetic fields. In examining the generally small response of cells to EMFs it is evident that fields represent a weak perturbation of the homeostatic state. Thus, unlike a chemical inhibitor that can irreversibly block a reaction, EMFs appear to represent a global interference or stress to which a cell can adapt without catastrophic consequences. There may be exceptions to this observation, however, such as the putative role of EMFs as promoters in the presence of a primary tumor initiator. Given the relatively weak intensities of the applied fields in most experiments, one would not expect a cell to respond at all; the nature of the response suggests that the cell is viewing EMFs as it would another subtle environmental change. The available data suggest that this small perturbation produces a small and measured reaction. However, the relation between the reaction and applied stimulus is neither simple nor linear. In many instances there is a threshold level below which no response is seen, followed by a response that rapidly reaches a saturation level, no longer increasing with further increasing field intensity or application time. In other instances there are frequency or intensity windows outside of which there is no response, or the response is in the opposite direction. Collectively, these effects and responses make model building extremely challenging.

As the number of experiments on EMF effects increases it is becoming increasingly evident that more than cursory consideration must be given to the cells being employed in the study. Many EMF experiments employ transformed rather than normal cells. One must question whether cells that are abnormal represent the best model systems for elucidating EMF interaction mechanisms. A better approach might be to use simpler, well-studied normal cells such as yeast or bacteria. The obvious advantage of employing these organisms to elucidate the transduction pathway(s) is that they are well characterized and, more importantly, an endless array of mutants is available to the investigator. Historically, the use of mutants has proved to be an essential tool for elucidation of cellular pathways.

Another potentially confounding factor emerging from EMF studies is that the age and state of the cell can profoundly affect the EMF bioresponse. This is seen, for example, in experiments with PHA-activated lymphocytes, in which the EMF response of the cell is directly correlated with the age of the donor. Related to these studies are other experiments indicating that cycling or metabolically active cells are more responsive to EMFs than quiescent ones.

The importance of subtle factors that affect EMF experiments is suggested in the data obtained from chick developmental studies discussed above. These data strongly indicate that eggs from a different egg-laying flock, although obtained from the same supplier, will differ in their responsiveness to EMF exposure. Eggs from a given breeding flock may either show susceptibility to a field in all experiments or show no effect at all. If correct, this might in part explain why the carefully replicated "Henhouse" experiment (Berman *et al.*, 1990) still showed differences between laboratories. In addition, it also reinforces the previously voiced caution that subtle and as yet undefined factors can impact on the final outcome of EMF experiments.

In examining the *in vivo* and *in vitro* response of cells to EMFs there appears to be a growing consensus among investigators that transcription is altered; direct posttranscriptional effects, if any, have yet to be determined. Although transcription alterations occur, no apparent disruption in routine physiological processes such as growth and division is immediately evident. What is usually observed is a transient perturbation followed by an adjustment by the normal homeostatic machinery of the cells. In view of the subtle response of cells to EMFs, it is essential that an investigator be aware of the following factors that can affect the outcome of replicative experiments.

1. Subtle differences in serum lots, concentrations, and sources can profoundly affect the growth of cells and their response of EMFs.
2. Using different cell strains even with apparently trivial differences must be carefully considered; aspects such as the cell density, growth conditions prior to use in an experiment, and the number of cells may impact on the final result.
3. Ambient EMFs from both the exposure apparatus and extraneous laboratory facilities, DC fields, and temperature have a pronounced, although subtle, effect on the outcome of an experiment. Although placing both control and exposed cells in the same chamber of an incubator has certain advantages, problems such as spillover fields and temperature gradients render this approach somewhat difficult.
4. On the basis of problems associated with exact replication of experiments, those attempting replication should spend time in the original investigator's laboratory to understand all of the subtleties of the experiment, prior to attempting replication in their own laboratory.

The data on EMF alterations of DNA replication are at best mixed; but in general, DNA does not appear to be significantly altered. If EMF exposure is associated with an increased risk of cancer, the absence of genotoxic effects would support the suggestion that the fields act in tumor promotion rather than initiation.

Finally, the site(s) and mechanisms of interaction remain to be elaborated. Although there are numerous studies and hypotheses that suggest the membrane represents the primary site of interaction, there are also several different studies showing that *in vitro* systems, including cell-free systems, are responsive to EMFs. The debate about potential hazards or therapeutic value of weak electromagnetic fields will continue until the mechanism has been clarified. As we have reiterated throughout this article, the attainment of this goal requires not only careful attention to detail in performing experiments, but is strongly dependent on the type and state of cells used by laboratories to address the question. The problem of how weak fields perturb cell function will be understood when the techniques of molecular biology, genetics, biochemistry, and biophysics are directed together to answer the question.

Acknowledgments

The authors acknowledge the National Institutes of Health and the Office of Naval Research for research support. We also acknowledge the excellent technical assistance of Pratima Tipness and Sandra Beyerlein.

References

Adair, R. (1991). Constraints on biological effects of weak extremely-low frequency electromagnetic fields. *Phys. Rev. A* **43,** 1039–1048.

Adair, R. (1992a). Criticism of Lednev's mechanism for the influence of weak magnetic fields on biological systems. *Bioelectromagnetics (N.Y.)* **13,** 231–235.

Adair, R. (1992b). Reply to "Comment on Constraints on biological effects of weak extremely-low-frequency electromagnetic fields." *Phys. Rev. A* **46,** 2185–2187.

Adair, R. (1994). Constraints of thermal noise on the effects of weak 60-Hz magnetic fields acting on biological magnetite. *Proc. Natl. Acad. Sci. U.S.A.* **91,** 2925–2929.

Adey, W. R. (1981). Tissue interactions with nonionizing electromagnetic fields. *Physiol. Rev.* **61,** 435–514.

Adey, W. R. (1989). The extracellular space and energetic hierarchies in electrochemical signalling between cells. *In* "Charge and Field Effects in Biosystems-2" (M. J. Allen, S. F. Cleary, and F. M. Hawkridge, eds.), pp. 263–290. Plenum, New York.

Adey, W. R. (1990). Nonlinear electrodynamics in cell membrane transductive coupling. *In* "Membrane Transport and Information Storage" (R. C. Aloria, C. C. Curtain, and L. M. Gordon, eds.), pp. 1–27. Alan R. Liss, New York.

Adey, W. R. (1993). Biological effects of electromagnetic fields. *J. Cell. Biochem.* **51,** 410–416.

Armstrong, P. F., Brighton, C. T., and Star, A. M. (1988). Capacitively coupled electrical stimulation of bovine growth plate chondrocytes grown in pellet form. *J. Orthop. Res.* **6,** 265–271.

Barnes, F. S. (1992). Some engineering models for interactions of electric and magnetic fields with biological systems. *Bioelectromagnetics (N.Y.)* **S1,** 67–86.

Bassen, H., Litovitz, T., Penafiel, M., and Meister, R. (1992). ELF *in vitro* exposure systems for inducing uniform electric and magnetic fields in cell culture media. *Bioelectromagnetics (N.Y.)* **13,** 183–198.

Battini, R., Monti, M. G., Moruzzi, M. S., Ferrari, S., Zaniol, P., and Barbiroli, B. (1991). ELF electromagnetic fields affect gene expression of regenerating rat liver following partial hepatectomy. *J. Bioelectr.* **10,** 131–139.

Bawin, S. M., and Adey, W. R. (1976). Sensitivity of calcium binding in cerebral tissue to weak environmental electric fields oscillating at low frequency. *Proc. Natl. Acad. Sci. U.S.A.* **73,** 1999–2003.

Bawin, S. M., Adey, W. R., and Sabbot, I. M. (1978). Ionic factors in release of $^{45}Ca^{2+}$ from chick cerebral tissue by electromagnetic fields. *Proc. Natl. Acad. Sci. U.S.A.* **75,** 6314–6318.

Bedlack, R. S., Jr., Wei, M., and Loew, L. M. (1992). Localized membrane depolarizations and localized calcium influx during electric field guided neurite growth. *Neuron* **9,** 393–403.

Berg, H. (1993). Electrostimulation of cell metabolism by low frequency electric and electromagnetic fields. *Bioelectrochem. Bioenerg.* **31,** 1–25.

Berman, E., Chacon, L., House, D., Koch, B. A., Leal, J., Lovstrup, S., Mantiply, E., Martin, A. H., Martucci, G. I., Mild, K. H., Monahan, J. C., Sandström, M., Shamsaifar, K., Tell, R., Trillo, M. A., Ubeda, A., and Wagner, P. (1990). Development of chick embryos in a pulsed magnetic field. *Bioelectromagnetics (N.Y.)* **11,** 169–187.

Blackman, C. F., Benane, S. G., Elder, J. A., House, D. E., Lampe, J. A., and Faulk, J. M. (1980). Induction of calcium-ion efflux from brain tissue by radiofrequency radiation: Effect of sample number and modulation frequency on the power-density window. *Bioelectromagnetics (N.Y.)* **1,** 35–43.

Blackman, C. F., Benane, S. G., House, D. E., and Joines, W. T. (1985a). Effects of ELF

(1-120 Hz) and modulated (50 Hz) RF fields on the efflux of calcium ions from brain tissue. *Bioelectromagnetics* (*N.Y.*) **6,** 1–11.

Blackman, C. F., Benane, S. G., Rabinowitz, J. R., House, D. E., and Joines, W. T. (1985b). A role for the magnetic field in the radiation-induced efflux of calcium ions from brain tissue *in vitro*. *Bioelectromagnetics* (*N.Y.*) **6,** 327–338.

Blackman, C. F., Benane, S. G., Elliott, D. J., House, D. E., and Pollock, M. M. (1988). Influence of electromagnetic fields on the efflux of calcium ions from brain tissue *in vitro:* A three-model analysis consistent with the frequency response up to 510 Hz. *Bioelectromagnetics* (*N.Y.*) **9,** 215–227.

Blackman, C. F., Kinney, L. S., House, D. E., and Joines, W. T. (1989). Multiple power-density windows and their possible origin. *Bioelectromagnetics* (*N.Y.*) **10,** 115–128.

Blackman, C. F., Benane, S. G., House, D. E., and Elliott, D. J. (1990). Importance of alignment between local DC magnetic field and an oscillating magnetic field in response of brain tissue *in vitro* and *in vivo*. *Bioelectromagnetics* (*N.Y.*) **11,** 159–167.

Blackman, C. F., Shawnee, G., Benane, S. G., and House, D. E. (1993a). Evidence for direct effect of magnetic fields on neurite outgrowth. *FASEB J.* **7,** 801–806.

Blackman, C. F., Benane, S. G., House, D. E., and Pollock, M. M. (1993b). Action of 50-Hz magnetic fields on neurite outgrowth in pheochromocytoma cells. *Bioelectromagnetics* (*N.Y.*) **14,** 273–286.

Blackman, C. F., Blanchard, J. P., Benane, S. G., and House, D. E. (1994). Empirical test of an ion parametric resonance model for magnetic field interactions with PC-12 cells. *Bioelectromagnetics* (*N.Y.*) **15,** 239–260.

Blanchard, J. P., and Blackman, C. F. (1994). Clarification and application of an ion parametric resonance model for magnetic field interactions with biological systems. *Bioelectromagnetics* (*N.Y.*) **15,** 217–238.

Blank, M. (1992). Na,K-ATPase function in alternating electric fields. *FASEB J.* **6,** 2434–2438.

Blank, M., and Goodman, R. (1988). An electrochemical model for the stimulation of biosynthesis by external electric fields. *Bioelectrochem. Bioenerg.* **19,** 569–580.

Blank, H., and Goodman, R. (1989). New and missing proteins in the electromagnetic and thermal stimulation of biosynthesis. *Bioelectrochem. Bioenerg.* **21,** 307–317.

Blank, M., and Soo, L. (1989). The effects of alternating currents on Na,K-ATPase function. *Bioelectrochem. Bioenerg.* **22,** 313–322.

Blank, M., and Soo, L. (1990). Ion activation of the Na,K-ATPase in alternating currents. *Bioelectrochem. Bioenerg.* **24,** 51–61.

Blank, M., and Soo, L. (1992). The threshold for alternating current inhibition of the Na,K-ATPase. *Bioelectromagnetics* (*N.Y.*) **13,** 329–333.

Blank, M., and Soo, L. (1993). The Na,K-ATPase as a model for electromagnetic field effects on cells. *Bioelectrochem. Bioenerg.* **30,** 85–92.

Blank, M., Soo, L., Lin, H., Henderson, A. S., and Goodman, R. (1992). Changes in transcription in HL-60 cells following exposure to alternating current from electric fields. *Bioelectrochem. Bioenerg.* **28,** 301–309.

Blank, M., Khorkova, O., and Goodman, R. (1993). Similarities in the proteins synthesized by *Sciara* salivary gland cells in response to electromagnetic fields and to heat shock. *Bioelectrochem. Bioenerg.* **31,** 27–38.

Bolognani, L., Del Monte, V., Francia, F., Venturelli, T., Volpi, N., and Costato, M. (1992a). Low-frequency electromagnetic pulsed field stimulation in yeast. *Electro- Magnetobiol.* **11,** 1–10.

Bolognani, L., Francia, F., Venturelli, T., Volpi, N., and Costato, M. (1992b). Fermentative activity of cold-stressed yeast and effect of electromagnetic pulsed fields. *Electro-Magnetobiol.* **11,** 11–17.

Borgens, R. (1982). What is the role of naturally produced electric current in vertebrate regeneration and healing. *Int. Rev. Cytol.* **76,** 245–298.

Bourguignon, G. J., and Bourguignon, L. Y. W. (1987). Electric stimulation of protein and DNA synthesis in human fibroblasts. *FASEB J.* **1,** 398–402.

Bracken, T. D. (1992). Experimental macroscopic dosimetry for extremely-low-frequency electric and magnetic fields. *Bioelectromagnetics (N.Y.)* **S1,** 15–26.

Brayman, A. A., Miller, M. W., Cox, C., Carstensen, E. L., and Schaedles, M. (1985). Absence of a 45 or 60 Hz electric field-induced respiratory effect in *Physarum polycephalum. Radiat. Res.* **104,** 242–261.

Brighton, C. T., and McClusky, W. P. (1987). Response of cultured bone cells to capacitively coupled electrical field: Inhibition of cAMP response to parathyroid hormone. *J. Orthop. Res.* **6,** 567–571.

Brighton, C. T., and Townsend, P. F. (1988). Increased c-AMP production after short-term capacitively coupled stimulation in bovine growth plate chondrocytes. *J. Orthop. Res.* **6,** 552–558.

Brighton, C. T., Jensen, L., Pollack, S. R., Tolin, B. S., and Clark, C. C. (1989). Proliferative and synthetic response of bovine growth plate chondrocytes to various capacitively coupled electrical fields. *J. Orthop. Res.* **7,** 759–765.

Byus, C. V., Lundak, R. L., Fletcher, R. M., and Adey, W. R. (1984). Alterations in protein kinase activity following exposure of cultured lymphocytes to modulated microwave fields. *Bioelectromagnetics (N.Y.)* **5,** 341–351.

Byus, C. V., Pieper, S. E., and Adey, W. R. (1987). The effects of low-energy 60-Hz environmental electromagnetic fields upon the growth-related enzyme ornithine decarboxylase. *Carcinogenesis (London)* **8,** 1385–1389.

Byus, C. V., Kartum, K., Pieper, S. E., and Adey, W. R. (1988). Ornithine decarboxylase activity in liver cells is enhanced by low-level amplitude modulated microwave fields. *Cancer Res.* **48,** 4222–4226.

Cadossi, R., Bersani, F., Cossarizza, A., Zucchini, P., Emilia, G., Torelli, G., and Franceschi, C. (1992). Lymphocytes and low frequency electromagnetic fields. *FASEB J.* **6,** 2667–2774.

Cain, C. D., Adey, W. R., and Luben, R. A. (1987). Evidence that pulsed electromagnetic fields inhibit coupling of adenylate cyclase. *J. Bone Miner. Res.* **2,** 437–441.

Cain, C. D., Thomas, D., and Adey, W. R. (1993). 60 Hz magnetic field acts as co-promoter in focus formation of C3H/10T1/2 cells. *Carcinogenesis (London)* **14,** 955–960.

Carson, J. J. L., Prato, F. S., Drost, D. J., Diesbourg, L. D., and Dixon, S. J. (1990). Time-varying magnetic fields increase cytosolic free Ca^{2+} in HL-60 cells. *Am. J. Physiol.* **28,** C687–C692.

Cohen, M. M., Kunska, A., Astemborski, J. A., and McCulloch, D. (1986a). The effect of low-level 60-Hz electromagnetic fields on human lymphoid cells. II. Sister-chromatid exchanges in peripheral lymphocytes and lymphoblastoid cell lines. *Muta. Res.* **172,** 177–184.

Cohen, M. M., Kunska, A., Astemborski, J. A., McCulloch, D., and Paskewitz, D. A. (1986b). Effect of low-level 60-Hz electromagnetic fields on human lymphoid cells. I. Mitotic rate and chromsome breakage in human peripheral lymphocytes. *Bioelectromagnetics (N.Y.)* **7,** 415–423.

Colacicco, G., and Pilla, A. (1983). Electromagnetic modulation of biological processes: Bicarbonate effect and mechanistic considerations in the Ca-uptake by embryonic chick tibia *in vitro. Z. Naturforsch.* **38,** 465–467.

Colacicco, G., and Pilla, A. (1984). Electromagnetic modulation of biological processes: Influence of culture media and significance of methodology in the Ca-uptake of embryonal chick tibia *in vitro. Calcif. Tissue Int.* **36,** 167–174.

Cook, M. R., Graham, C., Cohen, H., and Gerkovich, M. M. (1992). A replication study of human exposure to 60-Hz fields: Effects on neurobehavior measures. *Bioelectromagnetics (N.Y.)* **13,** 261–286.

Cooper, C. E. (1990). The steady state kinetics of cytochrome *c* oxidation by cytochrome oxidase. *Biochim. Biophys. Acta* **1017,** 187–203.

Cos, S., Blask, D. E., Lemus-Wilson, A., and Hill, A. B. (1991). Effects of melatonin on the cell cycle kinetics and "estrogen rescue" of MCF-7 human breast cancer cells. *J. Pineal Res.* **10,** 36–42.

Cossarizza, A., Monti, D., Bersani, F., Cantini, M., Cadossi, R., Sacchi, A., and Franceschi, C. (1989a). Extremely low frequency pulsed electromagnetic fields increase cell proliferation in lymphocytes from young and aged subjects. *Biochem. Biophys. Res. Commun.* **160,** 692–698.

Cossarizza, A., Monti, D., Bersani, F., Paganelli, R., Montagnani, G., Cadossi, R., Cantini, M., and Franceschi, C. (1989b). Extremely low frequency pulsed electromagnetic fields increase interleukin-2 (IL-2) utilization and IL-2 receptor expression in mitogen-stimulated human lymphocytes from old subjects. *FEBS Lett.* **248,** 141–144.

Cossarizza, A., Monti, D., Sola, P., Moschini, G., Cadossi, R., Bersani, F., and Franceschi, C. (1989c). DNA repair after γ irradiation in lymphocytes exposed to low-frequency pulsed electromagnetic fields. *Radiat. Res.* **118,** 161–168.

Cossarizza, A., Monti, D., Bersani, F., Cantini, M., Cadossi, R., Sacchi, A., and Franceschi, C. (1991a). Exposure to low-frequency pulsed electromagnetic fields increases mitogen-induced lymphocyte proliferation in Down's syndrome. *Aging* **3,** 241–246.

Cossarizza, A., Ortolani, C., Forti, E., Montagnani, G., Paganelli, R., Zanotti, M., Marini, M., Monti, D., and Franceschi, C. (1991b). Age-related expansion of functionally inefficient cells with markers of natural killer activity in Down's syndrome. *Blood* **77,** 1263–1270.

Dachà, M., Accorsi, A., Pierotti, C., Vetrano, F., Mantovani, R., Guido, G., Conti, R., and Nicolini, P. (1993). Studies on the possible biological effects of 50 Hz electric and/or magnetic fields: Evaluation of some glycolytic enzymes, glycolytic flux, energy and oxido-reductive potentials in human erythrocytes exposed *in vitro* to power frequency fields. *Bioelectromagnetics (N.Y.)* **14,** 383–391.

Delgado, M. R., Leal, J., Monteagudo, J. L., and Garcia, M. G. (1982). Embryological changes induced by weak, extremely low frequency electromagnetic fields. *J. Anat.* **134,** 533–551.

Deno, D. (1976). Transmission line fields. *IEEE Trans. Power Appar. Syst.* **PAS-95,** 1600–1611.

Dover, P. J., and McCaig, C. D. (1989). Enhanced development of striated myofibrils in *Xenopus* myoblasts cultured in an applied electric field. *Q. J. Exp. Physio. Cogn. Med. Sci.* **74,** 545–548.

Durney, C. H., Rushford, C. K., and Anderson, A. A. (1988). Resonant AC-DC magnetic fields: Calculated response. *Bioelectromagnetics (N.Y.)* **9,** 315–336.

Dutta, S. K., Subramaniam, A., Ghosh, B., and Parshad, R. (1984). Microwave radiation-induced calcium ion efflux from human neuroblastoma cells in culture. *Bioelectromagnetics (N.Y.)* **5,** 71–78.

Dutta, S. K., Ghosh, D. B., and Blackman, C. F. (1992). Dose dependence of acetylcholinesterase activity in neuroblastoma cells exposed to modulated radio-frequency electromagnetic radiation. *Bioelectromagnetics (N.Y.)* **13,** 317–322.

Edmonds, D. T. (1993a). Larmor precession as a mechanism for the detection of static and alternating magnetic fields. *Bioelectrochem. Bioenerg.* **30,** 3–12.

Edmonds, D. T. (1993b). Larmor precession as a mechanism for the biological sensing of small magnetic fields. *In* "Electricity and Magnetism in Biology and Medicine" (M. Blank, ed.), pp. 553–555. San Francisco Press, San Francisco.

Emilia, G., Torelli, G., Ceccherelli, G., Donelli, A., Ferrari, S., Zucchini, P., and Cadossi, R. (1985). Effect of low-frequency, low-energy pulsing electromagnetic fields on response to lectin stimulation of human normal and chronic lymphocyte leukemia lymphocytes. *J. Bioelectr.* **4,** 145–162.

Erickson, C. E., and Nuccitelli, R. (1984). Embryonic fibroblast motility and orientation can be influenced by physiological electric fields. *J. Cell Biol.* **98,** 296–307.

Farrell, J. M., Barber, M., Doinov, P., Krause, D., and Litovitz, T. A. (1993). Superposition of a temporally incoherent magnetic field suppresses the change in ornithine decarboxylase activity in developing chick embryos induced by a 60 Hz sinusoidal field. *In* "Electricity and Magnetism in Biology and Medicine" (M. Blank, ed.), pp. 342–344. San Francisco Press, San Francisco.

Feychting, M., and Ahlbom, A. (1993). Magnetic fields and cancer in children residing near Swedish high-voltage power lines. *Am. J. Epidemiol.* **138,** 467–481.

Fisher, S. J., Dulling, J., and Smith, S. D. (1986). Effect of a pulsed electromagnetic field on plasma membrane protein glycosylation. *J. Bioelect.* **5,** 253–267.

Fitzsimmons, R. J., Farley, J., Adey, W. R., and Baylink, D. J. (1986). Embryonic bone matrix formation is increased after exposure to a low amplitude capacitively coupled electric field *in vitro. Biochim. Biophys. Acta* **882,** 51–56.

Fitzsimmons, R. J., Farley, J., Adey, W. R., and Baylink, D. J. (1989). Frequency dependence of increased cell proliferation, *in vitro* in exposures to a low-amplitude, low-frequency electric field: Evidence for dependence on increased mitogen activity released into culture. *J. Cell. Physiol.* **139,** 586–591.

Fitzsimmons, R. J., Strong, D. D., Mohan, S., and Baylink, D. J. (1992). Low-amplitude, low-frequency electric field stimulated bone cell proliferation may in part be mediated by increased IGF-II release. *J. Cell. Physiol.* **150,** 84–89.

Foster, K. R. (1988). Health effects of low-level electromagnetic fields: Phantom or not-so-phantom risk. *Health Phys.* **62,** 429–435.

Frazier, M. E., Reese, J. A., Morris, J. E., Jostes, R. F., and Miller, D. L. (1990). Exposure of mammalian cells to 60-Hz magnetic or electric fields: Analysis of DNA repair of induced, single-strand breaks. *Bioelectromagnetics* (*N.Y.*) **11,** 229–234.

Frederiksen, J., Goodman, E. M., Greenebaum, B., and Marron, M. T. (1993). Altered RNA synthesis in a total synthetic system exposed to a sinusoidal magnetic field. *Bioelectromagn., 15th Annu. Meet,* Los Angeles, p. 45.

Galt, S., Sandbloom, and Hamnerius, Y. (1993a). Theoretical study of the resonant behavior of an ion confined to a potential well in a combination of AC and DC magnetic fields. *Bioelectromagnetics* (*N.Y.*) **14,** 299–314.

Galt, S., Sandbloom, J., Hamnerius, Y., Höjevik, P., Saalman, E., and Nordén, B. (1993b). Experimental search for combined AC and DC magnetic field effects on ion channels. *Bioelectromagnetics* (*N.Y.*) **14,** 315–327.

Garcia-Sagredo, J. M., Parado, L. A., and Monteagudo, J. L. (1990). Effect on SCE in human chromosomes *in vitro* of low-level pulsed magnetic fields. *Environ. Mol. Mutagen.* **16,** 185–188.

Garcia-Sagredo, J. M., and Monteagudo, J. L. (1991). Effect of low-level pulsed electromagnetic fields on human chromosomes *in vitro:* Analysis of chromosomal aberrations. *Hereditas* **115,** 9–11.

Goodman, E. M., Marron, M. T., and Greenebaum, B. (1979). Bioeffects of extremely low-frequency electromagnetic fields: Variance with intensity, waveform, and individual or combined electric and magnetic fields. *Radiat. Res.* **78,** 485–501.

Goodman, E. M., Marron, M. T., Sharpe, P. T., and Greenebaum, B. (1986). Pulsed magnetic fields alter the cell surface. *FEBS Lett.* **199,** 275–278.

Goodman, E. M., Marron, M. T., and Greenebaum, B. (1988). Electromagnetic energy and *Physarum. In* "Modern Bioelectricity" (A. Marino, ed.), pp. 393–425. Dekker, New York and Basel.

Goodman, E. M., Greenebaum, B., and Marron, M. T. (1993a). Altered protein synthesis in a cell-free system exposed to a sinusoidal magnetic field. *Biochim. Biophys. Acta* **1202,** 107–112.

Goodman, E. M., Greenebaum, B., and Frederiksen, J. (1993b). Effect of pulsed magnetic fields on human umbilical vein cells. *Bioelectrochem. Bioenerg.* **32,** 125–132.

Goodman, E. M., Greenbaum, B., and Marron, M. T. (1994). Magnetic fields alter translation in *E. coli. Bioelectromagnetics* (*N.Y.*) **15,** 77–84.

Goodman, R., and Henderson, A. S. (1986). Sine waves enhance cellular transcription. *Bioelectromagnetics* (*N.Y.*) **7,** 23–29.

Goodman, R., and Henderson, A. S. (1987). Stimulation of RNA synthesis in the salivary gland cells if *Sciara coprophila* by an electromagnetic signal used for treatment of skeletal problems in horses. *J. Bioelectr.* **6,** 37–47.

Goodman, R., and Henderson, A. S. (1988). Exposure of salivary glands to low-frequency electromagnetic fields alters polypeptide synthesis. *Proc. Natl. Acad. Sci. U.S.A.* **85,** 3928–3932.

Goodman, R., and Henderson, A. S. (1991). Transcription and translation in cells exposed to extremely low frequency electromagnetic fields. *Bioelectrochem. Bioenerg.* **25,** 335–355.

Goodman, R., Bassett, C. A. L., and Henderson, A. (1983). Pulsing electromagnetic fields induce cellular transcription. *Science* **220,** 1283–1285.

Goodman, R., Wei. L.-X., Xu, J.-C., and Henderson, A. S. (1989). Exposure of human cells to low-frequency electromagnetic fields results in quantitative changes in transcripts. *Biochim. Biophys. Acta* **1009,** 216–220.

Goodman, R., Weisbrot, D., Uluc, A., and Henderson, A. S. (1992a). Transcription in *Drosophila melanogaster* salivary gland cells is altered following exposure to low frequency electric and magnetic fields: Analysis of chromosomes 2R and 2L. *Bioelectrochem. Bioenerg.* **28,** 311–318.

Goodman, R., Weisbrot, D., Uluc, A., and Henderson, A. S. (1992b). Transcription in *Drosophila melanogaster* salivary gland cells is altered following exposure to low frequency electromagnetic fields: Analysis of chromosome 3R. *Bioelectromagnetics* (*N.Y.*) **13,** 111–118.

Goodman, R., Wei, L.-X., Bumann, J., and Henderson, A. S. (1992c). Exposure to electric and magnetic (EM) fields increases transcripts in HL-60 cells. Does adaptation to EM fields occur? *Bioelectrochem. Bioenerg.* **29,** 185–192.

Goodman, R., Bumann, J., Wei, L.-X., and Henderson, A. S. (1992d). Exposure of human cells to electromagnetic fields: Effect of time and field strength on transcript levels. *Electro-Magnetobiol.* **11,** 19–28.

Goodman, R., Chizmadzhev, Y., and Henderson, A. S. (1993). Electromagnetic fields and cells. *J. Cell. Biochem.* **51,** 436–441.

Grandolfo, M., Santini, M. T., Vecchia, P., Bonincontro, A., Camett, C., and Indovina, P. L. (1991). Non-linear dependence of the dielectric properties of chick embryo myoblast membranes exposed to sinusoidal 50 Hz magnetic field. *Int. J. Radiat. Biol.* **60,** 877–890.

Greene, J. J., Skowronski, W. J., Mullins, J. J., Nardone, R. M., Penafiel, M., and Meister, R. (1991). Delineation of electric and magnetic field effects of extremely low frequency electromagnetic radiation on transcription. *Biochem. Biophys. Res. Commun.* **174,** 742–749.

Greenbaum, B., Goodman, E. M., and Marron, M. T. (1975). Long-term effects of weak 45-75 Hz electromagnetic fields on the slime mold *Physarum polycephalum. USNS/URSI Annu. Meet.,* Boulder, CO, Vol. 1, pp. 449–459.

Grundler, W., Kaiser, F., Keilmann, F., and Walleczek, J. (1992). Mechanisms of electromagnetic information with cellular systems. *Naturwissenschaften* **79,** 551–559.

Gundersen, R. M., and Greenebaum, B. (1985). Low-voltage ELF electric field measurements in ionic media. *Bioelectromagnetics* (*N.Y.*) **6,** 157–168.

Gundersen, R. M., Greenebaum, B., and Schaller, M. (1986). Intracellular recording during magnetic field application to monitor neurotransmitter release events: Methods and preliminary results. *Bioelectromagnetics* (*N.Y.*) **7,** 271–282.

Guzelsu, N., Salkind, A. J., Shen, X., Patel, U., Thaler, S., and Berg, R. A. (1994). Effect of electromagnetic stimulation with different waveforms on cultured chick tendon fibroblasts. *Bioelectromagnetics* (*N.Y.*) **15,** 115–131.

Halle, B. (1988). On the cyclotron resonance mechanism for magnetic field effects on transmembrane ion conductivity. *Bioelectromagnetics* (*N.Y.*) **9,** 381–385.

Hamada, S. H., Witkus, R., and Griffith, R., Jr. (1989). Cell surface changes during electromagnetic field exposure. *Exp. Cell Biol.* **57,** 1–10.

Hart, F. X. (1992). Numerical and analytical methods to determine current density distributions produced in human and rat models by electric and magnetic fields. *Bioelectromagnetics* (*N.Y.*) **S1,** 27–42.

Hiraki, Y., Endo, N., Takigawa, M., Asada, A., Takahshi, H., and Suzuki, F. (1987). Enhanced responsiveness to parathyroid hormone and induction of functional differentiation of cultured rabbit costal chondrocytes by a pulsed electromagnetic field. *Biochim. Biophys. Acta* **931,** 94–100.

Iannacone, W. M., Pienkowski, D., Pollack, S. R., and Brighton, C. D. (1988). Pulsing electromagnetic field stimulation of the *in vitro* growth plate. *J. Orthop. Res.* **6,** 239–247.

Jaffe, L., and Poo, M. (1979). Neurites grow faster towards the cathode than the anode in a steady field. *J. Exp. Zool.* **209,** 115–128.

Jolley, W. B., Hinshaw, D. B., Knierim, K., and Hinshaw, D. B. (1983). Magnetic field effects on calcium efflux and insulin secretion in isolated rabbit Islets of Langerhans. *Bioelectromagnetics* (*N.Y.*) **4,** 103–105.

Kaune, W. T. (1992). Macroscopic dosimetry of power-frequency electric and magnetic fields. *Bioelectromagnetics* (*N.Y.*) **S1,** 11–14.

Kaune, W. T., King, A. J., Hungate, J., and Causey, S. C. (1984). System for the exposure of cell suspensions to power-frequency electric fields. *Bioelectromagnetics* (*N.Y.*) **5,** 117–129.

Kavaliers, M., Ossenkopp, K.-P., and Tysdale, D. M. (1991). Evidence for the involvement of protein kinase C in the modulation of morphine induced 'analgesia' and the inhibitory effects of exposure to 60-Hz magnetic fields in the snail *Cepaea nemoralis*. *Brain Res.* **554,** 65–71.

Khalil, A. M., and Qassem, W. (1991). Cytogenetic effects of pulsing electromagnetic field on human lymphocytes *in vitro:* Chromosome aberrations, sister-chromatid exchanges and cell kinetics. *Mut. Res.* **247,** 141–146.

Kirschvink, J. L. (1992a). Magnetite biomineralization in the human brain. *Proc. Natl. Acad. Sci. U.S.A.* **89,** 7683–7687.

Kirschvink, J. L. (1992b). Uniform magnetic fields and double-wrapped coil systems: Improved techniques for the design of bioelectromagnetic experiments. *Bioelectromagnetics* (*N.Y.*) **13,** 401–411.

Kirschvink, J. L. (1992c). Comment on "Constraints on biological effects of weak extremely low-frequency electromagnetic fields." *Phys. Rev. A* **46,** 2178–2184.

Kirschvink, J. L., Kobayashi-Kirschvink, A., Diaz-Ricci, J. C., and Kirschvink, S. J. (1992). Magnetite in human tissues: A mechanism for the biological effects of weak ELF magnetic fields. *Bioelectromagnetics* (*N.Y.*) **S1,** 101–114.

Knedlitschek, G., Nagy, M. N., Dertinger, H., and Schimmelpfeng, J. (1993). The action of alternating currents of different frequencies upon cyclic AMP and cell proliferation. *Bioelectromagn., 15th Annu. Meet.* Los Angeles, p. 97.

Kraus, J. D. (1984). "Electromagnetics," 3rd ed. McGraw-Hill, New York.

Krause, D., Skrowronski, W. J., Mullins, J. M., Nardone, R. M., and Greene, J. J. (1991). Selective enhancement of gene expression by 60 Hz electromagnetic radiation. *In* "Electromagnetics in Biology and Medicine" (C. T. Brighton and S. R. Pollack, eds.), pp. 133–138. San Francisco Press, San Francisco.

Lednev, V. V. (1991). Possible mechanism for the influence of weak magnetic fields on biological systems. *Bioelectromagnetics* (*N.Y.*) **12,** 71–76.

Lednev, V. V. (1993). Possible mechanism for the effect of weak magnetic fields on biological systems: Correction of the basic expression and its consequences. *In* "Electricity and Magnetism in Biology and Medicine" (M. Blank, ed.), pp. 550–552. San Francisco Press, San Francisco.

Lerchl, A., Nonaka, K. O., Stokkan, K.-A., and Reiter, R. J. (1990). Marked rapid alterations in nocturnal pineal serotonin metabolism in mice and rats exposed to weak intermittent magnetic fields. *Biochem. Biophys. Res. Commun.* **169,** 102–108.

Liboff, A. R. (1985). Cyclotron resonance in membrane transport. *In* "Interactions between Electromagnetic Fields and Cells" (A. Chiabrara, C. Nicolini, and H. P. Schwan, eds.), pp. 281–296. Plenum, London.

Liboff, A. R., and McLeod, B. R. (1988). Kinetics of channelized membrane ions in magnetic fields. *Bioelectromagnetics* (*N.Y.*) **9,** 39–51.

Liboff, A. R., and Parkinson, W. C. (1991). Search for ion-cyclotron resonance in a Na^+-transport system. *Bioelectromagnetics* (*N.Y.*) **12,** 77–83.

Liboff, A. R., Williams, T., Jr., Strong, D. M., and Wistar, R., Jr. (1984). Time-varying magnetic fields: Effect on DNA synthesis. *Science* **223,** 818–820.

Liboff, A. R., Rozek, R. J., Sherman, M. R., McLeod, B. R., and Smith, S. D. (1987). $^{45}Ca^{++}$ cyclotron resonance in human lymphocytes. *J. Bioelectr.* **6,** 13–22.

Liburdy, R. P. (1992). Calcium signalling in lymphocytes and ELF fields: Evidence for an electric field metric and a site of interaction involving the calcium ion channel. *FEBS Lett.* **301,** 53–59.

Liburdy, R. B., Sloma, T. R., Sokolić, R., and Yaswen, P. (1993). ELF magnetic fields, breast cancer, and melatonin: 60 Hz fields block melatonin's oncostatic action on breast cancer cell proliferation. *J. Pineal Res.* **14,** 89–97.

Lin, H., Goodman, R., and Shirley-Henderson, A. (1994). Specific region of the *c-myc* promoter is responsive to electric and magnetic fields. *J. Cell. Biochem.* **54,** 281–288.

Lindström, E., Lindström, P., Berglund, A., Mild, K. H., and Lundgren, E. (1993). Intracellular calcium oscillations induced in a T-cell line by weak 50 Hz magnetic field. *J. Cell. Physiol.* **156,** 395–398.

Lin-Liu, S., and Adey, W. R. (1982). Low frequency amplitude modulated microwave fields change calcium efflux rates from synaptosomes. *Bioelectromagnetics* (*N.Y.*) **3,** 309–322.

Lin-Liu, S., Adey, W. R., and Poo, M.-M. (1984). Migration of Cell-surface concanavalin A receptors in pulsed electric fields. *Biophys. J.* **45,** 1211–1218.

Litovitz, T. A., Mullins, J. M., and Krause, D. (1991). Effect of coherence time of the applied magnetic field on ornithine decarboxylase activity. *Biochem. Biophys. Res. Commun.* **178,** 862–865.

Litovitz, T. A., Montrose, C. J., and Wang, W. (1992). Dose-response implications of the transient nature of electromagnetic-field-induced bioeffects: Theoretical hypothesis and predictions. *Bioelectromagnetics* (*N.Y.*) **S1,** 237–246.

Litovitz, T. A., Farrell, J. M., Krause, D., Doinov, P., and Montrose, C. J. (1993a). Superimposing electromagnetic noise blocks the alteration of ornithine decarboxylase activity in developing chick embryos caused by a weak 60-Hz sinusoidal field. *Bioelectromagn., 15th Annu. Meet.,* Los Angeles, p. 95.

Litovitz, T. A., Krause, D., Penafiel, M., Elson, E. C., and Mullins, J. M. (1993b). The role of coherence time in the effect of microwaves on ornithine decarboxylase activity. *Bioelectromagnetics* (*N.Y.*) **14,** 395–403.

Litovitz, T. A., Montrose, C. J., and Doinov, P. (1993c). Spatial and temporal coherence affects the response of biological systems to electromagnetic fields. *In* "Electricity and Magnetism and Biology and Medicine" (M. Blank, ed.), pp. 339–341. San Francisco Press, San Francisco.

Litovitz, T. A., Montrose, C. J., Doinov, P., Brown, K. M., and Barber, M. (1994). Superimposing spatially coherent electromagnetic noise inhibits field-induced abnormalities in developing chick embryos. *Bioelectromagnetics* (*N.Y.*) **15,** 105–113.

Livingston, G. K., Witt, K. L., Gandhi, O. P., Chatterjee, I., and Roti Roti, J. L. (1991). Reproductive integrity of mammalian cells exposed to power frequency electromagnetic fields. *Environ. Mol. Mutagen.* **17,** 49–58.

Loew, L. M. (1992). Voltage-sensitive dyes: Measurement of membrane potentials induced by DC and AC electric fields. *Bioelectromagnetics* (*N.Y.*) **13** (S1), 179–190.

London, S. J., Thomas, D. C., Bowman, J. D., Sobel, E., Cheng, T.-C., and Peters, J. M. (1991). Exposure to residential electric and magnetic fields and risk of childhood leukemia. *Am. J. Epidemiol.* **134,** 923–937.

Luben, R. A. (1991). Effects of low-energy electromagnetic fields (pulsed and DC) on membrane signal transduction processes in biological systems. *Health Phys.* **61,** 15–28.

Luben, R. A., Cain, C. D., Chen, M. C. Y., Rosen, D. M., and Adey, W. R. (1982). Effects of electromagnetic stimuli on bone and bone cells *in vitro:* Inhibition of responses to parathyroid hormone by low-energy low-frequency fields. *Proc. Natl. Acad. Sci. U.S.A.* **79,** 4180–4184.

Lyle, D. B., Wang, X., Ayotte, R. D., Sheppard, A. S., and Adey, W. R. (1991). Calcium uptake by leukemic and normal T-lymphocytes exposed to low frequency magnetic fields. *Bioelectromagnetics* (*N.Y.*) **12,** 145–156.

Lyle, D. B., Doshi, J., Fuchs, T. A., Casamento, J. P., Sei, Y., and Swicord, M. L. (1992). Intracellular calcium signalling by human T-leukemic cells exposed to an induced 1 mV/cm 60 Hz, sinusoidal electric field. *World Congr. Electr. Magn. Biol. Med. 1st,* Orlando, FL, 1992, p. 13.

Malorni, W., Paradisi, S., Straface, E., Santini, M. T., and Donelli, G. (1992). An *in vitro* investigation on the subcellular effects of 50 Hz magnetic fields. *Proc. IRPA Int. Conf.* Montreal, Canada, May 17–22.

Markov, M. S., Wang, S., and Pilla, A. A. (1993). Effects of weak low frequency sinusoidal and DC magnetic fields on myosin phosphorylation in a cell-free preparation. *Bioelectrochem. Bioenerg.* **30,** 119–125.

Marron, M. T., Goodman, E. M., and Greenebaum, B. (1978). Effects of weak electromagnetic fields on *Physarum polycephalum:* Mitotic delay in heterokaryons and decreased respiration. *Experientia* **34,** 589–590.

Marron, M. T., Greenbaum, B., Swanson, J. E., and Goodman, E. M. (1983). Cell surface effect of 60-Hz electromagnetic fields. *Radiat. Res.* **94,** 217–220.

Marron, M. T., Goodman, E. M., Greenebaum, B., and Tipnis, P. (1986). Effects of sinusoidal 60-Hz electric and magnetic fields on ATP and oxygen levels in the slime mold *Physarum polycephalum. Bioelectromagnetics* (*N.Y.*) **7,** 307–314.

Marron, M. T., Goodman, E. M., Sharpe, P. T., and Greenebaum, B. (1988). Low frequency electric and magnetic fields have different effects on the cell surface. *FEBS Lett.* **230,** 13–16.

McCaig, C. D., and Dover, P. J. (1989). On the mechanism of oriented myoblast differentiation in an applied electric field. *Biol. Bull.* (*Woods Hole, Mass.*) **176,** 140–144.

McLauchlan, K. (1992). Are environmental magnetic fields dangerous? *Phys. World,* January, pp. 41–45.

McLeod, B. R., Pilla, A. A., and Sampsel, M. W. (1983). Electromagnetic fields induced by Helmholtz aiding coils inside saline-filled boundaries. *Bioelectromagnetics* (*N.Y.*) **4,** 357–370.

McLeod, B. R., Liboff, A. R., and Smith, S. D. (1992). Biological systems in transition: Sensitivity to extremely low-frequency fields. *Electro- Magnetobiol.* **11,** 29–42.

McLeod, K. J. (1992). Microelectrode measurements of low frequency electric field effects in cells and tissues. *Bioelectromagnetics (N.Y.)* **S1,** 161–178.

McLeod, K. J., Lee, R. C., and Ehrlich, H. P. (1987). Frequency dependence of electric field modulation of fibroblast protein synthesis. *Science* **236,** 1465–1468.

Miller, D. L., Miller, M. C., and Kaune, W. T. (1989). Addition of magnetic field capability to existing extremely-low-frequency electric field exposure systems. *Bioelectromagnetics (N.Y.)* **10,** 85–89.

Misakian, M., Sheppard, S. R., Krause, D., Frazier, M. E., and Miller, D. L. (1993). Biological, Physical and electrical parameters for *in vitro* studies with ELF magnetic and electric fields: A primer. *Bioelectromagnetics (N.Y.)* **S2,** 1–73.

Monti, M. G., Pernecco, L., Moruzzi, M. S., Battini, R., Zaniol, P., and Barbiroli, B. (1991). Effect of ELF pulsed electromagnetic fields on protein kinase C activation processes in HL-60 leukemia cells. *J. Bioelectr.* **12,** 119–130.

Mullins, J. M., Krause, D., and Litovitz, T. (1993). Simultaneous application of a spatially coherent noise field blocks response of cell cultures to a 60 Hz electromagnetic field. *In* "Electricity and Magnetism in Biology and Medicine" (M. Blank, ed.), pp. 345–346. San Francisco Press, San Francisco.

Mullins, R. D., Sisken, J. E., Ejase, H. A. N., and Sisken, B. F. (1993). Design and characterization of a system for exposure of cultured cells to extremely low frequency electric and magnetic fields over a wide range of field strengths. *Bioelectromagnetics (N.Y.)* **14,** 173–186.

Murray, J. C., and Farndale, R. W. (1985). Modulation of collagen production in cultured fibroblasts by a low-frequency pulsed magnetic field. *Biochim. Biophys. Acta* **838,** 98–105.

Nafzinger, J., Desjobert, H., Benamar, B., Guillosson, J., and Adolphe, M. (1993). DNA mutations and 50 Hz electromagnetic fields. *Bioelectrochem. Bioenerg.* **30,** 133–141.

Neidhardt, F. C., Vaughn, V., Phillips, T. A., and Bloch, P. L. (1983). Gene protein index of *Escherichia coli* K-12, Edition 7. *Microbiol. Rev.* **47,** 231–284.

Nossol, B., Buse, G., and Silny, J. (1993). Influence of weak static and 50 Hz magnetic fields on the redox activity of cytochrome-c oxidase. *Bioelectromagnetics (N.Y.)* **14,** 361–371.

Olcese, J., Reuss, S., and Vollrath, L. (1985). Evidence for the involvement in the visual system in mediating magnetic field effects on pineal melatonin synthesis in the rat. *Brain Res.* **333,** 382–384.

Paradisi, S., Donelli, G., Santini, M. T., Straface, E., and Malorni, W. (1993). A 50-Hz magnetic field induces structural and biophysical changes in membranes. *Bioelectromagnetics (N.Y.)* **14,** 247–255.

Parker, J. E., and Winters, W. (1992). Expression of gene-specific RNA in cultured cells exposed to rotating 60-Hz magnetic fields. *Biochem. Cell Biol.* **70,** 237–241.

Parkinson W. C., and Sulik, G. L. (1992). Diatom response to extremely low frequency magnetic fields. *Radiat. Res.* **130,** 319–330.

Parola, A. H., Porat, N., and Kiesow, L. A. (1993). Chicken embryo fibroblasts exposed to weak, time-varying magnetic fields share cell proliferation, adenosine deaminase activity, and membrane characteristics of transformed cells. *Bioelectromagnetics (N.Y.)* **14,** 215–228.

Patel, N. B., and Poo, M. (1982). Orientation of neurite growth by extracellular electric fields. *J. Neurosci.* **2,** 483–496.

Phillips, J. L. (1993). Effects of electromagnetic field exposure to gene transcription. *J. Cell. Biochem.* **51,** 381–386.

Phillips, J. L., and McChesney, L. (1991). Effect of 72 Hz pulsed magnetic field exposure on macromolecular synthesis in CCRF-CEM cells. *Cancer Biochem. Biophys.* **12,** 1–7.

Philips, J. L., Winters, W. L., and Rutledge, L. (1986a). *In vitro* exposure to electromagnetic fields: Changes in tumor cell properties. *Int. J. Radiat. Biol.* **49,** 463–469.

Phillips, J. L., Rutledge, L., and Winters, W. D. (1986b). Transferrin binding to two human colon carcinoma cell lines: Characterization and effect of 60 Hz electromagnetic fields. *Cancer Res.* **46,** 239–244.

Phillips, J. L., Haggren, W., Thomas, W. J., Ishida-Jones, T., and Adey, W. R. (1992). Magnetic field-induced changes in specific gene transcription. *Biochim. Biophys. Acta* **1132,** 140–144.

Pilla, A. A., Figueiredo, M., Nasser, P. R., Kaufman, J. J., Siffert, R. S., and Erickson, J. (1993). Broadband EMF acceleration on bone repair in a rabbit model is independent of magnetic component. *In* "Electricity and Magnetism in Biology and Medicine" (M. Blank, ed.), pp. 363–367. San Francisco Press, San Francisco.

Pines, J. (1993). Cyclins and their associated cyclin-dependent kinases in the human cell cycle. *Trends in Biochem. Sci.* **210,** 195–197.

Pollack, S. M., Reinbold, K. A., and Da Silva, O. L. (1992). Changes in the cytosolic calcium concentration of primary bone cell cultures due to electric fields at 10 mV/cm from 6 kHz to 600 kHz. *World Congr. Electr. Magn. Biol. Med., 1st,* Orlando FL, *1992,* p. 12.

Postow, E., and Swicord, M. L. (1989). Modulated fields and "window" effects. *In* "CRC Handbook of Biological Effects of Electromagnetic Fields" (C. Polk and E. Postow, eds.), pp. 426–460. CRC Press, Boca Raton, FL.

Ramoni, C., Dupuis, M. L., Grandolfo, M., Polichetti, A., and Vecchia, P. (1992). Modulation of the functional activity of NK cells by means of a sinusoidally varying magnetic field at 5-0 Hz. *Phys. Med.* **10,** 90–91.

Reiter, R. J. (1992). Changes in circadian melatonin synthesis in the pineal gland of animals exposed to extremely low frequency electromagnetic radiation: A summary of observations and speculation on their implications. *In* "Electromagnetic FIelds and Circadian Rhythmicity" (M. C. Moore-Ede, S. S. Campbell, and R. J. Reiter, eds.), pp. 13–27. Birkhauser, Boston.

Reiter R. J., and Richardson, B. A. (1992). Magnetic field effects on pineal indoleamine metabolism and possible biological consequences. *FASEB J.* **6,** 2283–2287.

Rodan, G. A., Bourret, L. A., and Norton, L. A. (1978). DNA synthesis in cartilage cells is stimulated by oscillating electric fields. *Science* **199,** 690–692.

Rodemann, H. P., Bayreuther, K., and Pfleiderer, G. (1989). The differentiation of normal and transformed human fibroblasts *in vitro* is influenced by electromagnetic fields. *Exp. Cell Res.* **182,** 610–621.

Rosenthal, M., and Obe, G. (1989). Effects of 50-Hz electromagnetic fields on proliferation and on chromosomal alterations in human peripheral lymphocytes untreated or pretreated with chemical mutagens. *Mutat. Res.* **210,** 329–335.

Rudolph, K., Wirtz-Justice, A., Krauchli, K., and Feer, H. (1988). Static magnetic fields decrease nocturnal pineal cAMP in rat brain. *Brain Res.* **446,** 159–160.

Russell, D. N., and Webb, S. J. (1981). Metabolic response of *Danaüs archippus* and *Saccharomyces cerevisiae* to weak oscillatory magnetic fields. *Int. J. Biometeorol.* **25,** 257–262.

Sakamoto, S., Hagino, N., and Winters, W. D. (1993). *In vivo* studies of the effect of magnetic field exposure on ontogeny of choline acetyltransferase in the rat brain. *Bioelectromagnetics* (*N.Y.*) **14,** 373–381.

Sandweiss, J. (1990). On the cyclotron resonance model of ion transport. *Bioelectromagnetics* (*N.Y.*) **11,** 203–205.

Savitz, D. A., Wachtel, H., Barnes, F., John, E. M., and Tvrdik, J. G. (1988). Case-control study of childhood cancer and exposure to 60 Hz magnetic fields. *Am. J. Epidemiol.* **128,** 21–38.

Scarfi, M. R., Bersani, F., Cossarizza, A., Monti, D., Castellani, G., Cadosi, R., Franceschetti, G., and Franceschi, C. (1991). Micronuclei formation in human lymphocytes exposed to 50 Hz AC electric fields. *Biochem. Biophys. Res. Commun.* **176,** 194–200.

Scarfi, M. R., Lioi, M. B., Zeni, O., Franceschetti, G., Franceschi, C., and Bersani, F. (1994). Lack of chromosomal aberration and micronucleus induction in human lymphocytes exposed to pulsed magnetic fields. *Muta. Res.* **306,** 129–133.

Schimmelpfeng, J., and Dertinger, H. (1993). The action of 50 Hz magnetic and electric fields upon cell proliferation and cyclic AMP content of cultured mammalian cells. *Bioelectrochem. Bioenerg.* **30,** 143–150.

Serpersu, E. H., and Tsong, T. Y. (1983). Stimulation of a ouabain-sensitive Rb^+ uptake in human erythrocytes with an external field. *J. Membr. Biol.* **74,** 191–201.

Serpersu, E. H., and Tsong, T. Y. (1984). Activation of electrogenic Rb^+ transport of (Na,K)-ATPase by an electric field. *J. Biol. Chem.* **259,** 7155–7162.

Smith, O. M., Goodman, E., Greenebaum, B., and Tipnis, P. (1991). An increase in the negative surface charge of U937 cells exposed to a pulsed magnetic field. *Bioelectromagnetics (N.Y.)* **12,** 197–202.

Smith, S. D., McLeod, B. R., Liboff, A. R., and Cooksey, K. (1987). Calcium cyclotron resonance and diatom mobility. *Bioelectromagnetics (N.Y.)* **8,** 215–228.

Stagg, R. B., Hardy, P. T., MacMurray, A., and Adey, W. R. (1992). Electric and magnetic field interactions with microsomal membranes: A novel system for studying calcium flux across membranes. *World Cong. Elect. Magn. Biol. Med. 1st,* Orlando, FL, 1992, p. 12.

Stegemann, S., Altman, K. I., Muhlensiepen, H., and Feinendegen, L. E. (1993). Influence of a stationary magnetic field on acetylcholinesterase in murine bone cells. *Radiat. Environ. Biophys.* **32,** 65–72.

Stump, R. F., and Robinson, K. R. (1983). *Xenopus* neural crest cell migration in an applied electrical field. *J. Cell Biol.* **97,** 1226–1233.

Sustachek, J. (1992). *In vivo* and *in vitro* transcriptional effects of electromagnetic field exposure on *E. coli* DNA-dependent RNA polymerase. Master's Thesis, University of Wisconsin-Parkside.

Takahashi, K., Kaneko, I., Date, M., and Fukada, E. (1986). Effects of pulsing electromagnetic fields on DNA synthesis in mammalian cells in culture. *Experientia* **43,** 331–332.

Teissie, J., and Tsong, T. Y. (1981). Voltage modulation of Na^+/K^+ transport in human erythrocytes. *J. Physiol. (Paris)* **77,** 1043–1053.

Tenforde T. S. (1992). Microscopic dosimetry of extremely-low-frequency electric and magnetic fields. *Bioelectromagnetics (N.Y.)* **S1,** 61–66.

Thomas, J. R., Schrot, R., and Liboff, A. R. (1986). Low-intensity magnetic fields alter operant behavior in rats. *Bioelectromagnetics (N.Y.)* **6,** 349–357.

Verma, M., Watson, B., Feelings, A., Coleman, M., and Dutta, S. K. (1986). Relationship of cell growth with neuron-specific acetylcholinesterase (AChE) and enolase activities in human neuroblastoma cells. *In* "Bioeffects of Electropollution" (S. K. Dutta and R. M. Millis, eds.), pp. 77–83. Information Ventures, Philadelphia.

Wachtel, H. (1992a). Bioelectric background fields and their implications for ELF dosimetry. *Bioelectromagnetics (N.Y.)* **S1,** 139–146.

Wachtel, H. (1992b). Methodological approaches to EMF dosimetry. *Bioelectromagnetics (N.Y.)* **S1,** 159–160.

Walleczek, J. (1992). Electromagnetic field effects on cells of the immune system: The role of calcium signalling. *FASEB J.* **6,** 3177–3185.

Walleczek, J., and Budinger, T. F. (1992). Pulsed magnetic field effects on calcium signalling in lymphocytes: Dependence on cell status and field intensity. *FEBS Lett.* **314,** 351–355.

Walleczek, J., and Liburdy, R. P. (1990). Nonthermal 60 Hz sinusoidal magnetic-field exposure enhances Ca^{2+} uptake in rat thymocytes: Dependence on mitogen activation. *FEBS Lett.* **271,** 157–160.

Walleczek, J., Miller, P. L., and Adey, W. R. (1992). Simultaneous dual-sample fluorometric detection of real-time effects of ELF electromagnetic fields on cytosolic free calcium and divalent cation flux in human leukemic T-cells (Jurkat). *World Congr. Electr. Magn. Biol. Med. 1st,* Orlando, FL, 1992, p. 12.

Weaver, J. C. (1992). Electromagnetic field dosimetry: Issues relating to background, noise, and interaction mechanisms. *Bioelectromagnetics* (*N.Y.*) **S1,** 115–118.

Weaver, J. C., and Astumian, R. D. (1990). The response of living cells to very weak electric fields: The thermal noise limit. *Science* **247,** 459–462.

Weaver, J. C., and Astumian, R. D. (1992). Estimates for ELF effects: Noise-based thresholds and the number of experimental conditions required for empirical searches. *Bioelectromagnetics* (*N.Y.*) **S1,** 119–138.

Wei, L.-X., Goodman, R., and Henderson, A. (1990). Changes in levels of c-myc and histone H2B following exposure of cells to low-frequency sinusoidal electromagnetic fields. *Bioelectromagnetics* (*N.Y.*) **11,** 297–311.

Weisbrot, D. R., Uluc, A., Henderson, A., and Goodman, R. (1993a). Transcription in *Drosophila melanogaster* salivary gland cells in response to electromagnetic fields and heat shock. *Bioelectrochem. Bioenerg.* **31,** 39–47.

Weisbrot, D. R., Khorkova, O., Lin, H., Henderson, H., and Goodman, R. (1993b). The effect of low frequency electric and magnetic fields on gene expression in *Saccharomyces cerevisiae*. *Bioelectrochem. Bioenerg.* **31,** 167–177.

Welker, H. A., Semm, P., Willig, J. C., Commentz, W., Wiltschko, L, and Vollrath, L. (1983). Effects of an artificial magnetic field on serotonin *N*-acetyltransferase and melatonin content of rat pineal gland. *Exp. Brain Res.* **50,** 426–432.

Westerhoff, H. V., Tsong, T. Y., Chock, P. B., Chen, Y. D., and Astumian, R. D. (1986). How enzymes can capture and transmit free energy from an oscillating electric field. *Proc. Natl. Acad. Sci. U.S.A.* **83,** 4734–4738.

Wilson, B. W., Chess, E. K., and Anderson, L. E. (1986). 60-Hz electric-field effects on pineal melatonin rhythms: Time course for onset and recovery. *Bioelectromagnetics* (*N.Y.*) **7,** 239–242.

Wolf, M., Levine, H., May, W. S., Cuatrecasas, P., and Sahyoun, N. (1985). A model for intracellular translocation of protein kinase C involving synergism between Ca^{2+} and phorbol esters. *Nature* (*London*) **317,** 546–549.

Yamaguchi, H., Ikehara, T., Hosokawa, K., Soda, A., Shono, M., Miyamoto, H., Kinouchi, Y., and Tasaka, T. (1992). Effects of time-varying electromagnetic fields on K^+ (Rb^+) fluxes and surface charge of HeLa cells. *Jpn. J. Physiol.* **42,** 929–943.

Yen-Patton, G.P.A., Patton, W. F., Beer, D. B., and Jacobson, B. S. (1988). Endothelial cell response to pulsed electromagnetic fields: Stimulation of growth rate and angiogenesis *in vitro*. *J. Cell. Physiol.* **134,** 37–46.

Yost, M. G., and Liburdy, R. P. (1992). Time-varying and static magnetic fields act in combination to alter calcium signal transduction in the lymphocyte. *FEBS Lett.* **296,** 117–122.

Zubay, G. (1993). "Biochemistry," 3rd ed., p. 678, fig. 24.17. Wm Brown, Dubuque, IA.

Index

A

B

C

D

F

N

O

P

T

U

V

Y

Z